INTERNATIONAL SERIES OF MONOGRAPHS IN
NATURAL PHILOSOPHY
GENERAL EDITOR: D. TER HAAR

VOLUME 48

INTRODUCTION TO THE THEORY OF MAGNETISM

INTRODUCTION TO THE THEORY OF MAGNETISM

By

D. WAGNER

Ruhr—Universität Bochum,
Institut für Theoretische Physik

Translated by
FERDINAND CAP

Translation edited by
D. ter Haar

PERGAMON PRESS

OXFORD · NEW YORK · TORONTO
SYDNEY · BRAUNSCHWEIG

Pergamon Press Ltd., Headington Hill Hall, Oxford
Pergamon Press Inc., Maxwell House, Fairview Park, Elmsford, New York 10523
Pergamon of Canada Ltd., 207 Queen's Quay West, Toronto 1
Pergamon Press (Aust.) Pty. Ltd., 19a Boundary Street, Rushcutters Bay, N.S.W. 2011, Australia
Vieweg & Sohn GmbH, Burgplatz 1, Braunschweig

Translated and revised from the German edition Einführung in die Theorie des Magnetismus *by Dieter Wagner*

First edition 1972
Library of Congress Catalog Card No. 76-183338

Printed in Hungary

08 016595 8

Contents

Preface to the German Edition

THE present book contains essentially the material of a lecture series on the theory of magnetism, which was held for students of physics at a medium advanced level. The set-up follows the natural subdivision of magnetic substances, i.e. diamagnetism was treated first, then paramagnetism, and finally we dealt with ferromagnetism, antiferromagnetism and ferrimagnetism. Naturally some parts of this field, magnetism, had to be dealt with in less detail than other parts, as the volume of the book was limited. Thus in the first part of the book we have put emphasis on the de Haas–van Alphen effect, which is very instructive as to the behaviour of electrons in crystals. In the second part the influence of the crystal field on the paramagnetism of atoms and connected problems (Kramers theorem, adiabatic demagnetisation, paramagnetic resonance, ...) have been dealt with more extensively. The major part of the last section is devoted to the exchange interaction and the resulting many-particle problem for a system of atomic spins. The properties of this system are discussed in several approximations. Within the framework of the present book it was, however, not possible to deal with the technologically interesting problems (such as hysteresis, magnetostriction). If the reader is interested in these problems, he can easily acquaint himself with this material by means of the literature cited.

I am indebted to Dr. Ballensiefen for his help with the preparation of the figures.

Cologne, November 1965

DIETER WAGNER

Foreword to the English Edition

THIS English edition is a substantially enlarged version of the German edition. It merely requires in a few places a slightly more advanced knowledge of quantum mechanics than that possessed by a student with an understanding of elementary quantum mechanics.

Abbreviations and Symbols

a_B	Bohr radius
$\boldsymbol{A}$	vector potential
B_s	Brillouin function
C	Curie constant
C	specific heat
C_m	magnetic specific heat
e	charge
e_0	elementary charge
E	energy
$f(E)$	Fermi distribution
F	free energy
g	spectroscopic splitting factor (tensor)
g'	magneto-mechanical factor
g_J	Landé factor
H, $\boldsymbol{H}$	magnetic field
H_A	anisotropy field
H_W	Weiss field
j, J; $\boldsymbol{j}$, $\boldsymbol{J}$	(quantum number of) total angular momentum
J, J_{ij}	exchange integral
k	Boltzmann constant
$\boldsymbol{k}$	wave vector
k_0	Fermi momentum
$\boldsymbol{K}$	Knight-shift constant
K	time reversal operator
K	anisotropy energy
l, L; $\boldsymbol{l}$, $\boldsymbol{L}$	(quantum number of) orbital angular momentum
m^*, m	electron (effective) mass
m, m_l, m_s, m_j, M_L, M_S, M_J	magnetic quantum numbers
m, M; $\boldsymbol{m}$, $\boldsymbol{M}$	magnetisation

p_{eff}; P	(effective) magnetic moment per atom in terms of μ_B
P	de Haas–van Alphen period
P	permutation operator
P	projection operator
s, S; $\boldsymbol{s}, \boldsymbol{S}$	spin (quantum number)
S'	effective spin
S	entropy
T	temperature
T_C, T_N	Curie point, Néel point
U	intra-atomic interaction
W, W_{ij}	Weiss factor
z	number of nearest-neighbour atoms
Z_{eff}; Z	(effective) atomic number
Z	partition function
$Z(E)$	number of states with energy $\leqslant E$
β	$1/kT$
γ	gyromagnetic factor
$\delta(x)$	δ-function
$\delta_{n, n'}$	Kronecker symbol
ζ, ζ_0	Fermi energy
ζ, $\zeta_{n, l}$	spin–orbit coupling constant
$\boldsymbol{\Theta}$	paramagnetic Curie point
Θ	Stoner's parameter
μ	chemical potential
$\boldsymbol{\mu}$	magnetic moment
μ_{B}^*; μ_{B}	(effective) Bohr magneton
μ_{nucl}	nuclear magneton
ϱ	density matrix, density operator
ϱ	density of states per atom
σ	relative magnetisation
$\boldsymbol{\sigma}$	Pauli matrices
φ_{cr}	crystal potential
φ_1	exchange interaction of quasi-particles
χ	susceptibility
ω_{L}^*; ω_{L}	(effective) Larmor frequency
ω_{c}	cyclotron resonance frequency
∇	Nabla operator
$\cong$	approximately equal to

Classification of Magnetic Substances

WHEN a substance is placed in a magnetic field $\boldsymbol{H}$, this field will induce inside the substance a field $\boldsymbol{B}$ which is given by

$$\boldsymbol{B} = \boldsymbol{H} + 4\pi\boldsymbol{M} = \mu\boldsymbol{H}.$$

Here $\boldsymbol{M}$ is the magnetisation of the substance and μ is its permeability. The magnetisation is due to the external field $\boldsymbol{H}$. It is an empirical fact that, to a good approximation, $\boldsymbol{M}$ (with the exception of some special cases) is a linear function of $\boldsymbol{H}$ and we can put

$$\boldsymbol{M} = \chi\boldsymbol{H}$$

where χ is called the magnetic susceptibility. Susceptibility is a parameter characteristic of the substance considered, which, in general, depends on its temperature. According to sign and order of magnitude of χ substances are subdivided into three large classes:

I. $\chi < 0$	Diamagnetic substances.
II. $\chi > 0$	Paramagnetic substances.
III. $\chi = \infty$	Ferromagnetic substances.

Class I is largest as almost all organic molecules belong to it, as well as all noble gases, several metals such as the noble metals, bismuth, zinc, mercury, ... and non-metals such as sulfur, iodine, silicon, ... (Bates, 1961). In principle also all superconductors belong to this class, since below the transition point in a superconductor $\boldsymbol{B} = 0$ (Meissner effect) so that $\chi = -1/4\pi$. Since, however, such a behaviour is due to quite different physical causes than normal diamagnetism, superconductors are not considered in this book. However, it should be mentioned that the "diamagnetic conception" of superconductors was important for the development of the theory of superconductivity (Bardeen, 1963).

The second class comprises all substances which possess a permanent magnetic moment, such as the transition elements of the periodic system and their compounds, as well as the rare earths, and the alkali metals, ...

A few substances belonging to Class II are ferromagnetic below a certain characteristic temperature, the Curie point; these substances represent Class III, which also comprises the antiferromagnetic and ferrimagnetic substances. The property of being ferromagnetic (antiferromagnetic, ferrimagnetic) is linked with definite crystalline structures so that this class comprises only solids.

The following treatment of the theories of magnetism is based on this classification of the substances.

I. Diamagnetism

1. The Bohr–van Leeuwen Theorem

In order to understand the fundamental effects of diamagnetism, one may be tempted first to develop a classical description and then to complete it with the necessary quantum-mechanical corrections. However, both diamagnetism and paramagnetism proved to be incomprehensible within the framework of an exact classical theory based on the magnetism of moving charges. This fact is sometimes called the Bohr–van Leeuwen theorem (Van Vleck, 1932a). If, for example, an electron with the charge e ($e = -e_0$; $e_0 = 4{\cdot}8\times 10^{-10}$ e.s.u.) is assumed to travel along a circular orbit (e.g. in a constant magnetic field), a magnetic moment $\boldsymbol{m}$ given by

$$\boldsymbol{m} = \frac{j}{c}\boldsymbol{A} = \frac{e}{2c}[\boldsymbol{r}\wedge\boldsymbol{v}]. \tag{1.1}$$

will correspond to the circular current $\boldsymbol{j}$ obtained in this way; here c is the velocity of light, $\boldsymbol{r}$ the radius vector of the electron, $\boldsymbol{v}$ its velocity, and $\boldsymbol{A}$ is a vector, the magnitude of which is the area enclosed by the current and the direction of which is given by the right-hand rule. When the electron is assumed to move in the x, y-plane, only the z component of moment $\boldsymbol{m}$ will be non-zero:

$$m_z = \frac{e}{2c}(x\dot{y} - y\dot{x}).$$

This formula was given only to show that there is a linear relation between the magnetic moment and the velocity of the charge. This linearity remains unchanged even if we have complex trajectories and many electrons. Consider, for example, a system of N electrons which are described by their coordinates $q_1 \ldots q_{3N}$ and the corresponding canonical momenta $p_1 \ldots p_{3N}$. For this system the z component of the magnetic moment will be given by

$$m_z = \sum_{i=1}^{3N} a_i^z(q_1 \dots q_{3N})\dot{q}_i, \tag{1.2}$$

where a_i^z depends only on the position coordinates, q_i. The thermal mean of (1.2) is obtained when this function is provided with a weight factor (the Boltzmann function) and integrated over the whole of phase space:

$$M = \int m_z \frac{e^{-\mathcal{H}/kT}}{Z} dq_1 \dots dq_{3N}\, dp_1 \dots dp_{3N}. \tag{1.3}$$

$\mathcal{H}$ is the Hamiltonian of the system generally given by

$$\mathcal{H} = \sum_{i=1}^{3N} \frac{1}{2m}\left(p_i - \frac{e}{c}A_i\right)^2 + eV(q_1, \dots, q_{3N});$$

here m is the mass of the charge carriers, A is the vector potential, and V is the interaction potential. Z is the classical partition function,

$$Z = \int e^{-\mathcal{H}/kT}\, dq_1 \dots dp_{3N},$$

k is Boltzmann's constant, and T is the absolute temperature. Substituting (1.2) in (1.3) we obtain

$$M = \frac{1}{Z}\sum_{i=1}^{3N} \int a_i^z(q_1, \dots, q_{3N})\dot{q}_i e^{-\beta\mathcal{H}}\, dq_1 \dots dp_{3N}; \qquad \beta = \frac{1}{kT}.$$

By virtue of the canonical equations of motion we have

$$\dot{q}_i = \frac{\partial \mathcal{H}}{\partial p_i}, \quad \dot{p} = -\frac{\partial \mathcal{H}}{\partial q_i},$$

and, integrating over the momenta, we obtain

$$\begin{aligned} M &= \frac{1}{Z}\int \sum_{i=1}^{3N} a_i^z(q_1 \dots q_{3N}) \frac{\partial \mathcal{H}}{\partial p_i} e^{-\beta\mathcal{H}}\, dq_1 \dots dp_{3N} \\ &= -\frac{1}{\beta Z}\int \sum_{i=1}^{3N} a_i^z(q_1 \dots q_{3N}) \frac{\partial}{\partial p_i} e^{-\beta\mathcal{H}}\, dq_1 \dots dp_{3N} \\ &= -\frac{1}{\beta Z}\int \sum_{i=1}^{3N} a_i^z(q_1 \dots q_{3N}) \left[e^{-\beta\mathcal{H}}\right]_{p_i=-\infty}^{p_i=+\infty} dq_1 \dots \\ &\qquad dq_{3N}\, dp_1 \dots dp_{i-1}\, dp_{i+1} \dots dp_{3N} = 0. \end{aligned}$$

Thus we see that **dia**magnetism is not a classical but a quantum-mechanical effect; when the susceptibility formula is derived in a classical way, the derivation is always based on assumptions involving quantum-mechanical results in a more or less implicit form.

2. Diamagnetism of Atoms and Ions

The classical derivation of diamagnetic susceptibility according to Langevin (1905) and Pauli (1920) is based upon the assumption of an electron moving in a fixed orbit (frequency ω_0) around a nucleus, i.e. it is based on a quantum-mechanical assumption. In this case the centrifugal force and the Coulomb force acting on the electron must be in equilibrium. When the nuclear charge is equal to Ze_0 we have

$$m\omega_0^2 r = \frac{Ze_0^2}{r^2},$$

r being the distance between electron and nucleus. Hence we obtain for the orbital frequency

$$\omega_0 = \pm\sqrt{\frac{Ze_0^2}{mr^3}}.$$

If now a magnetic field $\boldsymbol{H}$ is applied parallel to the z-axis, the Lorentz force

$$\boldsymbol{F} = \frac{e}{c}[\boldsymbol{v}\wedge\boldsymbol{H}]$$

will act on the electron in addition to the other forces. The condition of equilibrium for the forces in the plane perpendicular to the field is now given by

$$m\omega^2 r = \frac{Ze_0^2}{r^2} + \frac{e_0}{c}\,\omega r H,$$

where H is the magnitude of the magnetic field and ω is the orbital frequency of the electron, which has changed by ω_0.

This new frequency is given by

$$\omega = \omega_0\sqrt{1+\left(\frac{e_0 H}{2mc\omega_0}\right)^2} + \frac{e_0 H}{2mc}. \tag{1.4}$$

With field strengths achievable in practice, $e_0H/2mc$ is much smaller than ω_0 and we have in a good approximation

$$\omega = \omega_0 + \frac{e_0H}{2mc} = \omega_0 + \omega_L.$$

Thus the application of the magnetic field has changed the orbital frequency by ω_L, which is usually called the Larmor frequency. This additional Larmor frequency gives rise to an additional current and an additional magnetic moment induced by the field. As there is an additional number, $\omega_L/2\pi$, of revolutions of the electron, per unit time, the current due to the Larmor frequency is given by

$$j = \frac{e\omega_L}{2\pi}.$$

The magnetic moment m_z induced by it will then be

$$m_z = \frac{j}{c}A = \frac{e\omega_L}{2\pi c}A.$$

Here A is the circular area bounded by the electron's orbit in the x, y-plane: $A = \pi(x^2+y^2)$. Averaging over many identical electron orbits, we have

$$\overline{x^2} = \overline{y^2} = \overline{z^2} = \tfrac{1}{3}\overline{r^2},$$

r being the distance between electron and nucleus. Thus the magnetic moment of a diamagnetic substance consisting of N_0 identical atoms per cm^3, each possessing one electron, will be given by

$$M = -N_0 \frac{e_0^2 H}{6mc^2}\overline{r^2}.$$

If several electrons are bound to the nucleus, in addition to the electrostatic interaction between electrons and nucleus, there will also exist an electrostatic interaction between the electrons. Since this interaction is spherical-symmetric, the Larmor theorem remains unchanged, as can be easily verified: the orbital frequency of each electron will be changed by ω_L. Hence it follows that in the case of

N_0 identical atoms with Z electrons the mean magnetic moment is given by

$$M = -N_0 \frac{e_0^2}{6mc^2} \sum_{i=1}^{Z} \overline{r_i^2} \cdot H, \tag{1.5}$$

so that the susceptibility per unit volume is equal to

$$\chi_V = -N_0 \frac{e_0^2}{6mc^2} \sum_{i=1}^{Z} \overline{r_i^2}, \tag{1.6}$$

and the atomic or molar susceptibility will be given by

$$\chi_A = -N \frac{e_0^2}{6mc^2} \sum_{i=1}^{Z} \overline{r^2}. \tag{1.7}$$

N is the Avogadro number, equal to $6{\cdot}0022 \times 10^{23}$ mole^{-1}.

This calculation is based on the assumption of atoms which do not influence one another.

We have seen from the theorem by Bohr and van Leeuwen that a classical derivation of susceptibility will always be inconsistent. We therefore need a quantum-mechanical basis for eqn. (1.5).

The Hamiltonian of an atom with Z electrons in an external magnetic field is given by

$$\mathscr{H} = \frac{1}{2m} \sum_{i=1}^{Z} \left(\boldsymbol{p}_i - \frac{e}{c} A(\boldsymbol{r}_i) \right)^2 + \sum_{i=1}^{Z} U(r_i) + \frac{1}{2} \sum_{i \neq 1} V(|\boldsymbol{r}_i - \boldsymbol{r}_j|).$$

Here $U(r_i)$ is the interaction of the ith electron with the nucleus and $V(|\boldsymbol{r}_i - \boldsymbol{r}_j|)$ is the ordinary electrostatic interaction between an electron at $\boldsymbol{r}_i$ and an electron at $\boldsymbol{r}_j$. As these interactions are unimportant for what follows, we shall ignore them. The magnetic field is assumed to be constant, uniform, and parallel to the z-axis; the vector potential is then given by

$$A = \frac{H}{2}(-y, x, 0).$$

The expectation value of $\mathscr{H}$ is then obtained as

$$\overline{\mathscr{H}} = \sum_{i=1}^{Z} \frac{\overline{p_i^2}}{2m} - \frac{eH}{2mc} \sum_{i=1}^{Z} \overline{(xp_y - yp_x)_i} + \frac{e^2H^2}{8mc^2} \sum_{i=1}^{Z} \overline{(x^2+y^2)_i}.$$

Hence we obtain for the magnetic moment of the atom

$$\overline{m_z} = -\overline{\frac{\partial \mathcal{H}}{\partial H}} = \frac{e\hbar}{2mc}\sum_{i=1}^{Z}\overline{l_{zi}} - \frac{e^2 H}{4mc^2}\sum_{i=1}^{Z}\overline{(x^2+y^2)_i},$$

where

$$\hbar l_z = xp_y - yp_x$$

is the z-component of the orbital angular momentum of an electron, due to its motion around the nucleus. When the first term is different from zero, the atom has a permanent magnetic moment which is the cause of its paramagnetic behaviour, as this term as a rule exceeds the second, the diamagnetic term. As to the order of magnitude, in cgs units,

$$\hbar\overline{l_z} = m\overline{vr} \cong 10^{-27} \quad \text{and} \quad \frac{e_0}{c}H(\overline{x^2}+\overline{y^2}) = \frac{2}{3}\frac{e_0}{c}H\overline{r^2} \cong 10^{-36}H,$$

i.e. for practical field strengths and usual electron orbit radii, the diamagnetic term can be neglected in the expectation value for the magnetic moment. Thus, the diamagnetism of an atom can only be measured if the atom has no permanent magnetic moment. It is well known that only atoms with closed shells, i.e. noble gases, satisfy this condition. (Although the expectation value of the angular momentum vanishes also for s-electrons, for example, the diamagnetism of an atom with one s-electron is masked by the weak paramagnetism of the spin of this electron.) For such electron configurations we therefore obtain for the mean magnetic moment of the atom

$$M = -\frac{e_0^2 H}{6mc^2}\sum_{i=1}^{Z}\overline{r_i^2}, \tag{1.5}$$

where we can put $\overline{x^2} = \overline{y^2} = \frac{1}{3}\overline{r^2}$ because of the spherical symmetry of the Hamiltonian. (In the case of a crystal this would be generally inadmissible.) Hence we obtain for the molar or atomic susceptibility as before

$$\chi_A = -N\frac{e_0^2}{6mc^2}\sum_{i=1}^{Z}\overline{r_i^2} \tag{1.7}$$

which is the same as the previous formula. It is an advantage of the quantum-mechanical derivation, compared with the classical derivation by Langevin, that it reveals the conditions under which the diamagnetism of an atom can be observed, as the presence of closed shells cannot be understood within the framework of the classical concept. With the more exact derivation of eqn. (1.7), however, we did not achieve very much; the difficulties in the calculation of susceptibility consist of the determination of the expectation values of r^2. The simplest case to be considered is that of a hydrogenic atom with nuclear charge Z.

If the electron is in a state with principal quantum number n determining the energy of the electron, and the orbital quantum number, l, (because of spherical symmetry the magnetic quantum number plays no part), we obtain with the exact eigenfunctions of the hydrogen atom, according to Bethe and Salpeter (1957),

$$\overline{r^2_{n,l}} = \frac{a_B^2 n^2}{2Z^2}\{5n^2+1-3l(l+1)\},$$

where $a_B = \hbar^2/me_0^2 = 5 \cdot 29 \times 10^{-9}$ cm (Bohr radius); thus we obtain for the susceptibility

$$\chi_A = -N\frac{a_B^2 e_0^2}{6Z^2 mc^2} n^4 \left(\frac{5}{2} - \frac{3l(l+1)}{2n^2} + \frac{1}{2n^2}\right), \tag{1.8}$$

i.e. χ is proportional to the fourth power of the principal quantum number. When we apply (1.8) to the ground state of the hydrogen atom ($n = 1$, $l = 0$) we obtain

$$\chi_A = -2 \cdot 38 \times 10^{-6} \text{ cm}^3/\text{mole}.$$

Because of the electron spin this result cannot be verified by experiment. It is, however, known empirically that in larger atomic complexes the spin becomes saturated; it was therefore tried to draw conclusions from measurements on organic substances about the susceptibility of the hydrogen atom. In this way Pascal (1920) obtained

$$\chi_A = -2 \cdot 93 \times 10^{-6} \text{ cm}^3/\text{mole}.$$

Although this result is rather satisfactory, other experimenters arrived at quite different values so that this method is not well suited to check the theoretical result.

The calculation of $\overline{r^2}$ for atoms with two or more electrons immediately entails considerable difficulties, which are due to the mutual influence of the electrons. An electron at a distance r from the nucleus (charge Z) is not exposed to the action of the whole nuclear charge but to this charge minus the electron charge inside the sphere of radius r around the nucleus, i.e. to $Z_{eff}(r) = Z - Z_{el}(r)$, where according to the usual laws of electrostatics, $Z_{el}(r)$ is given by

$$Z_{el}(r) = 4\pi \int_0^r \varrho(r) r^2 \, dr,$$

where $\varrho(r)$ is the electron density at point r. (Here we tacitly assume a spherical-symmetric density distribution which strictly applies only to closed shells.) In a very rough approximation Z_{eff} can be replaced by a constant which will depend on the principal quantum number n and the orbital quantum number l (Van Vleck, 1932a):

$$Z_{eff} = Z - \sigma_{n,l}. \tag{1.9}$$

One could try to calculate the screening constants $\sigma_{n,l}$ (cf. Van Vleck, 1932a); in our rough approximation, however, it is better to take the screening constants as parameters to be obtained from, for example, the ground states of the atoms or ions, or their susceptibilities. In this case each electron in a shell with quantum numbers n and l moves in a Coulomb potential $Z_{eff}e_0/r$ and, according to (1.8), $\overline{r^2}$ is obtained for this electron in the form of

$$\overline{r^2_{n,l}} = \frac{a_B^2 n^2}{2(Z-\sigma_{n,l})^2}\left(5n^2+1-3l(l+1)\right).$$

As each energy level displays a $2(2l+1)$-fold degeneracy, we have for an atom or ion with closed shells

$$\chi_A = -N\frac{a_B^2 e_0^2}{6mc^2}\sum_{n=0}^{n_{max}}\sum_{l=0}^{l_n}\frac{(2l+1)\cdot n^2\cdot(5n^2+1-3l(l+1))}{(Z-\sigma_{n,l})^2}, \tag{1.10}$$

where the double sum is to be taken over all occupied states, i.e. the electron configuration of the atom or ion.

Another method of calculating $\overline{r^2}$ is based upon approximate analytical expressions for the radial part of the electron wave function which, besides the effective nuclear charge, also contains an effective principal quantum number n^*. (An effective quantum number is known to appear in the optical terms of the alkali metals; see, for example, Slater, 1960.)

Slater (1930; see also Angus, 1932) wrote the radial part in the form

$$\psi(r) \sim r^{n^*-1} e^{-\frac{Z-s_{n,l}}{n^*}\frac{r}{a_B}} \tag{1.11}$$

and gave simple rules as to the attribution of n^* and $s_{n,l}$ to the electron configurations in the atom (Slater, 1930; Angus, 1932). The form of (1.11) corresponds to the asymptotic behaviour of the hydrogen eigenfunctions for large distances from the nucleus, see, for example, the text by Landau and Lifshitz (1958).

Using (1.11) we then obtain

$$\overline{r^2_{n,l}} = \frac{a_B^2(n^*)^2}{(Z-s_{n,l})^2}\left(n^*+\frac{1}{2}\right)(n^*+1)$$

and for the molar susceptibility

$$\chi_A = -N\frac{a_B^2 e_0^2}{3mc^2}\sum_{n=1}^{n_{\max}}\sum_{l=0}^{l_n}\frac{(n^*)^2\left(n^*+\frac{1}{2}\right)(n^*+1)}{(Z-s_{n,l})^2}. \tag{1.12}$$

The methods sketched in this way are more or less semi-empirical methods as the final formulas contain parameters obtained by choosing them in agreement with other data on the atom. Theoretically $\overline{r^2}$ can be calculated in three ways. One of them is the Thomas–Fermi method for statistical treatment of the atom (Gombás, 1956). The other two ways are the Hartree method and the Hartree–Fock method. Both are based on a variational principle for the electron wave functions with the approximation of replacing the interaction potential between the atomic electrons and all other electrons by its mean value averaged over all angles, where the still unknown wave functions are used in the averaging. In this way, we obtain from the Schrödinger equation for the many-electron problem an integro-differential equation which can only be solved numerically (Slater, 1960; Hartree, 1957). Unlike the Hartree method, the Hartree–Fock method takes the Pauli principle into account; for most atoms it is the best and most exact way of determining the physical quantities of an atom (Slater, 1960).

TABLE 1

Element	Exper. (Landolt-Börnstein, 1950)	Eqn. (1.10) (Van Vleck, 1932a)	Eqn. (1.12) (Slater, 1930)	Thomas–Fermi (Gombás, 1956)	Hartree (Stoner, 1929)	Hartree–Fock (Hartree and Hartree, 1938)
He	−1·93	−1·54 −11·85	−1·64	−	−1·90	
Ne	−7·2	−5·7 −	−5·6	−12·8	−8·6	
Ar	−19·4	−13·6 −21·5	−18·5	−18·2	−24·8	−20·6
Kr	−28	−17·2 −42	−31	−28·3		
Xe	−43	−25·4 −66	−47	−47·3		

The table contains data on the atomic susceptibility $\chi_A \times 10^6$ mole/cm³ for noble gases. The first values of the third column are the susceptibilities obtained with the theoretical screening constants; the second values of this column were obtained by means of empirical screening constants. The values of the fourth column were calculated according to Slater's rules.

Table 1 gives the molar susceptibilities of the noble gases as obtained by means of various methods and compared with the experimental results. With almost all elements the agreement between theoretical and experimental values is rather good.

The He atom is a special case since for this atom the wave function is known relatively exactly (Slater, 1960; Hylleraas, 1929; Kinoshita, 1957; Pekeris, 1958). The accuracy is essentially better than using the Hartree–Fock method. Using the results by Pekeris (1959; Stewart, 1963), for helium a value of $\chi_A = -1{\cdot}8905 \times 10^{-6}$ is obtained, while the experimental value obtained by Havens (1933) is

$$\chi_A = -(1{\cdot}906 \pm 0{\cdot}006) \times 10^{-6}$$

so that in this case we can speak of a very good agreement.

A comparison between theory and experiment is more difficult in the case of ions with noble-gas electron configurations (closed shells). First of all, the susceptibilities of these ions can only be measured in solutions or crystals, e.g. Na^+Cl^-, as these ions do not exist in the gaseous phase; secondly, there is a decisive uncertainty to be taken into account, which is due to the fact that the noble-gas configuration of ions is essentially less stable than that of the noble gases. Their susceptibility will therefore depend on the partner of the ion to be measured, on the structure of the crystal, or on the solvent. It is therefore difficult to conclude from the susceptibilities measured, the susceptibilities of the free ions which were calculated theoretically (Myers, 1952).

TABLE 2

Element	Exper. (Myers, 1952)	Eqn. (1.10) (Pauling, 1927)	Eqn. (1.12) Slater (see Myers, 1952)	Thomas–Fermi (Gombás, 1956)	Hartree (Hartree, 1928)	Hartree–Fock (Hartree and Hartree, 1938)
F^-	− 9·4; −12·1	− 8·1	− 8·1	−24·4	−17·0	
Cl^-	−18·3; −26·5	−29	−25·2	−34·6	−41·3	−30·4
Br^-	−30; −54	−54	−39·2	−51·2		
J^-	−47·7; −55·4	−80	−58·5	−63·7		
Li^+	+ 0·1; − 6·7	− 0·63	− 0·7	—	− 0·7	− 0·7
Na^+	− 5·25; −12·5	− 4·2	− 4·1	− 7·3	− 5·6	− 4·9
K^+	−13·4; −21·1	−16·7	−14·1	−14·4	−17·3	−18·3
Rb^+	−20·4 ; −28·2	−35	−25·1	−26·8	−29·5	
Cs^+	−35·1; −44·3	−55	−38·7	−37·2	−47·5	
Mg^{++}	− 1·6; −13·6	− 3·2	− 3·1	− 5·8	− 4·2	− 3·74
Ca^{++}	− 4·5; −18·5	−13·3	−11·1	− 9·0	−13·1	
Sr^{++}	−14; −28·5	−28	−21·0	−21·0		
Ba^{++}	−25·3; −41·1	−46	−32·6	−30·5		

The table contains data on the atomic susceptibilities ($\chi_A \times 10^6$ mole/cm^3) of some ions with noble-gas configurations. The second column gives the largest and smallest experimental values (after Myers). In the fifth column the susceptibility of Mg^{++} was calculated by Hirone (1935), and that of Ca^{++}, Sr^{++}, and Ba^{++} by Jensen (1936). The value for Mg^{++} in the seventh column was calculated by Yost (1940).

Naturally the values of χ compiled in Table 2 show a considerable spread. Given the spread in the experimental values we cannot say much about the agreement between theory and experiment. However, all theoretical methods yield a qualitative agreement.

3. The Fermi Gas

Passing over from substances whose atoms have a noble-gas configuration to the alkali metals, we see that the alkali ions also have noble-gas configurations; however, each of them has also an electron which, in the metallic state, is more or less separated from the atom. The appearance of free charges is a characteristic phenomenon of conductors and semiconductors. Like the ions they contribute to the magnetic behaviour of the substance. This behaviour depends partly on the temperature; in the following we shall therefore consider the thermodynamic properties of such an electron gas as far as they are essential for the magnetism. We shall neglect the electron–electron Coulomb interaction: this corresponds to the old Sommerfeld treatment of the electron gas in metals. (The influence of temperature on the diamagnetism of atoms and ions could be neglected because the noble-gas configuration is very stable and cannot be changed by temperature effects.) We now determine the (Helmholtz) free energy F of the electron gas with a given number of electrons and given energy:

$$F = E - TS. \tag{1.13}$$

Here E is the energy, T is the temperature, and S is the entropy. When the electrons are assumed to possess a discrete spectrum of states (marked by a subscript i), the energy is given by

$$E = \sum_i E_i \cdot f(E_i), \tag{1.14}$$

where E_i is the energy of the ith state and $f(E_i)$ is the mean number of electrons in this state. As in quantum mechanics, the subscript i characterises a whole set of variables determining the properties of the electron, e.g. momentum and spin in the case of free electrons without an external field. According to the Pauli principle $f(E_i) = f_i$ can then be at most equal to 1 since for particles of half-odd-integral spin each state can be occupied only once. (This formulation of the Pauli principle applies only to a free Fermi gas. As to a more general formulation we refer to the text book literature (e.g. Landau and Lifshitz, 1958; Becker and Sauter, 1959); see also Section III.1.) The entropy can be obtained from Boltzmann's principle

$$S = k \ln W \tag{1.15}$$

where k is the Boltzmann constant and W is the statistical weight of the state of the entire system. In order to calculate W we consider the g_j states pertaining to a very small energy interval about the energy E_j; of these g_j states n_j are assumed occupied so that $g_j - n_j$ are free. (The size of the interval depends only on the requirement that the numbers g_j and n_j are sufficiently large.) The statistical weight W_j of this group of states with the approximate energy E_j will then be given by

$$W_j = \frac{g_j!}{(g_j - n_j)!n_j!},$$

since a permutation of the occupied or non-occupied states yields the same physical state because of the indistinguishability of the electrons. As $W = \Pi_j W_j$ we have

$$S = k \sum_j \ln W_j.$$

Using Stirling's formula $n! \cong \sqrt{2\pi n}\, e^{-n} n^n$ we obtain for S

$$S = k \sum_j (g_j \ln g_j - (g_j - n_j) \ln (g_j - n_j) - n_j \ln n_j),$$

where only the dominant terms are taken into account. We are only interested in the state with the largest entropy for given particle number and energy; S must therefore be varied with respect to the n_j in order to obtain the most probable distribution $\overline{n_j}$. The variation must be subjected to the two constraints saying that the number of electrons and the energy of the total system must remain unchanged:

$$\delta\left(N-\sum_j n_j\right) = 0 \qquad \delta\left(E-\sum_j n_j E_j\right) = 0,$$

where δ denotes the variation. These constraints can be taken into account by means of the Lagrange method, and then we have to determine the extremum of

$$S' = k\sum_j \left(g_j \ln g_j-(g_j-n_j)\ln (g_j-n_j)-n_j \ln n_j\right)$$
$$+\alpha\left(N-\sum_j n_j\right)+\beta\left(E-\sum_j n_j E_j\right)$$

where α and β are the corresponding Lagrangian multipliers. From $\partial S'/\partial n_j = 0$ we obtain immediately

$$\overline{n_j} = \frac{g_j}{e^{\alpha+\beta E_j}+1}.$$

Since the occupation probability f_i depends only on the energy, we obtain for the mean number of electrons in state E_i:

$$f_i = \frac{\overline{n_i}}{g_i} = \frac{1}{e^{\alpha+\beta E_i}+1}. \tag{1.16a}$$

We put $\alpha = -\zeta/kT$; this parameter is obtained from the constraint $\sum f_i = N$. The other parameter is obtained from the consideration that at high temperatures the electron gas must behave like a classical gas, i.e. f_i must become a Boltzmann distribution. Obviously this is only possible if the exponential term in the denominator of (1.16a) is considerably larger than 1, i.e. if $f_i \sim \exp(-\beta E_i)$, and hence we obtain $\beta = 1/kT$. We thus have for the so-called Fermi function

$$f(E_i) = \frac{1}{e^{\beta(E_i-\zeta)}+1} \tag{1.16b}$$

where ζ is the Fermi energy. The Fermi function displays characteristic deviations from the usual Boltzmann distribution. In metals the electron energies, and thus also ζ, are of the order of a few eV, while kT at room temperature amounts to about 1/40 eV. If $E_i < \zeta$, f_i is virtually equal to 1 and drops to zero near $E = \zeta$ over a distance of a few times kT (see, for example, Kittel, 1962; Wilson, 1958; ter Haar, 1966). At $T = 0, f(E)$ is a step function: $f(E) = 1$ if $E \leqslant \zeta$, and $f(E) = 0$ if $E \geqslant \zeta$; the derivative $\partial f/\partial E$ is thus a δ-function, i.e. $\partial f/\partial E = -\delta(E-\zeta)$ so that

$$\int_0^\infty \frac{\partial f}{\partial E} dE = -1,$$

if $\zeta > 0$.

At finite temperatures, i.e. as long as $kT \ll \zeta$, this behaviour of f remains essentially unchanged. In the case of metals this is true for all temperatures achievable in practice. The quantity ζ is determined from the first constraint $\sum f_i = N$. When we consider free electrons without an external field we have $E_i = \hbar^2 k_i^2/2m$, $\hbar \boldsymbol{k}_i$ denoting the electron momentum; $\boldsymbol{k}_i$ is the wave vector. Formally the wave vectors are assumed to be discrete; as here the E_i are independent of the spin, the equation for ζ reads as follows: $\sum_{k,s} f(E_k) = N$; s indicates summation over the two directions of spin and yields a factor 2. When we now change in the usual way from summation to integration

$$\sum_k \rightarrow \frac{V}{(2\pi)^3} \int d^3\boldsymbol{k}$$

(V being the electron gas volume), we obtain for $T = 0$

$$2 \cdot \frac{4\pi V}{(2\pi)^3} \int_0^{k_0} k^2 dk = \frac{V}{3\pi^2} k_0^3 = N.$$

Since $\hbar^2 k_0^2/2m = \zeta_0$ is the Fermi energy at the temperature $T = 0$ we obtain from this

$$\zeta_0 = \frac{h^2}{2m} \left(\frac{3N}{8\pi V} \right)^{2/3}; \tag{1.17}$$

here $h = 2\pi\hbar$.

In the case of typical metals ζ_0 is of the order of 10 eV. We often put $\zeta_0 = kT_0$, thus defining the so-called degeneracy temperature of the Fermi gas. As we may verify easily from (1.17), ordinary metals have degeneracy temperatures between 10^4 and 10^5 °K. Several metals and semiconductors whose electron density is very low may have a particularly low degeneracy temperature so that in this case a Boltzmann distribution is obtained instead of the Fermi distribution. Bismuth, for example, has a degeneracy temperature of 210°K. At finite temperatures the Fermi energy of ordinary metals depends slightly on temperature. From the definition of ζ we obtain easily (see II.1)

$$\zeta = \zeta_0\left(1-\frac{\pi^2}{12}\left(\frac{T}{T_0}\right)^2\ldots\right).$$

A magnetic field H has little influence on the Fermi energy since, because of the spin of the electrons, their energy levels are only shifted by the small amount of $\pm\mu_B H$ ($\mu_B = e_0\hbar/2mc$ is the Bohr magneton). Another simple calculation (Wilson, 1958) yields

$$\zeta = \zeta_0\left(1-\frac{\pi^2}{12}\left(\frac{T}{T_0}\right)^2-\frac{1}{4}\left(\frac{\mu_B H}{\zeta_0}\right)^2\right), \tag{1.18}$$

in which terms of higher order are neglected. We see that in good conductors the corrections are always very small. The form of the Fermi distribution (1.16b) permits a transformation of the expression for the entropy S; using (1.16a) and the most probable population numbers $\bar{n}_j$, we obtain from the above expression for the entropy

$$\begin{aligned} S &= -k\sum_i (f_i \ln f_i + (1+f_i)\ln(1+f_i)) \\ &= k\sum_i (\ln(1+e^{-\beta(E_i-\zeta)}) + \beta(E_i-\zeta)f_i) \\ &= k\left(\sum_i \ln(1+e^{-\beta(E_i-\zeta)})\right) + \beta E - \beta N\zeta. \end{aligned}$$

Hence we obtain for the free energy F according to (1.13)

$$F = N\zeta - k\sum_i \ln(1+e^{-\beta(E_i-\zeta)}). \tag{1.19}$$

4. The Landau Levels

Since the Bohr–van Leeuwen theorem also holds for a Fermi gas in a magnetic field, the problem must be subjected to a quantum-mechanical treatment. Again we assume the field in the z direction and choose the vector potential defining the field $\boldsymbol{H} = \text{curl}\,\boldsymbol{A}$ in such a way that it has only a y-component, i.e. $\boldsymbol{A} = (0, xH, 0)$. The Schrödinger equation of an electron in this field (disregarding the spin) will then read

$$\frac{1}{2m}\left(\boldsymbol{p}-\frac{e}{c}\boldsymbol{A}\right)^2\psi+V(\boldsymbol{r})\psi = E\psi.$$

$V(\boldsymbol{r})$ is the potential energy of the electron in the periodic lattice potential; in a very rough approximation the influence of the lattice can be taken into account when the mass of the free electron is replaced by a so-called effective mass m^*. This approximation will be used in the following as it does not affect the main behaviour of an electron in a magnetic field. So we have the simple Schrödinger equation

$$\frac{1}{2m^*}\left(\boldsymbol{p}-\frac{e}{c}\boldsymbol{A}\right)^2\psi = E\psi$$

which is to be solved; in an explicit form it reads

$$\left(\frac{p^2}{2m^*}-\frac{e}{m^*c}(\boldsymbol{A}\cdot\boldsymbol{p})+\frac{e^2}{2m^*c^2}A^2\right)\psi = E\psi.$$

Substituting in it $\boldsymbol{p} = \hbar\nabla/i$ (∇ = nabla or gradient operator) and $\boldsymbol{A} = (0, xH, 0)$ we obtain

$$\left(\frac{\partial^2}{\partial x^2}+\frac{\partial^2}{\partial z^2}\right)\psi+\left(\frac{\partial}{\partial y}-i\frac{eHx}{\hbar c}\right)^2\psi+\frac{2m^*E}{\hbar^2}\psi = 0. \quad (1.20)$$

Putting $\psi = e^{i(k_y y+k_z z)}\cdot u(x)$ we can transform it to

$$u''(x)+\left(\frac{2m^*E_1}{\hbar^2}-\left(k_y-\frac{eHx}{\hbar c}\right)^2\right)u(x) = 0, \quad (1.21)$$

where $E_1 = E - \hbar^2 k_z^2/2m^*$. Using the substitutions

$$x_0 = \frac{\hbar c}{eH} k_y, \qquad \omega_L^* = \frac{e_0 H}{2m^* c} \tag{1.22}$$

we finally obtain the following differential equation for $u(x)$:

$$-\frac{\hbar^2}{2m^*} u''(x) + \frac{m^*}{2} (2\omega_L^*)^2 (x - x_0)^2 u(x) = E_1 u(x). \tag{1.23}$$

This is the well-known Schrödinger equation of a harmonic oscillator which oscillates with frequency $2\omega_L^*$ about x_0. The eigenvalues of (1.22) are therefore (see, for example, Becker and Sauter, 1959):

$$E_{1,n} = 2\hbar\omega_L^* \left(n + \frac{1}{2}\right) = \frac{e_0 \hbar}{2m^* c} H(2n+1), \tag{1.24}$$

where n is an arbitrary, positive integer. The energy eigenvalues of the electron are

$$E(n, k_z) = \mu_B^* H(2n+1) + \frac{\hbar^2 k_z^2}{2m^*}, \tag{1.25}$$

where $\mu_B^* = e_0 \hbar / 2m^* c$ is the effective Bohr magneton. Thus we see that the electron motion in the z-direction is not influenced by the field, whereas the motion perpendicular to the direction of the field will be quantised. These discrete energy levels (1.24) are called Landau levels (Landau, 1930). According to Peierls (1933) their influence on the magnetisation can be studied by means of a two-dimensional model for the electron gas. The electrons are assumed to move for $T = 0$ in the x, y-plane, within a rectangular area bounded by the edges L_1 and L_2. The magnetic field is assumed perpendicular to it, in the z-direction; the possible energy states are then given exactly by (1.24). As shown in the following (see (1.28)), each of these levels comprises $L_1 L_2 e_0 H / 2\pi\hbar c$ states, disregarding the spin. It is now easy to see in which way the energy of this gas and thus the magnetisation vary with the field. When H is first chosen so strong that all electrons are in the lowest level (1.24), i.e. in the energy state $E_0 = \mu_B^* H$, the energy increase will be proportional to the field strength as the latter is raised; thus the magnetisation remains constant. When the field strength is reduced, a point will be reached at which precisely all the N electrons are in the energy state E_0. This occurs at the field strength H_0, where

$$N = 2\frac{L_1 L_2 e_0}{2\pi\hbar c} H_0 .$$

We have here taken into account that, because of the spin, each state is doubly occupied. Of course the total energy of the electron gas will now be equal to $N\mu_B^* H_0$. If the field strength is further reduced some electrons must go over to the next level, $E_1 = 3\mu_B^* H$, so that the total energy will first grow as the field strength is reduced. When the field strength further decreases a point is reached where also the second level is filled; this is the case at precisely $H_1 = H_0/2$. Both levels contain the same number of states. The total energy at the field strength H_1 is therefore given by

$$\frac{N}{2} E_0 + \frac{N}{2} E_1 = N\mu_B^* H_0 ,$$

i.e. it is equal to the energy of the electrons when the ground state was just filled. We can now assume the field further to decrease and the same process will be repeated. The energy becomes a "periodic" function of the field. The nth level will be completely filled at $H_n = H_0/(n+1)$. If the field strength lies within the limits $H_{n-1} \geqslant H \geqslant H_n$, we obtain for the total energy

$$E = N\mu_B^* H \left(2n+1-n(n+1)\frac{H}{H_0}\right),$$

so that we obtain the curve illustrated in Fig. 1. From the energy we immediately obtain the magnetisation; we have

$$M = -\frac{\partial E}{\partial H}$$

at $T = 0$, and we obtain the field-dependence of magnetisation as shown in Fig. 2. At the points H_n the magnetisation cannot be differen-

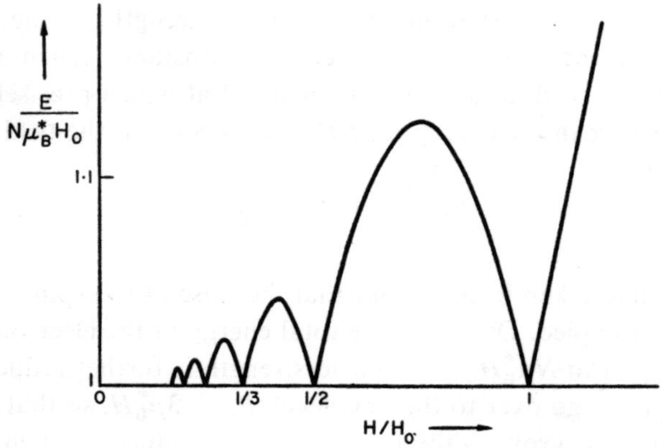

FIG. 1.
The energy of a two-dimensional free electron gas as a function of the magnetic field at $T = 0$.

tiated so that there the susceptibility χ cannot be defined. Apart from these discontinuities, χ increases as the field strength drops.

The three-dimensional treatment of this problem considered in the following will show that the essential features of the two-dimensional model remain unchanged. The fact that the susceptibility oscillates with the field strength is called the de Haas–van Alphen effect.

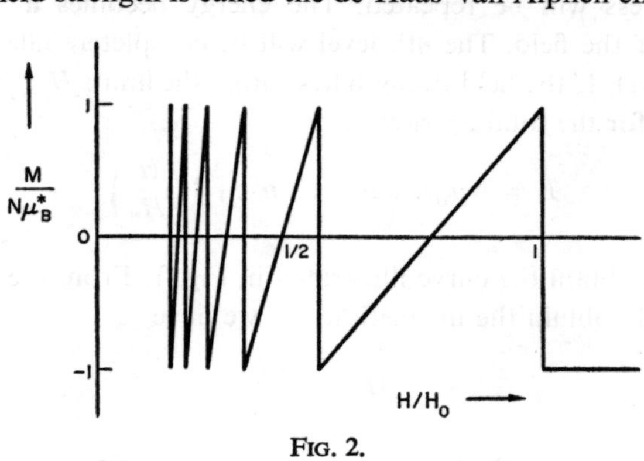

FIG. 2.
Relative magnetisation of a two-dimensional free electron gas as a function of the magnetic field at $T = 0$.

5. Diamagnetism of Electrons

Knowing the energy levels of an electron in a magnetic field, we can now calculate the Gibbs free energy of the electrons and thus their magnetisation, on the basis of the relation $M = -\partial F/\partial H$. The eigenvalues (1.25), however, are incomplete, since each electron has an intrinsic magnetic moment equal to the Bohr magneton μ_B so that, instead of (1.25), the eigenvalues are given by

$$E_{\pm}(n, k_z) = \frac{\hbar^2 k_z^2}{2m^*} + \mu_B^* H(2n+1) \pm \mu_B H. \tag{1.26}$$

The upper (lower) sign indicates that the electron spin is parallel (antiparallel) to the field. Equation (1.26) reveals the difference between the static and the dynamic properties of the electron. The static property is the spin which cannot be influenced by the lattice potential (at least not in the free-electron approximation). The magnetic moment connected with the spn is therefore equal to μ_B. The dynamic property is the quantised motion of the electron in the lattice potential so that the magnetic moment connected with it contains the effective mass m^*. Owing to the first term on the right-hand side of (1.26), the energy levels lie very close to one another so that the sum in (1.19) can be replaced by an integral. Separating the two spin components in (1.19) we obtain for F

$$F = N\zeta - kT \int_0^\infty \ln\left(1 + e^{-\beta(E-\zeta)}\right) dZ_+ - kT \int_0^\infty \ln\left(1 + e^{-\beta(E-\zeta)}\right) dZ_-.$$

$Z_{\pm}(E)$ is the number of states up to the energy E with parallel or antiparallel spin. Applying the equality $dZ_+ = (dZ_+/dE)dE$ the integrals can be transformed through integration by parts:

$$\int_0^\infty \ln\left(1 + e^{-\beta(E-\zeta)}\right) dZ = \int^\infty \ln\left(1 + e^{-\beta(E-\zeta)}\right) \frac{dZ}{dE}\, dE$$

$$= Z(E) \ln\left(1 + e^{-\beta(E-\zeta)}\right)\Big]^\infty + \beta \int^\infty Z(E) f(E)\, dE,$$

the lower limit of the integrals over the energy is given by $Z(E) = 0$, and $f(E)$ is the Fermi distribution. Thus we obtain

$$F = N\zeta - \int^{\infty} Z_{+}(E) f(E)\, dE - \int^{\infty} Z_{-}(E) f(E)\, dE. \tag{1.27}$$

In order to calculate the two integrals we must know the number Z of states as a function of the energy. In order to count them conveniently it is necessary to have appropriate boundary conditions for the solutions of the Schrödinger equation (1.20). The electron gas is assumed to be contained in a volume defined by the boundary conditions

$$-\frac{L_1}{2} \leqslant x \leqslant +\frac{L_1}{2},$$

in the x-direction,

$$-\frac{L_2}{2} \leqslant y \leqslant +\frac{L_2}{2}$$

in the y-direction, and

$$-\frac{L_3}{2} \leqslant z \leqslant +\frac{L_3}{2}$$

in the z-direction. We impose periodic boundary conditions for the eigensolutions of (1.10) with respect to the y-and z-axes, i.e.

$$\psi_{k_y, k_z, n}(x, y, z) = \psi_{k_y, k_z, n}(x, y+L_2, z+L_3).$$

No such periodicity can be required for the x-axis since the eigenfunctions of the harmonic oscillator have no periodic properties and are defined along the whole of the x-axis. This deviation from the physically correct boundary conditions, which is $\psi = 0$ on the surface of the parallelepepid considered, is inessential as long as the radii of the circles corresponding to the Landau levels are small compared to the linear dimensions of the volume containing the electrons. Owing to the periodic boundary conditions for $\psi_{k_y, k_z, n}(x, y, z)$, the parameters k_y and k_z are integral multiples of $2\pi/L_2$ and $2\pi/L_3$, respectively. As the eigenvalues (1.26) are independent of k_y, there are *a priori* as many states for each eigenvalue as we have k_y-values. This number is obtained from the range of k_y-values divided by $2\pi/L_2$. According to

(1.22) we have

$$(k_y)_{\substack{\max\\ \min}} = \pm\frac{e_0 H L_1}{2\hbar c},$$

and

$$\frac{(k_y)_{\max}-(k_y)_{\min}}{2\pi/L_2} = \frac{L_1 L_2 e_0 H}{2\pi\hbar c} \tag{1.28}$$

is thus the degeneracy of each level (1.26). In order to obtain the number $Z(E)$ we have to multiply (1.28) by the number of possible values of k_z, the energies (1.26) (upper sign) assuming all values up to E. With fixed E and n we have according to (1.26)

$$(k_z)_{\substack{\max\\ \min}} = \pm\frac{\sqrt{2m^*}}{\hbar}\sqrt{E-(2n+1)\mu_B^* H-\mu_B H},$$

so that for fixed n the number of different k_z-values is

$$\frac{(k_z)_{\max}-(k_z)_{\min}}{2\pi/L_3} = \frac{\sqrt{2m^*}L_3}{\pi\hbar}\sqrt{E-(2n+1)\mu_B^* H-\mu_B H},$$

and finally

$$Z_\pm(E) = 2\frac{\sqrt{2m^*}L_1L_2L_3}{(2\pi\hbar)^2 c}e_0 H\sum_{n=0}^{n_{\max}}\sqrt{E-(2n+1)\mu_B^* H\mp\mu_B H} \tag{1.29}$$

where n can assume all values consistent with the root being real. Using the substitutions

$$V = L_1L_2L_3,\quad \varepsilon = \frac{E}{2\mu_B^* H},\quad \varepsilon_0 = \frac{\zeta}{2\mu_B^* H},\quad \Theta = \frac{kT}{2\mu_B^* H}$$

we can write for the free energy F

$$F = N\zeta - \frac{4(m^*)^{3/2}V(\mu_B^* H)^{5/2}}{\pi^2\hbar^3}\left(\int^{\infty} f(\varepsilon)\sum_{n=0}^{n_{\max}}\left(\varepsilon-n-\frac{1}{2}-\frac{1}{2}\frac{m^*}{m}\right)^{1/2}d\varepsilon\right.$$

$$\left.+\int^{\infty} f(\varepsilon)\sum_{n=0}^{n_{\max}}\left(\varepsilon-n-\frac{1}{2}+\frac{1}{2}\frac{m^*}{m}\right)^{1/2}d\varepsilon\right),$$

where the lower limits are again to be chosen according to the rule mentioned above, and

$$f(\varepsilon) = \frac{1}{e^{(\varepsilon-\varepsilon_0)/\Theta}+1}.$$

Since the derivative of $f(\varepsilon)$, denoted by $f'(\varepsilon)$, has the nature of a δ-function (see above), it is convenient to transform the integrals in F once more through integration by parts. We obtain

$$\begin{aligned} F &= N\zeta + A \int_{-\infty}^{+\infty} d\varepsilon f'(\varepsilon)\left\{\sum_{n=0}^{n_{\max}} \left(\varepsilon - n - \frac{1}{2}\left(1+\frac{m^*}{m}\right)\right)^{3/2} \right. \\ &\quad \left. + \sum_{n=0}^{n_{\max}} \left(\varepsilon - n - \frac{1}{2}\left(1-\frac{m^*}{m}\right)\right)^{3/2}\right\} \qquad (1.30) \\ &= N\zeta + A \int_{-\infty}^{+\infty} d\varepsilon \left\{f'\left(\varepsilon+\frac{m^*}{2m}\right) + f'\left(\varepsilon - \frac{m^*}{2m}\right)\right\} \sum_{n=0}^{n_{\max}} \left(\varepsilon - n - \frac{1}{2}\right)^{3/2}. \end{aligned}$$

As $f'(\varepsilon)$ is essentially zero except near $\varepsilon = \varepsilon_0$, the lower limit of integration has here been shifted to $-\infty$. The constant A is defined by

$$A = \frac{8(m^*)^{3/2}\cdot V\cdot(\mu_B^* H)^{5/2}}{3\pi^2\hbar^3}.$$

It is expedient to replace the sum over n by an integral

$$\begin{aligned} \sum_{n=0}^{n_{\max}} (\varepsilon - n - \tfrac{1}{2})^{3/2} &= \int_0^{\infty} (\varepsilon - x)^{3/2}\, dx \{\delta(x-\tfrac{1}{2}) + \delta(x - \tfrac{3}{2}) \\ &\quad + \ldots + \delta(x - x_{\max})\}, \qquad (1.31) \end{aligned}$$

i.e. we integrate the function $(\varepsilon - x)^{3/2}$, which is now continuous, and the δ-functions select the correct arguments over which we sum. The sum of the δ-functions at all half-odd-integral points of the entire x-axis can be easily transformed to a Fourier series:

$$\sum_{v=-\infty}^{+\infty} \delta\left(x - \frac{v}{2}\right) = \sum_{l=-\infty}^{+\infty} e^{2\pi i l(x-1/2)}.$$

Since in this sum only the δ-functions at the half-odd-integral points in the interval $(0, \varepsilon)$ are needed, the limits of integration in (1.31) must be chosen appropriately.

We have

$$\sum_{n=0}^{n_{\max}} (\varepsilon-n-\tfrac{1}{2})^{3/2} = \sum_{l=-\infty}^{+\infty} (-)^l \int_0^{\varepsilon} dx(\varepsilon-x)^{3/2}\, e^{2\pi ilx}.$$

In the case of $l = 0$ the integral is trivial:

$$\int_0^{\varepsilon} (\varepsilon-x)^{3/2}\, dx = \tfrac{2}{5}\, \varepsilon^{5/2}.$$

In the case of $l \neq 0$ we integrate twice by parts and obtain easily

$$\int_0^{\varepsilon} (\varepsilon-x)^{3/2}\, e^{2\pi ilx}\, dx = -\frac{\varepsilon^{3/2}}{2\pi il}+\frac{3\varepsilon^{1/2}}{8\pi^2 l^2}-\frac{3e^{2\pi il\varepsilon}}{8\pi^2 l^2}\int_0^{\sqrt{\varepsilon}} e^{-2\pi ilu^2}\, du.$$

In the subsequent summation over l the first term of this integral vanishes and, according to (1.30), we obtain

$$F = N\zeta+A\int_{-\infty}^{+\infty} d\varepsilon \left\{f'\left(\varepsilon+\frac{m^*}{2m}\right)+f'\left(\varepsilon-\frac{m^*}{2m}\right)\right\}$$
$$\times\left\{\frac{2}{5}\,\varepsilon^{5/2}-\frac{1}{16}\,\varepsilon^{1/2}-\frac{3}{4\pi^2}\sum_{l=1}^{\infty}\frac{(-)^l}{l^2}\,\mathrm{Re}\, e^{2\pi il\varepsilon}\int_0^{\sqrt{\varepsilon}} e^{-2\pi ilu^2}\, du\right\}, \quad (1.32)$$

where we made use of the fact that

$$\sum_{l=-\infty}^{+\infty}\frac{(-)^l}{l^2} = -\frac{\pi^2}{6}$$

(see, for example, Magnus and Oberhettinger, 1948); Re denotes the real part. To begin with, consider the first two integrals in (1.32):

$$\int_{-\infty}^{+\infty}\left\{f'\left(\varepsilon+\frac{m^*}{2m}\right)+f'\left(\varepsilon-\frac{m^*}{2m}\right)\right\}\left\{\frac{2}{5}\,\varepsilon^{5/2}-\frac{1}{16}\,\varepsilon^{1/2}\right\} d\varepsilon =$$
$$\int_{-\infty}^{+\infty} d\varepsilon f'(\varepsilon)\left\{\frac{2}{5}\left(\varepsilon-\frac{m^*}{2m}\right)^{5/2}+\frac{2}{5}\left(\varepsilon+\frac{m^*}{2m}\right)^{5/2}-\frac{1}{16}\left(\varepsilon-\frac{m^*}{2m}\right)^{1/2}\right.$$
$$\left.-\frac{1}{16}\left(\varepsilon+\frac{m^*}{2m}\right)^{1/2}\right\}.$$

Since $f'(\varepsilon)$ is a kind of δ-function around $\varepsilon = \varepsilon_0$, in the integrand ε can always be assumed large with respect to 1 so that we can expand in a series. Continuing this expansion until the term $\varepsilon^{1/2}$ we obtain

$$\int_{-\infty}^{+\infty} d\varepsilon f'(\varepsilon)\{\ldots\} = -\frac{4}{5}\,\varepsilon_0^{5/2}+\frac{1}{8}\,\varepsilon_0^{1/2}\left(1-\frac{3m^{*2}}{m^2}\right), \qquad (1.33)$$

since, because of the δ-function character of $f'(\varepsilon)$, the slowly varying powers are taken at the point $\varepsilon = \varepsilon_0$ and can be put in front of the integral sign.

The integral over u in (1.32) is an error integral. Since $\varepsilon \gg 1$, it can be expanded in a rapidly converging series. We have (Magnus and Oberhettinger, 1948):

$$\int_0^{\sqrt{\varepsilon}} e^{-2\pi i l u^2}\,du = \frac{e^{-i\pi/4}}{2\sqrt{2l}}-\frac{e^{-2\pi i l \varepsilon}}{4\pi i l\sqrt{\varepsilon}}\left\{1-\frac{1}{4\pi i l\varepsilon}+\frac{3}{(4\pi i l\varepsilon)^2}\cdots\right\}.$$

Only the first term on the right-hand side is essential here. A comparison with (1.33) shows at once that the other terms are of the same kind as those omitted in (1.33) as terms of higher order. So they must be omitted here also. Thus we have the following expression for the free energy:

$$F = N\zeta - A\left\{\frac{4}{5}\,\varepsilon_0^{5/2}-\frac{1}{8}\,\varepsilon_0^{1/2}\left(1-\frac{3(m^*)^2}{m^2}\right)\right.$$

$$\left.+\frac{3}{4\sqrt{2}\pi^2}\sum_{l=1}^{\infty}\frac{(-)^l}{l^{5/2}}\cos\left(\pi l\frac{m^*}{m}\right)\operatorname{Re}\int_{-\infty}^{+\infty} d\varepsilon f'(\varepsilon)\, e^{2\pi i l\varepsilon - i\pi/4}\right\}. \qquad (1.34)$$

The last integral cannot be evaluated by considering $f'(\varepsilon)$ as a δ-function, since in the neighbourhood of $\varepsilon = \varepsilon_0$ the exponential function $\exp(2\pi i l\varepsilon)$ itself is a rapidly varying function. It is possible, however, to calculate this integral exactly. After a few simple transformations

we have

$$\int_{-\infty}^{+\infty} d\varepsilon\, e^{2\pi i l\varepsilon - i\pi/4} f'(\varepsilon) = -\frac{1}{4}\, e^{2\pi i l\varepsilon_0 - i\pi/4} \int_{-\infty}^{+\infty} \frac{e^{2\pi i l\Theta\eta}}{\cosh^2 \frac{\eta}{2}}\, d\eta.$$

The integral can be calculated by means of the method of residues; since l is positive, the path of integration can be closed in the upper half-plane. The integrand has poles at $\eta_n = (2n+1)\pi i$, $n = 0, 1, \ldots,$ and the residue at η_n is equal to $-8\pi i l\Theta \exp(-\pi^2 l(2n+1)\Theta)$, as can be verified easily. We therefore have

$$\int_{-\infty}^{+\infty} e^{2\pi i l\varepsilon - i\pi/4} f'(\varepsilon)\, d\varepsilon = -4\pi^2 l\Theta\, e^{2\pi i l\varepsilon_0 - i\pi/4 - 2\pi^2 l\Theta} \cdot \sum_{n=0}^{\infty} e^{-4\pi^2 l\Theta n}$$

$$= -\frac{2\pi^2 l\Theta e^{2\pi i l\varepsilon_0 - i\pi/4}}{\sinh 2\pi^2 l\Theta}.$$

Substituting this in (1.34) and using the abbreviations given, we finally arrive at

$$F = N\zeta - N\zeta_0^{-3/2}\left\{\frac{2}{5}\zeta^{5/2} + \frac{1}{4}(\mu_B^* H)^2 \zeta^{1/2}\left(\frac{3(m^*)^2}{m^2} - 1\right)\right.$$

$$\left. -\frac{3}{2}(\mu_B^* H)^{3/2} kT \sum_{l=1}^{\infty} \frac{(-)^l}{l^{3/2}} \cos\left(\pi l \frac{m^*}{m}\right) \cdot \frac{\cos\left(\frac{\pi}{4} - \frac{\pi l\zeta}{\mu_B^* H}\right)}{\sinh \frac{\pi^2 l kT}{\mu_B^* H}}\right\}. \quad (1.35)$$

In the second term in the braces ζ is replaced by ζ_0, as this term, compared with the first one, is a correction. The Fermi energy, which was undetermined so far, is now obtained from the equation $\partial F/\partial\zeta = 0$ which, as can be seen from the general formula (1.19) for the free energy, is equivalent to the condition $\sum_i f_i = N$. Thus we obtain from (1.35):

$$\zeta = \zeta_0 \left\{ 1 - \frac{\pi kT}{\zeta_0} \left(\frac{\mu_B^* H}{\zeta_0} \right)^{1/2} \sum_{l=1}^{\infty} \frac{(-)^l}{l^{1/2}} \cos\left(\pi l \frac{m^*}{m} \right) \frac{\sin\left(\frac{\pi}{4} - \frac{\pi l \zeta_0}{\mu_B^* H} \right)}{\sinh \frac{\pi^2 l kT}{\mu_B^* H}} \right\},$$

where the second term in the braces obviously is a small correction to ζ_0. We therefore replaced in it ζ by ζ_0. (The corrections to the Fermi energy at finite temperatures and field strengths mentioned in the discussion of the Fermi gas cannot arise here, since in the integral (1.33) responsible for it the derivative of the Fermi distribution was considered to be a δ-function and we did not, as would have been more correct, account for the finite width of the region where the Fermi function decreases at $E = \zeta$.)

The magnetisation can now be obtained from the thermodynamic relation $M = -\partial F/\partial H$. This differentiation is independent of the dependence of Fermi energy on the magnetic field, since the corresponding terms cancel because of the condition $\partial F/\partial \zeta = 0$. In the differentiation of the sum in (1.35) the derivative of the cosine contributes most to the magnetisation; all other derivatives are smaller by a factor kT/ζ_0 or $\mu_B^* H/\zeta_0$. Omitting them, we obtain from (1.35)

$$M = \frac{N}{2} \frac{\mu_B^{*2} H}{\zeta_0} \left\{ \frac{3m^{*2}}{m^2} - 1 \right.$$

$$\left. + 3\pi \frac{kT}{\mu_B^* H} \left(\frac{\zeta_0}{\mu_B^* H} \right)^{1/2} \sum_{l=1}^{\infty} \frac{(-)^l}{l^{1/2}} \cos\left(\pi l \frac{m^*}{m} \right) \frac{\sin\left(\frac{\pi}{4} - \frac{\pi l \zeta_0}{\mu_B^* H} \right)}{\sinh \frac{\pi^2 l kT}{\mu_B^* H}} \right\}. \quad (1.36)$$

Thus the susceptibility $\chi = M/H$ is constituted of a constant component χ_0, which is independent of temperature and magnetic field, and a component χ_{osc}, which depends on T and H. We have

$$\chi_0 = \frac{3}{2} N \frac{\mu_B^2}{\zeta_0}\left(1 - \frac{m^2}{3m^{*2}}\right) = \chi_p + \chi_L. \tag{1.37}$$

The first term, χ_p, is the Pauli spin paramagnetism. The fact that it derives from the spin is proven by the fact that it contains the Bohr magneton and not the effective magneton. The properties of χ_p will be discussed in detail in the next chapter. The second term in (1.37), χ_L, is the Landau–Peierls diamagnetism. For free electrons ($m^* = m$) its absolute magnitude is one-third of that of the spin paramagnetism. It describes the orientation of the quantised orbital angular momenta in the external magnetic field. In ordinary metals the effective mass is sometimes considerably different from the electron mass so that, as, for example, in bismuth, the diamagnetic term in (1.37) exceeds the paramagnetic term. Since the concept of free electrons, to which (1.36) and (1.37) apply, is suitable to a rather good approximation for the alkali metals, we shall discuss χ_0 for these metals in the following. In Table 3 we have compiled the parameters for the alkali metals. The non-dimensional quantity r_s determines the electron density according to the relation $V = N\cdot 4\pi/3(a_B r_s)^3$. Thus r_s is the radius of the alkali atom in terms of the Bohr radius a_B. Using this parameter we can write the Fermi energy for free electrons

$$\zeta_0 = \frac{\hbar^2}{2m}\left(\frac{3N}{8\pi V}\right)^{2/3} = \frac{e_0^2}{2a_B}\cdot\frac{1}{(\alpha r_s)^2} \quad \text{where} \quad \alpha = \left(\frac{4}{9\pi}\right)^{1/3} \tag{1.38}$$

and $e_0^2/2a_B$ is the Rydberg constant equal to 13·605 eV.

We see from Table 3 that Li and Cs deviate most strongly from the values given for free electrons. The values of the electron effective masses given by Ham (1962) tend to reduce both the Fermi energy and the orbital angular momentum compared to the values for free electrons. Therefore, according to (1.37), the value of χ_L becomes smaller.

The introduction of an effective mass is disputable; strictly speaking, it is only justified if the electron energies in the crystal (zero mag-

TABLE 3

	$\alpha \times 10^8$ (cm) ($T = 0\,^\circ$K)	$r_0 \times 10^8$ (cm)	r_s	ζ_0 (eV)	$T_0 \times 10^{-4}$ (°K)	$-\chi_L^0 \times 10^6$	$\frac{m^*}{m}$	$-\chi_L \times 10^6$
Li	3·491	1·718	3·247	4·752	5·515	3·409	1·66	2·05
Na	4·225	2·080	3·931	3·242	3·763	4·985	1·00	4·99
K	5·225	2·573	4·862	2·120	2·460	7·624	1·09	6·99
Rb	5·585	2·750	5·197	1·856	2·153	8·710	1·21	7·20
Cs	6·045	2·977	5·625	1·584	1·838	10·205	1·76	5·80

In the first column of the table the lattice constants of the alkali metals are given according to Ham (1962). The second column contains the radii of the atoms. Since the alkali metals are body-centred cubic crystals, they contain two atoms per elementary cell and therefore $2(4\pi/3)r_0^3 = a^3$. Again r_s is the atomic radius in atomic units, ζ_0 is the free-electron Fermi energy calculated from (1.38), and T_0 is the corresponding degeneracy temperature, $\zeta_0 = kT_0$. The symbol χ_L^0 denotes the diamagnetic atomic susceptibility calculated according to (1.37) with $m^* = m$; m^*/m are the effective masses calculated by Ham (1962) and χ_L is the diamagnetic susceptibility of the electrons corrected according to (1.37).

netic field) can be represented by $\varepsilon_k = \hbar^2 k^2/2m^*$, i.e. if the surface of constant energy in momentum space (**k**-space), the Fermi surface, remains spherical. This, however, does not even apply to the alkali metals and still less to the other metals (see, for example, Kittel, 1962; Wilson, 1958; Seitz, 1940; Ziman, 1962; Ham, 1962). In the simplest case we have a quadratic dependence of the electron energies on the wave vector, that is,

$$\varepsilon_k = \frac{\hbar^2}{2} \sum_{i,j=1}^{3} \frac{k_i k_j}{m_{ij}}$$

which defines the so-called mass tensor. The constant-energy surfaces are then ellipsoids or hyperboloids whose principal axes are determined by the type of crystal structure. Therefore the experiments, as far as they depend on the orientation with respect to the crystal axes, aim at determining the mass tensor. However, the quadratic dependence of the electron energies is also generally but an approximation which, as we know for some semiconductors (Ziman, 1962), does not hold true. When we speak in the following, in spite of this fact, of an isotropic effective mass, we always consider quantities somehow averaged over the Fermi surface. It is thus possible that different experiments do yield different effective masses if the averaging over the Fermi surface has been different. In spite of this, it is not entirely unreasonable to stick to this concept, since in measurements of scalar quantities, such as the specific heat of electrons, or in experiments with polycrystalline materials, one is forced to determine an isotropic effective mass.

Another problem connected with the diamagnetic susceptibility is Sommerfeld's treatment of metal electrons. It always uses plane waves for the electrons and not, as should be the case, Bloch functions for the electrons in the crystal (Wilson, 1958): $\psi_k = e^{i(k \cdot r)} u_k(\boldsymbol{r})$ where $u_k(\boldsymbol{r})$ is a lattice-periodic function. If we assume, for example, that the conduction electrons are bound relatively strongly to the lattic atoms, the $u_k(\boldsymbol{r})$ will behave approximately like an atomic eigenfunction in the immediate proximity of an atom (tight-binding approximation; see Wilson, 1958). It is easy to verify that this corresponds to a large effective mass (see also Mott and Jones, 1936; Sachs, 1963), as is obvious, since stronger binding corresponds to lower electron mobility. In the limiting case of $m^* \to \infty$, however, the Landau diamagnetism vanishes according to (1.37). An exact formula, however, should be reduced in this limiting case to the general formula (1.7) for the diamagnetism of atoms. The Landau diamagnetism can therefore be considered only a first term of a series expansion for the exact formula (Wilson, 1958; Kjeldaas and Kohn, 1957). According to Kjeldaas and Kohn (1957), a calculation with the help of Bloch functions shows that the susceptibility of lithium is strongly changed; the authors obtained for Li: $\chi_L' = 0{\cdot}37\chi_L$. For sodium, however, they obtained but a small deviation from (1.37) as was to be expected, since the electrons in sodium, and also in potassium, may be considered to behave like free electrons in a relatively good approximation ($m^* \cong m$).

Another deviation from Sommerfeld's electron model is obtained when we consider the electron–electron Coulomb interaction. Hitherto this problem is only dealt with (Stephen, 1962) for electron gases of high densities ($r_s < 1$) so that the methods applied cannot be used in the case of metals.

The oscillating component of susceptibility

$$\chi_{osc} = \frac{3\pi N\mu_B^{*2}}{2\zeta_0}\cdot\frac{kT}{\mu_B^* H}\cdot\left(\frac{\zeta_0}{\mu_B^* H}\right)^{1/2} \times \sum_{l=1}^{\infty}\frac{(-)^l}{l^{1/2}}\cos\left(\pi l\frac{m^*}{m}\right)\frac{\sin\left(\frac{\pi}{4}-\frac{\pi l\zeta_0}{\mu_B^* H}\right)}{\sinh \pi^2 l\frac{kT}{\mu_B^* H}} \tag{1.39}$$

describes the de Haas–van Alphen effect (1930), i.e. the oscillations of the magnetic susceptibility with the field strength. From (1.39) we see at once that the spin contribution to the susceptibility is not simply an additive contribution to the total susceptibility, since the factor $\cos(\pi l m^*/m)$ in the sum can be traced back to the fact that we took the spin in the Landau levels (1.26) into account. It is therefore generally impossible to calculate the orbital susceptibility and the spin susceptibility separately. The series in (1.39) converges rapidly (because of the sinh in the denominator), even at low temperatures. As a rule it is possible in most cases to break off after the first term ($l = 1$) of the series and neglect the other terms. For practically achievable field strengths the argument of the sinh is large; when we put, for example, $\mu_B^* = \mu_B$ and $T = 1°K$, we obtain from $\pi^2 kT = \mu_B^* H$ a field strength of $H = 1{\cdot}5\times10^5$ oersteds. Such field strengths have only recently been reached in studies of the de Haas–van Alphen effect (Shoenberg, 1957, 1960). Hence it follows that the effect can only be measured at low temperatures and high field strengths. These conditions are less stringent in the case when the effective Bohr magneton is large with respect to μ_B, i.e. when $m^* \ll m$. This will happen only if parts of the Fermi surface deviate strongly from the spherical shape, so that large curvatures and thus small effective masses can appear in the $\boldsymbol{k}$ space.

The de Haas–van Alphen effect is characterised by the periodic behaviour of susceptibility when the magnetic field is varied. We see from (1.39) that the oscillation period is given by

$$P \equiv \Delta\frac{1}{H} = \frac{2\mu_B^*}{\zeta_0}. \tag{1.40}$$

The period is therefore independent of temperature, a fact always verified by experiment (Shoenberg, 1957). In our approximation of an isotropic effective mass m^*, P will also be independent of m^*, since μ_B and ζ_0 contain m^* in the same way. Using the data given in Table 3 we obtain, for example, for rubidium $P = 6{\cdot}24 \times 10^{-9}$ oe^{-1} which deviates from the values measured by at most 1% (Ham, 1962). The determination of the oscillations is essentially easier if the periods are long. According to (1.40) this requires a low Fermi energy or,

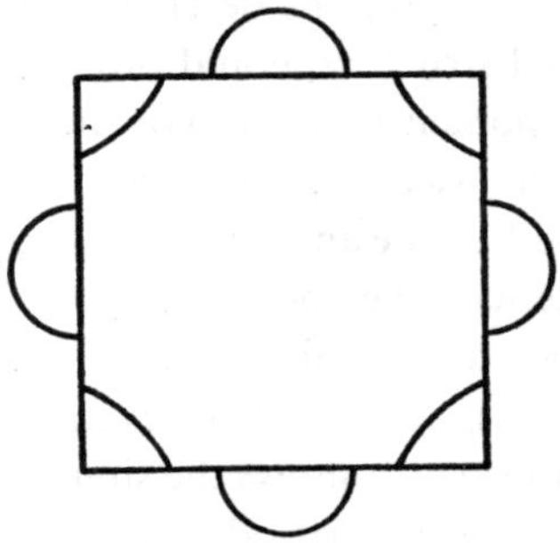

FIG. 3.
Schematic representation of the transition of electrons to the next higher Brillouin zone. In the higher zone the Fermi surface is strongly curved so that it contains (few) electrons with small effective masses.

more exactly, a lower electron density per atom. This is the case only for multivalent metals where most of the valence electrons have filled almost completely the first or several lower Brillouin zones, and only a small fraction of electrons comes into the higher Brillouin zone, as shown schematically in Fig. 3. These electrons usually have a very small effective mass since in this region the Fermi surface is strongly curved. It is thus understandable that the de Haas–van Alphen effect was first observed for one of these multi-valent metals, namely bismuth. Bismuth has small effective masses ($\cong 10^{-2}\, m$) and a low effective electron density, about 10^{-5} electrons per atom (Shoenberg, 1957). Accordingly periods P between 2 and 8×10^{-5} oe^{-1} were found which exceed the periods to be expected for alkali metals by a factor of 10^4.

From this we understand that for low field strengths (10^4 oe) the de Haas–van Alphen effect involves only small parts of the Fermi surface, those parts which contain electrons with small effective masses. The shape of the Fermi surface, however, depends on the structure of the crystal; therefore the de Haas–van Alphen effect is generally highly anisotropic, i.e. it depends as to amplitude and period on the orientation of the magnetic field with respect to the crystallographic axes of the single crystal investigated. Thus, for example, Sydoriak and Robinson (1949), studying the de Haas–van Alphen effect of zinc, which has a hexagonal structure, could observe this effect if the magnetic field had a component parallel to the hexagonal axis, and found no effect when the field was perpendicular to this axis. (Fig. 4). Since this behaviour is based on the anisotropy of the Fermi surface, Blackman and Landau extended the above theory to ellipsoidal Fermi surfaces; this enabled them to describe correctly the direction-dependence of the de Haas–van Alphen effect of several metals (see Shoenberg, 1957). Besides this, Fig. 4 also shows the simplest de Haas–van Alphen effect: there exists a constant period P and the amplitude of oscillations varies towards smaller field strengths, as $\exp(-\pi^2 kT/\mu_B^* H)$; this agrees with eqn. (1.39) where the parameters m^* and P contained in it must now be considered quantities which depend on the orientation of the field with respect to the crystal axes. Complex patterns may appear, however, if the Fermi surface is not simply connected. If it consists of, for example, three separated ellipsoids, the contributions to the susceptibility of these three ellipsoids are superimposed; since in general the orientations of the ellipsoids with respect to the field are different, this superposition takes place with three different periods, P_1, P_2, P_3, which may result in the formation of very complex patterns which are difficult to analyse (Shoenberg, 1957). The concept of ellipsoidal Fermi surfaces can only be applied to a few metals and semiconductors. We therefore need a more general theory of quantised electron orbits and, in particular, of the de Haas–van Alphen period P. Since the problem of electrons in a crystal and a magnetic field cannot be solved exactly, we cannot expect a more general theory to yield exact results. Onsager (1952) developed a method for the determination of the de Haas–van Alphen period which is applicable to arbitrary, simply connected Fermi surfaces. Later on, Onsager's result was verified by Lifshitz and Kosevich (1956) by means of a calculation similar to that given above. Consider a wave packet (momentum $\hbar \boldsymbol{k}$) built up of Bloch functions. For such wave packets the classical equations of motion hold approximately; in particular,

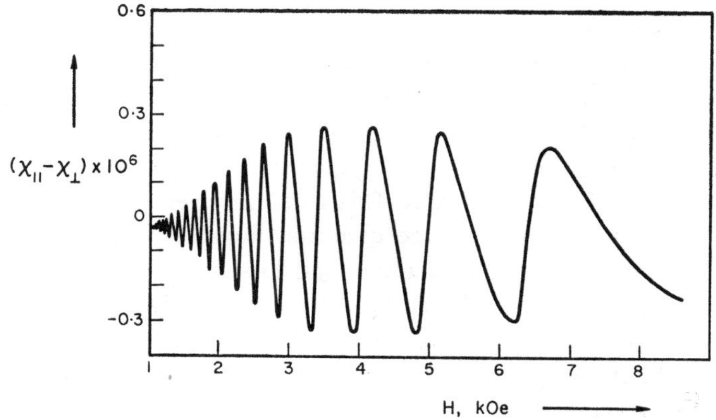

FIG. 4.

The difference of the principal susceptibility per gram for zinc at 4·5°K as a function of the magnetic field, after Sydoriak and Robinson (1949). The susceptibility perpendicular to the hexagonal axis, $\chi_{\perp}$, is independent of the magnetic field.

the variation with time of the wave-packet momentum in a magnetic field is given by (Ziman, 1962):

$$\hbar \dot{\boldsymbol{k}} = \frac{e}{c}[\dot{\boldsymbol{r}} \wedge \boldsymbol{H}]. \tag{1.41}$$

On the right-hand side is the Lorentz force, and $\dot{\boldsymbol{r}} = \boldsymbol{v}(\boldsymbol{k})$ is the velocity of the wave packet. The magnetic field is assumed to be constant and parallel to the z-axis. In this case eqn. (1.41) indicates the presence of a close relationship between the motion of the electron (wave packet) in coordinate space and in momentum space. To within a factor $eH/\hbar c$, the time derivative of $\boldsymbol{k}$ corresponds to the time derivative of the coordinate vector and they are perpendicular to one another. Equation (1.41) can be integrated with respect to time and, after a suitable choice of the origins in coordinate space and momentum space, i.e. of the constants of integration, we obtain

$$\hbar \boldsymbol{k} = \frac{e}{c}[\boldsymbol{r} \wedge \boldsymbol{H}]. \tag{1.42}$$

The spatial motion of the electron takes place in a plane perpendicular to the direction of the field, along a closed trajectory, and, according to (1.42), in $\boldsymbol{k}$ space in a plane $k_z = 0$, or $k_z = \text{const}$, according to (1.41). When the electron travels along closed curves in these planes, the curves can be quantised according to the Bohr–Sommerfeld quantisation rule. This rule, however, yields the sought quantum states only for large quantum numbers, that is, near the classical limiting case (see Landau and Lifshitz, 1958). In our case, however, this condition obviously seems to be satisfied as, for example, in the above-mentioned case of rubidium, at a field strength of 10^5 oe, about the lowest 1000 Landau levels are involved ($2\mu_B^* Hn \leqslant \zeta$). In its most general form for periodic orbits the Bohr–Sommerfeld rule reads

$$\oint p\, dq = (n+\varphi)h,$$

where p is the momentum conjugate to the position coordinate; integration is carried out over a closed orbit, φ is an indefinite constant ($0 \leqslant \varphi \leqslant 1$) which is equal to $\frac{1}{2}$ in the case of a linear harmonic oscillator. The momentum conjugate to the local variable $\boldsymbol{r}$ is given by $\hbar\boldsymbol{k}+e\boldsymbol{A}/c$ and the quantisation rule will read

$$\oint \left(\left[\hbar\boldsymbol{k}+\frac{e}{c}\boldsymbol{A}\right]\cdot d\boldsymbol{r}\right) = (n+\varphi)h. \tag{1.43}$$

Here $d\boldsymbol{r}$ is the line element of the electron trajectory in coordinate space. The first integral can be transformed by means of the integrated equation of motion (1.42):

$$\oint \hbar(\boldsymbol{k}\cdot d\boldsymbol{r}) = \frac{e}{c}\oint ([\boldsymbol{r}\wedge\boldsymbol{H}]\cdot d\boldsymbol{r}) = -\frac{e}{c}\oint (\boldsymbol{H}\cdot[\boldsymbol{r}\wedge d\boldsymbol{r}]) = -\frac{e}{c}\boldsymbol{H}\cdot 2\boldsymbol{S},$$

where S is the projection of the surface bounded by the electron's orbit onto the x, y-plane. The second contour integral is obtained by means of the Stokes integral theorem

$$\oint (\boldsymbol{A} \cdot \boldsymbol{dr}) = \int (\text{curl}\, \boldsymbol{A} \cdot \boldsymbol{dS}) = HS = \boldsymbol{\Phi},$$

and represents the magnetic flux through the surface $\boldsymbol{S}$. Thus we have the quantisation rule

$$\Phi(n, k_z) = HS(n, k_z) = (n+\varphi)\frac{hc}{e_0},$$

or, since the spatial dimensions differ from the dimensions in the momentum space only by the factor $eH/\hbar c$,

$$A(n, k_z) = 2\pi(n+\varphi)\frac{\hbar c}{e_0 H}\cdot\left(\frac{e_0 H}{\hbar c}\right)^2 = 2\pi(n+\varphi)\frac{e_0 H}{\hbar c}, \quad (1.44)$$

where $A(n, k_z)$ is the area circumscribed by the orbit in k-space, in agreement with the quantisation rule. Equation (1.44) determines implicitly the levels $E(n, k_z)$ allowed for the electron. Two neighbouring levels which differ in n by 1 have an energy difference ΔE, which for large n can be calculated according to (1.44)

$$\Delta E\frac{dA}{dE} = A(n, k_z) - A(n-1, k_z) = \frac{2\pi e_0 H}{\hbar c}$$

or

$$\Delta E \equiv \hbar\omega_c = \frac{e_0 \hbar H}{m_c c} \quad (1.45)$$

where $m_c = (\hbar^2/2\pi)\, dA/dE$.

The effective mass defined in this way is called the cyclotron mass; it is generally different from the effective masses introduced above. For spherical-symmetric Fermi surfaces ($A = \pi k^2$) and free electrons with mass m^* ($E(k) = \hbar^2 k^2/2m^*$), $m_c = m^*$, and we obtain from (1.45) the energy difference of two neighbouring Landau levels known from (1.25)

$$\Delta E = 2\mu_B^* H.$$

The de Haas–van Alphen period to be determined is obtained by means of the following consideration: the periodic oscillations of the susceptibility are due to a successive draining of the quantised levels with increasing field strength, as we have seen from the two-dimensional model discussed above. In the three-dimensional case, an energy state $E(n, k_z)$ is drained at a certain n precisely at the moment when the field strength has reached a value ($H = H_n$) at which the Fermi surface is still just touched, i.e. there still exist k_z values so that the con-

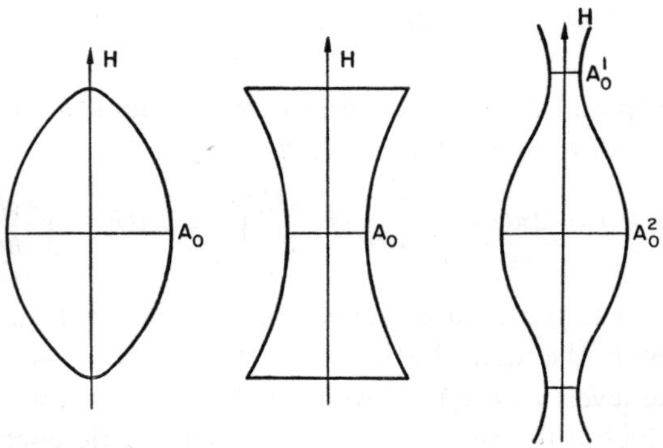

FIG. 5.

Extremal cross-sections A_0 through the Fermi surface which are parallel to the external field.

dition $E(n, k_z) = \zeta$ is satisfied. If the field strength is raised above this value, the level considered will be empty. It has its last contact with the Fermi surface (which is assumed to be convex) when the maximum cross-section within the Fermi surface, perpendicular to the direction of field, is reached; see Fig. 5. Let us denote this cross-section by A_0; it is determined by the geometry of the Fermi surface and the direction of the field. The next level is completely drained at the higher field strength H_{n-1}; the difference determines the period P. We therefore obtain according to (1.44) for the de Haas–van Alphen period

$$P \equiv \Delta \frac{1}{H} = \frac{2\pi e_0}{\hbar c A_0}. \tag{1.46}$$

In the case of a concave Fermi surface (see Fig. 5), the period P is obtained when we ask for the field strength at which a certain level just begins to be drained. Obviously we thus obtain the same formula for the period, namely (1.46), with the only difference that A_0 is now the minimum cross-section possible within the Fermi surface for a given field direction. Thus A_0 in (1.46) is understood to be in arbitrary Fermi surfaces the extremal cross-section, which is perpendicular to the direction of field. This is the most important result of Onsager's considerations; in principle it permits the determination of the structure of the Fermi surface of many metals with the help of the de Haas–van Alphen effect. In the case of a spherically symmetric electron distribution in momentum space we have

$$A_0 = \pi k_0^2 = \frac{2\pi m^*}{\hbar^2}\zeta,$$

so that we arrive again at eqn. (1.40) for P.

Hitherto we have not taken into account in the model the influence of lattice vibrations and impurity atoms on the electrons. In an extreme case, collisions with the lattice atoms or with impurity atoms may prevent the electrons from travelling along closed orbits. If τ be the collision time of an electron, the energy uncertainty which is due to it will be given by $\Delta E \cdot \tau \geqslant \hbar$. Of course, the behaviour of the electron will only be influenced by the quantisation of its levels, if this uncertainty ΔE is small compared with the energy difference between two quantised states, that is, $\Delta E \ll \hbar\omega_c$. This condition requires high purity of the metal and low temperatures to make the collision times as large as possible. According to Dingle (1952a; Kahn and Frederikse, 1959), the influence of the collisions can be taken into account by introducing an additional temperature T' in the sinh of eqn. (1.39) for the susceptibility. The so-called Dingle temperature is defined by $kT' = 2\hbar/\pi\tau$.

So far we have restricted our discussion to the one-electron model, i.e. we have neglected the electron–electron Coulomb interaction when considering the de Haas–van Alphen oscillations. This interaction, however, seems not to exert an essential influence on the effect. According to Kohn (1961), in a two-dimensional model at least, the period P is not influenced by this interaction.

Of course the quantised states in the magnetic field also influence the other physical properties and give rise to oscillations in other quantities which can be measured. These effects are known best in the transport phenomena, i.e. electric resistance (Shubnikow–de Haas effect), thermal conductivity, Hall effect,..., which will not be considered in this book. There are several review articles on these effects (see, for example, Ziman, 1962; Kahn and Frederikse, 1959; Pippard, 1960).

6. Cyclotron Resonance

Classical free electrons in a magnetic field travel on circular orbits in planes normal to the magnetic field; the orbital frequency is given according to eqn. (1.4), with $\omega_0 = 0$, by the double Larmor frequency, which was denoted by ω_c (cyclotron frequency) in the previous section and defined by eqn. (1.45), with $m_c = m$. When a variable electric field is applied perpendicular to the constant magnetic field, and the energy absorbed by the electrons is observed, a resonance effect is obtained in the classical description if the electric field has the frequency ω_c. This effect is called diamagnetic or cyclotron resonance. It is based on the fact that at this frequency the electric field constantly loses energy which is spent to increase the radius of the circular orbit perpendicular to the magnetic field. Of course, this process only takes place until the electron loses all its energy or part of it in a collision.

The classical equation of motion for an electron reads

$$m\dot{\boldsymbol{v}} = e\left(\boldsymbol{E} + \left[\frac{\boldsymbol{v}}{c} \wedge \boldsymbol{H}\right]\right) - m\frac{\boldsymbol{v}}{\tau}. \qquad (1.47)$$

The last terms on the right-hand side represent the total effect of the collisions in terms of the collision time τ. The magnetic field is assumed to be in the z-direction, the electric field to be in the x, y-plane and to be left- or right-handedly circularly polarised; $\boldsymbol{E} = \boldsymbol{E}_0 (\cos \omega t, \pm \sin \omega t, 0)$.

With $z \equiv v_x + iv_y$ we can rewrite (1.47):

$$\dot{z} = \frac{eE_0}{m} e^{\pm i\omega t} - i\omega_c z - \frac{z}{\tau}.$$

A solution of this equation will be given by $z = z_0 \exp(\pm i\omega t)$ where

$$z_0 = \frac{eE_0\tau}{m} \frac{1 - i(\omega \pm \omega_c)\tau}{1 + (\omega \pm \omega_c)^2\tau^2}.$$

The work done by the field is equal to $A = (\boldsymbol{g} \cdot \boldsymbol{E})$, $\boldsymbol{g}$ being the electron current: $\boldsymbol{g} = ne\boldsymbol{v}$. Using the solution of (1.47) we obtain

$$A(\omega) = \frac{\sigma_0 E_0^2}{1 + (\omega \pm \omega_c)^2\tau^2}.$$

For A we took a time-average and for σ_0 we applied the well-known formula for d.c. conductivity: $\sigma_0 = ne^2\tau/m$. The energy absorbed has thus a marked resonance at ω_c for an appropriate polarisation of the electric field; however, only if $\omega\tau \gg 1$ and $\omega_c\tau \gg 1$. This means that the mean free path of the carriers must be sufficiently large.

A classical treatment can be applied to both electrons and ions. The phenomenon of selective absorption has therefore been known a long time from radiowave absorption measurements in the ionosphere whose cyclotron frequency is determined by the magnetic field of the earth. These experiments can also be simulated in the laboratory and thus one can measure, for example, the resonance absorption by ions (Wobschall *et al.*, 1963).

It is interesting to consider these effects in the case of a solid. In the approximation of an effective isotropic mass, eqn. (1.47) remains unchanged if m is replaced by m_c and $\boldsymbol{v}$ is understood to be the drift velocity of the charge carriers, so that the relation $\boldsymbol{g} = ne\boldsymbol{v}$ also remains valid. In this way absorption measurements yield the cyclotron mass m_c. In general, the cyclotron mass m_c will depend on the direction of the magnetic field with respect to the crystal axes, since the Fermi surface is not spherically symmetric. If, for example, the constant energy surfaces are ordinary ellipsoids

$$E(\boldsymbol{k}) = \frac{\hbar^2}{2}\left(\frac{k_1^2}{m_1}+\frac{k_2^2}{m_2}+\frac{k_3^2}{m_3}\right),$$

and the cosines of direction of the magnetic field with respect to the three principal axes of the ellipsoid are given by α_1, α_2 and α_3, we can calculate the surface A in (1.45) when we intersect the surfaces $\alpha_1 k_1 + \alpha_2 k_2 + \alpha_3 k_3 = \text{const}$ with the ellipsoid $E(\boldsymbol{k}) = \zeta$. A few simple steps of calculation to obtain m_c from (1.45) yield

$$m_c = \sqrt{\frac{m_1 \cdot m_2 \cdot m_3}{\alpha_1^2 m_1 + \alpha_2^2 m_2 + \alpha_3^2 m_3}}, \tag{1.48}$$

so that we obtain the parameters m_1, m_2, m_3 from the directional dependence of the cyclotron resonance. If the Fermi surface consists of several non-connected ellipsoids, this will generally result in the appearance of several resonance absorption lines, Fig. 6 (see Lax and Mavroides, 1960). Besides the de Haas–van Alphen effect, cyclotron resonance is a perfect means of determining the Fermi surface.

This simple description must be completed by a few corrections, if it is to apply to a solid. The above conditions $\omega\tau \gg 1$ and $\omega_c\tau \gg 1$ indicate that, as in the case of the de Haas–van Alphen effect, very pure materials and low temperatures are necessary conditions. Metals, for which Dorfman (1951) and Dingle (1952b) first suggested cyclotron resonance, are primarily unsuited for these considerations as, owing to the skin effect, an electric field cannot penetrate into metals. Thus only semiconductors and semi-metals (bismuth) are left; cyclotron resonance was therefore observed first in a semiconductor, namely germanium (Lax and Mavroides, 1960). Since at low temperatures the semiconductors are virtually non-conductors, the carrier (electrons or holes) concentration must be raised sufficiently by, for example, light absorption causing electron transitions from the valence band to the higher conduction band. These carriers are polarised by the external field so that in semiconductors and semi-metals the polarisation current of the charges must be taken into account in the calculation (for metals, however, it is possible as a rule to neglect the polarisation current as compared to the conduction current). Of course the rest of the medium becomes also slightly polarised. We therefore have to replace in (1.47) the external field $\boldsymbol{E}$ by the internal field $\boldsymbol{E}_{\text{eff}}$ which is given by

$$\boldsymbol{E}_{\text{eff}} = \boldsymbol{E} - L\boldsymbol{P}$$

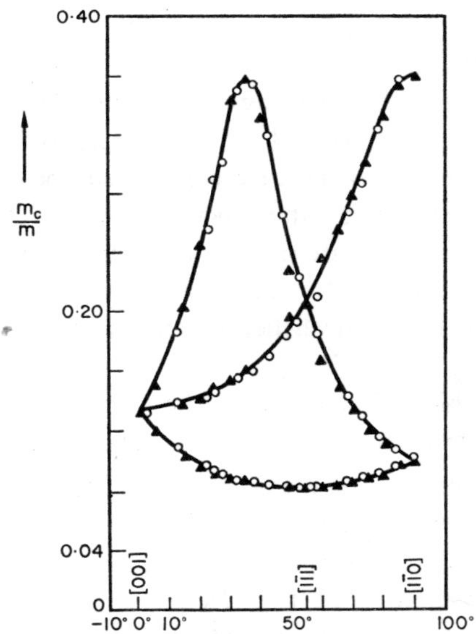

FIG. 6.

The cyclotron mass of germanium at 4°K as a function of the magnetic field orientation, after Dresselhaus, Kip, and Kittel (1955). The field coincides with the (110)-plane and the angles are measured from the [001]-axis. Germanium has cubic symmetry and the ellipsoids are rotationally symmetric along the space diagonals in $\boldsymbol{k}$ space. The theoretical curves are obtained from (1.48) with $m_1 = m_2 = 0{\cdot}082m$ and $m_3 = 1{\cdot}58m$.

(see, for example, Kittel, 1962); here L is the de-electrisation factor of the medium ($L = 4\pi/3$ for spheres) and $\boldsymbol{P}$ is the polarisation vector. It consists of the component of the medium and the dipole moment of the charges: $\boldsymbol{P} = \chi_D \boldsymbol{E}_{\text{eff}} + ne\boldsymbol{r}$. These polarisation effects or magnetoplasma effects change the resonance frequency† in semiconductors or semi-metals (Dresselhaus *et al.*, 1955); they will be discussed below by way of an example.

In a quantum-mechanical consideration, cyclotron resonance is due to transitions between Landau levels caused by the electric field. In the free electron approximation, these transitions are proportional to the square of the matrix element $< n'|(\boldsymbol{E} \cdot \boldsymbol{r})|n >$, where the eigenfunctions of (1.20) are described by Dirac's bra- and ket-formalism; n is the quantum number of the Landau level (1.25). From the well-known properties of the eigenfunctions of the harmonic oscillator, we obtain the selection rule $n'-n = \pm 1$, so that the resonance frequency is $\omega = \omega_c$ as in the case considered above. The quantum theory of cyclotron resonance in semiconductors is much more complicated, since their bands are partly degenerate. The most important result is that the Landau levels, e.g. in germanium, are no longer equidistant. This result has been verified by experiment (Lax and Mavroides, 1960).

In metals the skin effect prevents the penetration of the electric field and thus the ordinary cyclotron diamagnetic resonance. The classical depth of penetration of the field is given by $d = c/\sqrt{2\pi\sigma_0\omega}$ (see, for example, Becker and Sauter, 1957); because of the conditions $\omega\tau \gg 1$ and $\omega_c\tau \gg 1$, however, the skin depth is here essentially smaller than the mean free path l of the electrons (anomalous skin effect), so that essentially only the electrons travelling more or less parallel to the metal surface are appreciably influenced by the field and thus contribute to the conductivity. This gives rise to an effective conductivity used in the equation for d, which is given approximately by $d\sigma_0/l$, as only the fraction d/l of all electrons is constantly in the skin layer between two subsequent collisions. Within the range of the anomalous skin effect the penetration depth is therefore larger than in the classical case (Pippard, 1960). Typical numbers for copper, for example, are $\tau = 10^{-10}$ sec at liquid-helium temperatures and since

† The resonance frequency is easy to obtain from (1.47):

$$\omega_{res} = \frac{\omega_c}{2} \pm \sqrt{\frac{\omega_c^2}{4} + \frac{ne^2}{m}\,\frac{L}{1+\chi_D L}}\,.$$

copper has a Fermi energy of 7 eV, the electron velocity is about 10^8 cm/sec and their mean free path is $l = v\tau = 10^{-2}$ cm. In the frequency range of cyclotron resonance, at about 10^{11} sec^{-1}, the skin depth obtained for copper is 10^{-5}cm; this is in the range of an extremely anomalous skin effect. Azbel and Kaner (1956, 1957) indicated the possibility of resonance absorption in metals in the range of the anomalous skin effect, which is based on the following assumptions: if a magnetic field is applied parallel to the surface of the metal, there will exist a large number of electrons moving almost exclusively in a plane perpendicular to the magnetic field, so that they have a small velocity component parallel to the direction of the field. In a plane perpendicular to the latter, they move along circular orbits with radius $r = v/\omega_c = 10^{-3}$ cm. These electrons are therefore only for a very short time in the skin layer in which a microwave field perpendicular to the magnetic field can act upon them. Since electrons which were at $t = 0$ in the skin layer are again in it after a time of $2\pi/\omega_c$, or an integral multiple of it, resonance absorption may occur if the microwave field is applied at the cyclotron frequency ω_c, or an integral multiple of ω_c, since in the time in which the microwave field is not in phase with the electrons, these electrons are inside the metal, far below the skin depth. Cyclotron resonance in metals is thus characterised by an appearance of resonance absorption at $\omega = n\omega_c$; $n = 1, 2, 3, \ldots$ (see Fig. 7). A similar effect can be observed in semiconductors (Lax and Mavroides, 1960) if the Fermi surface is irregularly distorted; there, however, this effect is due to other reasons. In metals, resonance absorption is a consequence of the special orientation of the magnetic field and the condition for the anomalous skin effect $d \ll r \ll l$. The latter restriction indicates that electrons which do not move in a plane normal to the magnetic field cannot be involved in resonance absorption, since their velocity components in this plane, which determines the radius of the circular orbits, $r = v/\omega_c$, are too small, so that these electrons are more or less always inside the penetration depth d of the electric field.

The resonance frequencies measured depend again on the orientation of the magnetic field with respect to the crystal, so that also in the case of metals, cyclotron resonance is a good means of determining the effective masses, or, more generally, the band structure of these metals. The effective masses measured in this way are in good agreement with the mass parameters determined with the help of other methods (Lax and Mavroides, 1960).

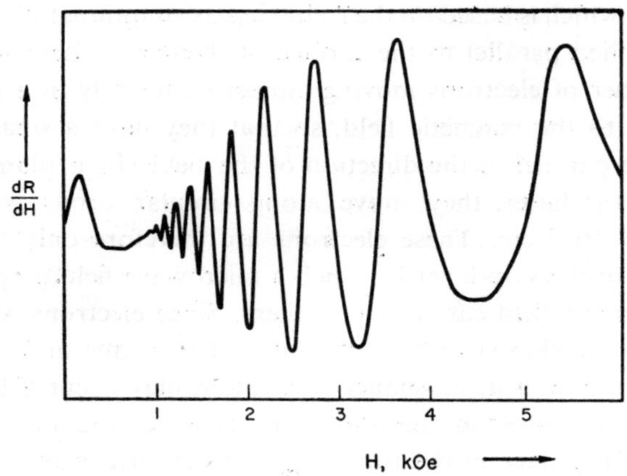

FIG. 7.

Cyclotron resonance in copper at 4°K after Kip *et al.* (1961). The graph shows the field dependence of the derivative of the absorption with respect to the magnetic field.

Finally we should like to discuss a particular magnetoplasma effect, namely the reflection of electromagnetic waves from a non-absorbing medium in a magnetic field. Such a medium may be considered a semiconductor. Assume the medium to be hit by a monochromatic wave of frequency ω and perpendicular or almost perpendicular incidence; its reflection coefficient is then given by

$$R = \left(\frac{n-1}{n+1}\right)^2,$$

n being the refractive index of the medium. If the medium has a dielectric constant ε and permeability μ, the Maxwell equations will read

$$\operatorname{curl} \boldsymbol{E} = -\frac{\mu}{c}\dot{\boldsymbol{H}}, \qquad \operatorname{curl} \boldsymbol{H} = \frac{4\pi}{c}\boldsymbol{g} + \frac{\varepsilon}{c}\dot{\boldsymbol{E}},$$

or, after eliminating $\boldsymbol{H}$,

$$\operatorname{curl}\operatorname{curl} \boldsymbol{E} = -\frac{\mu\varepsilon}{c^2}\left(\frac{4\pi}{\varepsilon}\dot{\boldsymbol{g}} + \ddot{\boldsymbol{E}}\right).$$

In addition to them we have the expression for the conduction current $\boldsymbol{g} = ne\boldsymbol{v}$ and eqn. (1.47) without the collision term, which is unimportant for this problem. These equations determine the possible waves propagated in the medium and its refractive index. We are interested only in those waves which are polarised at right angles to the magnetic field, as only these waves can be influenced by the magnetic field. Assuming the solutions to have the form $\boldsymbol{E} = \boldsymbol{E}_0 \exp[i\omega t - i(\boldsymbol{k}\cdot\boldsymbol{r})]$, we obtain two interrelated equations for the two components of the electric field perpendicular to $\boldsymbol{H}$; from the fact that the determinant of the set of equations must vanish, we obtain an equation which determines $\boldsymbol{k}$ as a function of the given frequency, i.e. we obtain the refractive index $k = n\omega/c$. Two cases may arise:

(a) the wave vector $\boldsymbol{k}$ is parallel to the external field $\boldsymbol{H}$. In the way described above we obtain for the refractive index

$$n_{\parallel}^2 = \varepsilon\mu\left(1 - \frac{\omega_p^2}{\omega}\frac{1}{\omega \pm \omega_c}\right) \quad \text{where} \quad \omega_p^2 = \frac{4\pi n e^2}{m_c \varepsilon}.$$

It should be mentioned that in the expression for the plasma frequency ω_p, the quantity n denotes the electron density.

(b) The wave vector $\boldsymbol{k}$ is perpendicular to the magnetic field. We then obtain in the same way

$$n_{\perp}^2 = \varepsilon\mu\left(1 - \frac{\omega_p^2}{\omega^2}\frac{\omega^2 - \omega_p^2}{\omega^2 - \omega_c^2 - \omega_p^2}\right).$$

The reflection coefficient shows a distinct dependence on the magnetic field; in particular, R vanishes at a certain frequency which, for given ε, μ, and electron density, depends only on the cyclotron mass. For indium antimonide, which has a spherical-symmetric Fermi surface for the electrons, Lax and Wright (1960) obtained $m_c = (0{\cdot}041 \pm 0{\cdot}002)m$ by means of this method, which is in good agreement with the results of other measurements (Lax and Mavroides, 1960).

Finally it should be mentioned that, according to Kohn (1961), the electron–electron Coulomb interaction does not influence the cyclotron resonance frequency.

II. Paramagnetism

1. Paramagnetism of Free Electrons

In Section I.5 we calculated the magnetic moment of a free electron gas, which is produced by the intrinsic magnetic moment of the electrons. According to (1.37) the susceptibility of an electron gas is described to first approximation by

$$\chi_p = \frac{3}{2} N \frac{\mu_B^2}{\zeta_0}. \tag{2.1}$$

It is a characteristic feature of this Pauli spin paramagnetism that it is independent of the temperature or, if terms of higher order are taken into account, that χ_p of ordinary metals is a very weak function of T. Equation (2.1) is generally derived in a very simple way. Neglecting the quantised orbital motion of the electrons the number $N_\pm$ of electrons whose magnetic moment μ_B is parallel or antiparallel to the external field H is calculated. According to Section I.3 this number is given by

$$N_\pm = \int_0^\infty f(E \mp \mu_B H) \frac{dZ}{dE}\, dE. \tag{2.2}$$

Here $Z(E)$ is again the number of states up to the energy E, and $f(E)$ is the Fermi distribution (1.16). The magnetisation of the electron gas will then be given by

$$M = \mu_B(N_+ - N_-). \tag{2.3}$$

The electron energies are influenced only insignificantly by the external field; we can therefore expand $N_\pm$ in a series breaking off after terms linear in H and obtain

$$M = -2\mu_B^2 H \int_0^\infty \frac{dZ}{dE} \frac{\partial f}{\partial E}\, dE = 2\mu_B^2 \left(\frac{dZ}{dE}\right)_{E=\zeta} \cdot H,$$

again taking into account that if

$$\zeta \gg kT, \qquad \frac{\partial f}{\partial E} \cong -\delta(E-\zeta).$$

Hence we obtain for the spin susceptibility

$$\chi_{\mathrm{P}} = 2\mu_{\mathrm{B}}^2 \left(\frac{dZ}{dE}\right)_{E=\zeta}. \tag{2.4}$$

For free particles, we can obtain $Z(E)$ from the general principles of quantum statistics. According to those, the six-dimensional phase-space of a particle is divided into cells of size h^3, and to each cell precisely one state is attributed. The function $Z(E)$ is then given by the volume of the phase space containing all states with energies $\leqslant E$ divided by h^3.

For free electrons with momentum $\boldsymbol{p} = \hbar \boldsymbol{k}$ we have $E = p^2/2m$ and therefore

$$Z(E) = \frac{V}{\hbar^3} \frac{4\pi}{3} p^3, \tag{2.5}$$

where V is the volume in ordinary space. Hence we obtain the density of states

$$\frac{dZ}{dE} = \frac{3}{4} N \frac{\sqrt{E}}{\zeta_0^{3/2}}, \tag{2.6}$$

when we introduce the Fermi energy ζ_0 of the electrons at $T = 0$ according to (1.17). Substituting this in (2.4), we arrive once more at (2.1), if the small difference between ζ and ζ_0 is neglected.

If we take into account the temperature-dependent correction terms of lowest order, we obtain from (2.3)

$$\chi_{\mathrm{P}} = -2\mu_{\mathrm{B}}^2 \int_0^\infty \frac{\partial f}{\partial E}\, dE \left\{ Z'(\zeta) + (E-\zeta)\, Z''(\zeta) + \frac{1}{2}(E-\zeta)^2 Z'''(\zeta) \right\},$$

when we expand $Z(E)$ in a Taylor series in the neighbourhood of $E = \zeta$. Since $\partial f/\partial E$ is essentially different from zero only in the neighbourhood of $E = \zeta$, the lower limit of integration can be shifted to $-\infty$. The second integral will then vanish as $\partial f/\partial E$ is an even function of $E-\zeta$. Introducing the variable $y = \beta(E-\zeta)$ and integrating by parts we obtain for the third integral

$$\frac{1}{2}\int_{-\infty}^{+\infty}(E-\zeta)^2\frac{\partial f}{\partial E}\,dE = -2(kT)^2\int_0^\infty \frac{y\,dy}{e^y+1} = -\frac{\pi^2}{6}(kT)^2.$$

Neglecting terms of higher order in kT/ζ_0 we therefore have

$$\chi_p = 2\mu_B^2\left\{Z'(\zeta)+\frac{\pi^2}{6}(kT)^2 Z'''(\zeta_0)\right\}.$$

In the first term the difference between ζ and ζ_0 can no longer be neglected; we have

$$Z'(\zeta) \cong Z'(\zeta_0)+(\zeta-\zeta_0)\,Z''(\zeta_0).$$

On the other hand, at finite temperatures and $H = 0$ the Fermi energy is obtained from the equation

$$N = 2\int_0^\infty f(E)\,Z'(E)\,dE \cong -2\int_{-\infty}^{+\infty} Z(E)\frac{\partial f}{\partial E}\,dE$$

$$= 2Z(\zeta)+\frac{\pi^3}{3}Z''(\zeta)\cdot(kT)^2,$$

if again only the lowest correction terms are taken into account. Since in this approximation $Z(\zeta)$ must be replaced by

$$Z(\zeta) = Z(\zeta_0)+(\zeta-\zeta_0)\,Z'(\zeta_0)$$

we finally have

$$N = 2Z(\zeta_0)+2(\zeta-\zeta_0)\,Z'(\zeta_0)+\frac{\pi^2}{3}(kT)^2 Z''(\zeta_0).$$

Equating quantities of the same order of magnitude

$$N = 2Z(\zeta_0), \quad \zeta-\zeta_0 = -\frac{\pi^2}{6}(kT)^2\frac{Z''(\zeta_0)}{Z'(\zeta_0)},$$

we see that, using (2.5), the first equation yields (1.17) for ζ_0, and the second equation, together with (2.6), yields the temperature correction of the Fermi energy of (1.18). Thus we obtain for the temperature dependence of susceptibility in the lowest approximation

$$\chi_p(T) = \chi_p(0)\left\{1-\frac{\pi^2}{6}(kT)^2\left(\left(\frac{Z''(\zeta_0)}{Z'(\zeta_0)}\right)^2-\frac{Z'''(\zeta_0)}{Z'(\zeta_0)}\right)\right\}. \quad (2.7)$$

The factor $\chi_p(0)$ is to be calculated from (2.4). The derivation of these formulas is based only on the assumption that we are dealing with free electrons with an isotropic distribution in momentum space. In this generalisation we have the advantage of working with state densities. Since for free electrons with $E = p^2/2m$ eqn. (2.6) is valid, we have

$$\chi_p(T) = \chi_p(0)\left\{1-\frac{\pi^2}{12}\left(\frac{kT}{\zeta_0}\right)^2\right\}.$$

Thus we see that the temperature dependence of ordinary metals is weak; it will be neglected in the following.

In real metals the density of states will generally deviate from the value for free electrons. This variation is most simply described by introducing an effective mass. For alkali metals this will be a reasonable approximation; in the case of multivalent metals, however, the effective masses will be considerably anisotropic and thus also the susceptibility will be anisotropic.

In Table 4 the susceptibilities are compiled for the alkali metals; the effective masses given in Table 3 result partly in a strong increase of Pauli paramagnetism compared with the case of free electrons (where $m^* = m$).

It is difficult to account for the influence of the electron–electron interaction; only within the framework of Landau's phenomenological theory of Fermi fluids (Landau, 1956, 1957; Silin, 1958; Abrikosov *et al.*, 1965; Nozières, 1964; Pines and Nozières, 1966) does this problem become relatively simple. This theory is based upon the fact that, even in the case of interacting Fermi particles, there still exist particle-type excitations if their energy is near the Fermi energy μ and if the

TABLE 4

	χ_P	χ_P^*	χ_P^{exch}	χ_P^{corr}	$(\chi_P)_{\text{exp}}$	χ_L	χ_{ion}	$(\chi_{\text{tot}})_{\text{theor}}$	$(\chi_{\text{tot}})_{\text{exp}}$
Li	10	17	150	27·8	26·6±1·3	−2	− 0·7	25	25
Na	15	15	43	19·6	25·8±2·6	−5	− 4·9	9·7	14
K	23	25	195	31·7		−7	−18·3	9·4	18
Rb	26	32	—	41·5		−7·2	−25	9·0	17
Cs	31	54	—	76		−5·8	−39	31	35

The table shows the atomic susceptibilities of the alkali metals; the values given are to be multipled by 10^{-6}. The first column contains the values for the Pauli spin paramagnetism according to eqn. (2.1); in the second, third, and fourth columns values are given which are obtained when the effective mass (Table 3), the exchange energy (2.10), and the correlation energy (Silverstein, 1969) are taken into account. The fourth column is to be compared with the experimental values (5th column) after Schumacher and Vehse (1960, 1963). The two following columns give the Landau diamagnetism (Table 3) and the susceptibility of the alkali ions (Table 2). For the ions Li^+, Na^+, and K^+ the Hartree–Fock values are given, for Rb^+ and Cs^+ the values obtained by Slater. The 7th column gives the susceptibilities for the metal itself obtained from these values; they are to be compared with the experimental values (last column).

temperature is low. These so-called quasi-particles obey Fermi statistics, since the interaction cannot influence the fermionic nature of the particles as long as interaction may be considered as a perturbation, that is, in the case of normal systems (Abrikosov *et al.*, 1965; Nozières, 1964). Moreover, at $T = 0$ the distribution function of the quasi-particles has a discontinuity at the same point in momentum space as the distribution function of free electrons, i.e. at $p_0 = \hbar k_0$. This is plausible, if the interaction is assumed not to introduce essential alterations in the spectrum of the system, in the sense that at $T = 0$ there are still exactly N states occupied, and none of them has been split off from the continuum of the free particle states, e.g. as a bound state. This means that the correspondence between the free particle states and the quasi-particle states is assumed to be unambiguous. However, since the counting of momenta depends only on the geometry of the system considered (cf. Section I.5), which is independent of interaction, the limiting momentum of the occupied states at $T = 0$ is $p_0 = \hbar k_0$, just as in a system without interaction. The Fermi energy μ (chemical potential), however, differs from the value of ζ of the free particle system.

In the case of thermal equilibrium the distribution function of quasi-particles near the Fermi energy is therefore given by

$$f_0(\varepsilon_0) = \frac{1}{e^{\beta(\varepsilon_0-\mu)}+1}.$$

Here ε_0 is the equilibrium energy of the quasi-particles which is a function of momentum $\boldsymbol{p}$ and spin s if a magnetic field is present; if there is no spin–orbit coupling, ε_0 can always be assumed to be a diagonal matrix in the spin space. Since quasi-particles are not free particles, ε_0 will also depend on the distribution of the other quasi-particles and thus be a function of temperature.

Let us now assume this system to be perturbed by the application of external fields or a change of temperature. Such a change will cause an excitation of the quasi-particles, and the distribution function will differ from the equilibrium distribution f_0 by $\delta f(\boldsymbol{p}, \boldsymbol{s}, \boldsymbol{r})$. Here, in general, δf will no longer be diagonal in the spin and, if the perturbation is non-uniform, it will also be a function of the spatial coordinates. Thus the total energy of the system will vary by

$$\delta E = \int \frac{d^3\boldsymbol{r}\, d^3\boldsymbol{p}}{(2\pi\hbar)^3} \operatorname{Tr}_s \varepsilon(\boldsymbol{p}, \boldsymbol{s}, \boldsymbol{r})\, \delta f(\boldsymbol{p}, \boldsymbol{s}, \boldsymbol{r}).$$

The trace over the spin is due to the fact that δE is a scalar. This equation defines the quasi-particle energy ε as a functional derivative of the energy with respect to the density in the phase space. We also may consider $\varepsilon(\boldsymbol{p}, \boldsymbol{s}, \boldsymbol{r})$ as a local Hamilton operator or the Hamiltonian for the quasi-particles. It follows that this definition of ε will apply only in the case of weak inhomogeneities, i.e. slowly varying perturbations δf, when in the classical limiting case the non-commuting quantities $\boldsymbol{p}$ and $\boldsymbol{r}$ can be replaced by suitable averages (Landau and Lifshitz, 1958), as was done in the present case.

If δf is small, ε can be expanded in a series of δf and, breaking off after the linear terms, we have

$$\begin{aligned}\varepsilon(\boldsymbol{p}, \boldsymbol{s}, \boldsymbol{r}) &= \varepsilon_0(\boldsymbol{p}, \boldsymbol{s})+\delta\varepsilon(\boldsymbol{p}, \boldsymbol{s}, \boldsymbol{r})\\ &= \varepsilon_0(\boldsymbol{p}, \boldsymbol{s})+\mathrm{Tr}_{s'}\int \frac{d^3r'\,d^3p'}{(2\pi\hbar)^3}\,\phi(\boldsymbol{p}, \boldsymbol{s}, \boldsymbol{r}; \boldsymbol{p}', \boldsymbol{s}', \boldsymbol{r}')\,\delta f(\boldsymbol{p}', \boldsymbol{s}', \boldsymbol{r}').\end{aligned}$$

The second term of the right-hand side obviously describes the interaction of the quasi-particles, ϕ being the interaction of two quasi-particles. It is therefore symmetric with respect to the two particles:

$$\phi(\boldsymbol{p}, \boldsymbol{s}, \boldsymbol{r}; \boldsymbol{p}', \boldsymbol{s}', \boldsymbol{r}') = \phi(\boldsymbol{p}', \boldsymbol{s}', \boldsymbol{r}'; \boldsymbol{p}, \boldsymbol{s}, \boldsymbol{r}).$$

The function ϕ was introduced by Landau as a result of phenomenological considerations; in perturbation theory it proved to be a renormalised two-particle interaction (Abrikosov, *et al.*, 1965; Nozières, 1964). In the following we shall assume the interaction ϕ to be local, i.e. proportional to $\delta\,(\boldsymbol{r}-\boldsymbol{r}')$, so that we obtain for ε

$$\varepsilon(\boldsymbol{p}, \boldsymbol{s}, \boldsymbol{r}) = \varepsilon_0(\boldsymbol{p}, \boldsymbol{s})+\mathrm{Tr}_{s'}\int \frac{d^3p'}{(2\pi\hbar)^3}\,\phi(\boldsymbol{p}, \boldsymbol{s}; \boldsymbol{p}', \boldsymbol{s}')\,\delta f(\boldsymbol{p}', \boldsymbol{s}', \boldsymbol{r}).$$

Of course this approximation is only reasonable in the case of short-range forces; however, it is not difficult to take long-range forces into account (Pines and Nozières, 1966). For time-dependent perturbations Landau postulates the validity of a Liouville or Boltzmann equation for the quasi-particle distribution function f. When collisions are neglected, it reads $df/dt = 0$. If f were a purely quantum-mechanical quantity, we would have

$$\frac{df}{dt} = \frac{\partial f}{\partial t}+\frac{i}{\hbar}\,[H, f]_-,$$

where $\mathscr{H}$ is the Hamiltonian of the system. As $\boldsymbol{p}$ and $\boldsymbol{r}$ are classical quantities (which is not true for the spin), we have to pass over from the commutator $[\mathscr{H}, f]_-$ to the Poisson brackets in the well-known way, replacing $\mathscr{H}$ by $\varepsilon(\boldsymbol{p}, \boldsymbol{s}, \boldsymbol{r}, \boldsymbol{t})$ (Landau and Lifshitz, 1958):

$$\frac{\partial f}{\partial t}+\left(\frac{\partial f}{\partial \boldsymbol{r}}\cdot\dot{\boldsymbol{r}}\right)+\left(\frac{\partial f}{\partial \boldsymbol{p}}\cdot\dot{\boldsymbol{p}}\right)+\frac{i}{\hbar}\,[\varepsilon, f]_- = 0.$$

Since f is a matrix with respect to the spin, the products must (similarly as for the case of the quantum-mechanical current) be replaced by symmetrised products (Silin, 1958):

$$\frac{\partial f}{\partial t}+\frac{1}{2}\left[\left(\frac{\partial f}{\partial \boldsymbol{r}}\cdot\dot{\boldsymbol{r}}\right)+\left(\dot{\boldsymbol{r}}\cdot\frac{\partial f}{\partial \boldsymbol{r}}\right)\right]+\frac{1}{2}\left[\left(\frac{\partial f}{\partial \boldsymbol{p}}\cdot\dot{\boldsymbol{p}}\right)+\left(\dot{\boldsymbol{p}}\cdot\frac{\partial f}{\partial \boldsymbol{p}}\right)\right]+\frac{i}{\hbar}[\varepsilon,f]_-=0$$

($\partial f/\partial \boldsymbol{r} \equiv \nabla f, \ldots$). The commutator involves only the spin. In addition to this we have the classical equations of motion

$$\dot{\boldsymbol{r}} = \boldsymbol{v} = \frac{\partial \varepsilon}{\partial \boldsymbol{p}}, \qquad \dot{\boldsymbol{p}} = -\frac{\partial \varepsilon}{\partial \boldsymbol{r}},$$

where $\boldsymbol{v}$ is the velocity of the quasi-particle. Finally we expand the Liouville equation in a series ending with the terms linear in $\delta f (f = = f_0 + \delta f)$, and obtain

$$\frac{\partial \delta f}{\partial t}+\frac{1}{2}\left[\left(\frac{\partial \delta f}{\partial \boldsymbol{r}}\cdot\boldsymbol{v}\right)+\left(\boldsymbol{v}\cdot\frac{\partial \delta f}{\partial \boldsymbol{r}}\right)\right]-\frac{1}{2}\left[\left(\frac{\partial f_0}{\partial \boldsymbol{p}}\cdot\frac{\partial \delta\varepsilon}{\partial \boldsymbol{r}}\right)+\left(\frac{\partial \delta\varepsilon}{\partial \boldsymbol{r}}\cdot\frac{\partial f_0}{\partial \boldsymbol{p}}\right)\right]$$
$$+\frac{i}{\hbar}[\delta\varepsilon, f_0]_- + \frac{i}{\hbar}[\varepsilon_0, \delta f)]_- = 0. \tag{2.8}$$

This equation describes the dynamic behaviour of the system, for example, the spin waves (III.5).

In the case of zero magnetic field, f_0 and $\delta\varepsilon$ are diagonal and the kinetic equation is reduced to

$$\frac{\partial \delta f}{\partial t}+\left(\boldsymbol{v}\cdot\frac{\partial \delta f}{\partial \boldsymbol{r}}\right)-\left(\frac{\partial \delta\varepsilon}{\partial \boldsymbol{r}}\cdot\frac{\partial f_0}{\partial \boldsymbol{p}}\right)=0.$$

Integrating this equation over all momenta we obtain the continuity equation

$$\frac{\partial \delta n}{\partial t}+\operatorname{div}(\delta \boldsymbol{j}(\boldsymbol{r},t))=0,$$

with the density variation δn:

$$\delta n(\boldsymbol{r},t) = \mathrm{Tr}_s \int \frac{d^3p}{(2\pi\hbar)^3}\,\delta f(\boldsymbol{p},\boldsymbol{s},\boldsymbol{r},t),$$

and the current connected with it:

$$\delta \boldsymbol{j}(\boldsymbol{r}, t) = \mathrm{Tr}_s \int \frac{d^3p}{(2\pi\hbar)^3} \Big\{ \boldsymbol{v}(\boldsymbol{p})\, \delta f(\boldsymbol{p}, \boldsymbol{s}, \boldsymbol{r}, t) - \frac{\partial f_0}{\partial \boldsymbol{p}} \mathrm{Tr}_{s'} \int \frac{d^3p'}{(2\pi\hbar)^3} \phi(\boldsymbol{p}, \boldsymbol{s}; \boldsymbol{p}', \boldsymbol{s}')\, \delta f(\boldsymbol{p}', \boldsymbol{s}', \boldsymbol{r}, t) \Big\}$$

$$= \mathrm{Tr}_s \int \frac{d^3p}{(2\pi\hbar)^3} \Big\{ \boldsymbol{v}(\boldsymbol{p}) - \mathrm{Tr}_{s'} \int \frac{d^3p'}{(2\pi\hbar)^3} \frac{\partial f_0}{\partial \boldsymbol{p}'} \phi(\boldsymbol{p}, \boldsymbol{s}; \boldsymbol{p}', \boldsymbol{s}') \Big\}\, \delta f(\boldsymbol{p}, \boldsymbol{s}, \boldsymbol{r}, t),$$

where the expression of the last line was obtained when making use of the symmetry of ϕ. The terms in braces obviously represent the current of a quasi-particle with momentum $\boldsymbol{p}$. In the case of zero interaction this current would be equal to $\boldsymbol{p}/m$; since in translation–invariant systems the current is a conserved quantity (independent of the strength of the interaction), we must have

$$\boldsymbol{p}/m = \boldsymbol{v}(\boldsymbol{p}) - \mathrm{Tr}_{s'} \int \frac{d^3p'}{(2\pi\hbar)^3} \frac{\partial f_0}{\partial \boldsymbol{p}'} \phi(\boldsymbol{p}, \boldsymbol{s}; \boldsymbol{p}', \boldsymbol{s}').$$

If for momenta near the Fermi momentum p_0 we define an effective mass of the quasi-particles

$$\boldsymbol{v}(\boldsymbol{p}) = \frac{\boldsymbol{p}}{m^*}, \qquad |\boldsymbol{p}| \cong p_0,$$

we obtain for quasi-particles with momentum $|\boldsymbol{p}| \cong p_0$

$$\frac{1}{m} = \frac{1}{m^*} - \mathrm{Tr}_{s'} \int \frac{d^3p'}{(2\pi\hbar)^3} \left(\frac{\boldsymbol{p}}{p_0^2} \cdot \frac{\partial f_0}{\partial \boldsymbol{p}'} \right) \phi(\boldsymbol{p}, \boldsymbol{s}; \boldsymbol{p}', \boldsymbol{s}').$$

In rotationally invariant systems the scalar function ϕ may depend only on the relative orientation of the spins, i.e. we shall have in general

$$\phi(\boldsymbol{p}, \boldsymbol{s}; \boldsymbol{p}', \boldsymbol{s}') = \varphi_0(\boldsymbol{p}, \boldsymbol{p}') + (\boldsymbol{s} \cdot \boldsymbol{s}')\, \varphi_1(\boldsymbol{p}, \boldsymbol{p}'),$$

where the second term is of the nature of an exchange interaction, see (III.1); when taking the trace, this term vanishes. Therefore, when we transform to polar coordinates, we obtain

$$\frac{1}{m} = \frac{1}{m^*} - 2\int \frac{d^3\boldsymbol{p}'}{(2\pi\hbar)^3}\left(\frac{\boldsymbol{p}}{p_0^2}\cdot\frac{\partial f_0}{\partial \boldsymbol{p}'}\right)\varphi_0(\boldsymbol{p},\boldsymbol{p}')$$

$$= \frac{1}{m^*} + \frac{2p_0}{(2\pi\hbar)^3}\int \cos\vartheta\,\varphi_0(\boldsymbol{p}_0,\boldsymbol{p}')\,d^2\Omega_{p'=p_0},$$

since f_0 depends only on $\boldsymbol{p}$ and $\partial f_0/\partial \boldsymbol{p}$ is non-vanishing only at $p = p_0$. The function $\varphi_0(\boldsymbol{p}_0, \boldsymbol{p}_0')$ depends only on the angle ϑ between $\boldsymbol{p}_0$ and $\boldsymbol{p}_0'$ and the rest of the integral is taken over the Fermi surface.

Just like the crystal potential, the interaction between the fermions changes the particle mass. The effective mass defined in this way is immediately connected with the specific heat and is therefore also called thermal mass.

This is obvious when we assume the temperature of the system to rise from T to $T+\delta T$; the distribution function f_0 will then vary by

$$\delta f = f_0(T+\delta T) - f_0(T) = \frac{\partial f_0}{\partial T}\,\delta T,$$

which causes an energy variation of the system equal to

$$\delta E = V\,\mathrm{Tr}_s \int \frac{d^3\boldsymbol{p}}{(2\pi\hbar)^3}\,\varepsilon_0(\boldsymbol{p})\,\delta f,$$

so that we obtain for the specific heat at constant volume

$$C_v = \left(\frac{\partial E}{\partial T}\right)_v = 2V\int \frac{d^3\boldsymbol{p}}{(2\pi\hbar)^3}\,\varepsilon_0(\boldsymbol{p})\,\frac{\partial f_0}{\partial T}.$$

This is the same expression as obtained for the model of free electrons, with the only exception that $\varepsilon_0 = p^2/2m^*$ near $\varepsilon = \mu$. Therefore we have

$$C_v = \gamma T = \frac{m^*}{m}\gamma_0 T; \qquad \gamma_0 = \frac{2}{3}\pi^2 k^2 Z'(\mu);$$

γ_0 is Sommerfeld's constant of the free particle system, $Z'(\mu)$ is the density of states, and k is Boltzmann's constant.

It is now easy to obtain the susceptibility of the system. In a constant magnetic field $\delta\boldsymbol{H}$ the quasi-particles obtain an additional energy

$$\delta\varepsilon(\boldsymbol{p}, \boldsymbol{s}) = -2\mu_{\mathrm{B}}(\boldsymbol{s}\cdot\delta\boldsymbol{H}) + \mathrm{Tr}_{s'}\int \frac{d^3p'}{(2\pi\hbar)^3}\,\phi(\boldsymbol{p}, \boldsymbol{s}; \boldsymbol{p}', \boldsymbol{s}')\,\delta f(\boldsymbol{p}', \boldsymbol{s}').$$

The first term is the Zeeman energy of the quasi-particle, $\boldsymbol{s}$ are the angular momentum matrices in units of $\hbar$; the second term describes the variation of interaction caused by the change in occupation of states. The change in occupation of the quasi-particle states takes place near the Fermi energy μ (which may be assumed to be constant to a good approximation, see (1.18)). Similarly, as with (2.2), this leads us to

$$\delta f = f_0(\varepsilon) - f_0(\varepsilon_0) = \frac{\partial f_0}{\partial \varepsilon_0}\,\delta\varepsilon,$$

so that we obtain

$$\delta\varepsilon(\boldsymbol{p}, \boldsymbol{s}) = 2\mu_{\mathrm{B}}(\boldsymbol{s}\cdot\delta\boldsymbol{H}) + \mathrm{Tr}_{s'}\int \frac{d^3p'}{(2\pi\hbar)^3}\,\phi(\boldsymbol{p}, \boldsymbol{s}; \boldsymbol{p}'\, \boldsymbol{s}')\,\frac{\partial f_0}{\partial \varepsilon_0(\boldsymbol{p}')}\,\delta\varepsilon(\boldsymbol{p}', \boldsymbol{s}').$$

As $\delta\varepsilon$ is obviously a linear function of $\delta\boldsymbol{H}$, we are led to the Ansatz

$$\delta\varepsilon = -\gamma(p)\,(\boldsymbol{s}\cdot\delta\boldsymbol{H}).$$

Using the splitting of ϕ into φ_0 and φ_1 given above, we see that when forming the trace the term containing φ_0 vanishes and only the exchange–interaction term is left. When we also take into account that

$$\mathrm{Tr}_{s'}\,(\boldsymbol{s}\cdot\boldsymbol{s}')\,(\boldsymbol{s}'\cdot\delta\boldsymbol{H}) = \tfrac{1}{2}(\boldsymbol{s}\cdot\delta\boldsymbol{H}),$$

we obtain for $\gamma(\boldsymbol{p})$

$$\gamma(\boldsymbol{p}) = 2\mu_{\mathrm{B}} + \frac{1}{2}\int \frac{d^3p'}{(2\pi\hbar)^3}\,\varphi_1(\boldsymbol{p}, \boldsymbol{p}')\,\frac{\partial f_0}{\partial \varepsilon_0(\boldsymbol{p}')}\,\gamma(\boldsymbol{p}').$$

Again we are interested only in momenta $p \cong p_0$ and $p' \cong p_0$ and we obtain

$$\gamma(p_0) = \frac{2\mu_B}{1+\frac{1}{2}Z'(\mu)\bar{\varphi}_1},$$

where

$$\bar{\varphi}_1 = \frac{1}{4\pi}\int \varphi_1(\boldsymbol{p}_0, \boldsymbol{p}_0')\, d^2\Omega_{p'=p_0}.$$

Just as in the case of (2.3), the magnetisation is obtained from the difference of the occupied states parallel and antiparallel to the field $\delta\boldsymbol{H}$ (which is assumed to be parallel to the z-axis)

$$\begin{aligned} \boldsymbol{M} &= \mu_B(\delta N_+ - \delta N_-) = \mu_B V(\delta n_+ - \delta n_-) \\ &= \mu_B V\,\mathrm{Tr}_s \int \frac{d^3\boldsymbol{p}}{(2\pi\hbar)^3} 2s_z\, \delta f(\boldsymbol{p}, s) = -\mu_B V \int \frac{d^3\boldsymbol{p}}{(2\pi\hbar)^3} \frac{\partial f_0}{\partial\varepsilon_0(\boldsymbol{p})}\gamma(\boldsymbol{p})\,\delta H, \end{aligned}$$

so that, in contrast to (2.4), we obtain for the susceptibility

$$\tilde{\chi}_p = \frac{2\mu_B^2 Z'(\mu)}{1+\frac{1}{2}Z'(\mu)\bar{\varphi}_1} = \frac{\chi_p}{1+\frac{1}{2}Z'(\mu)\bar{\varphi}_1}, \tag{2.10a}$$

where χ_p was substituted from (2.4). $\tilde{\chi}_p$ is the susceptibility of a Fermi fluid corrected for the exchange interaction. Equation (2.10a) shows that the simple proportionality of specific heat and susceptibility of a free electron gas will no longer hold true; the thermal mass contains the function φ_0 and the susceptibility φ_1. As to the values of φ_0 and φ_1, nothing can be said within the framework of Landau's theory; they must be chosen in agreement with experiment or perturbation-theoretical calculations (cf. Silverstein, 1963; Herring, 1966). Comparison with experiment (Table 9 on p. 127) shows that the deviation from the free-electron model may be considerable in some cases (Pt). This is easy to understand when we take into account that the electrons, because of the repulsive character of Coulomb interaction, will tend to arrange their spins in parallel to avoid mutual approach (see below), i.e. to keep the system's energy as low as possible. Formally this can be described by a negative $\bar{\varphi}_1$ in (2.10a) which, in its turn, raises the susceptibility in agreement with the data given in Table 9.

Since Landau's theory is a phenomenological theory, it depends on a series of parameters and/or functions which can be chosen in agreement with experiment. However, this does not answer the question of the exact influence of Coulomb interaction in an electron gas. We shall here discuss it within the framework of the Hartree–Fock approximation. In this approximation the electrons are assumed to be independent of one another and each electron moves in the mean potential of the other electrons. When φ and ψ are two orthonormalised electron eigenfunctions in the Hartree–Fock approximation, the normalised wave function of the two-electron system allowed by Pauli's principle is given by

$$\frac{1}{\sqrt{2}}(1-P)\,\varphi(r_1)\,\psi(r_2),$$

where P is the permutation of the two "coordinates" r_1 and r_2. Here, for the time being, we use $r_{1,2}$ to describe position and spin (i.e. the complete set of observables) of an electron. The two-particle density $\varrho(r_1, r_2)$ for the two electrons is equal to the absolute square of the wave function

$$\begin{aligned}\varrho(r_1, r_2) &= \mathrm{Re}\,\big(\varphi^*(r_1)\,\psi^*(r_2)\,(1-P)\,\varphi(r_1)\,\psi(r_2)\big)\\ &= |\varphi(r_1)|^2\,|\psi(r_2)|^2 - \mathrm{Re}\;\varphi(r_1)\,\psi^*(r_1)\,\psi(r_2)\,\varphi^*(r_2),\end{aligned}$$

since $P\varphi(r_1)\,\psi(r_2) = \varphi(r_2)\,\psi(r_1)$.

In the translationally invariant electron gas considered here (where the positive ionic charge is assumed to be spatially constant), we have $|\varphi|^2 = |\psi|^2 = 1/V$, V being the volume of the electron gas. We therefore have

$$\varrho(r_1, r_2) = \frac{1}{V^2} - \mathrm{Re}\,\varphi(r_1)\,\psi^*(r_1)\,\psi(r_2)\,\varphi^*(r_2).$$

Neglecting spin–orbit coupling, the functions φ and ψ are products of functions of spin and spatial coordinates. The second term (the exchange density) will then be non-zero only if φ and ψ have the same spin function, i.e. if both electron spins are parallel. When the electrons have the momentum $\hbar\boldsymbol{k}$ and $\hbar\boldsymbol{k}'$, respectively, their eigenfunctions in the Hartree–Fock approximation will be $\left(1\sqrt{V}\right) e^{i(\boldsymbol{k}\cdot\boldsymbol{r})}$ and $\left(1\sqrt{V}\right) e^{i(\boldsymbol{k}'\cdot\boldsymbol{r})}$ and we obtain for the density of electrons with parallel spins

$$\varrho(\boldsymbol{r}_1, \boldsymbol{r}_2) = \frac{1}{V^2} - \frac{1}{V^2} \operatorname{Re} e^{i([\boldsymbol{k}-\boldsymbol{k}']\cdot[\boldsymbol{r}_1-\boldsymbol{r}_2])}.$$

Summing over all occupied states ($k < k_0$) and the two possible directions of spin, we obtain the so-called static pair distribution function for two different electrons with parallel spin in the Hartree–Fock approximation

$$G_{\mathrm{p}}^{\mathrm{HF}}(\boldsymbol{r}_1 - \boldsymbol{r}_2) = \frac{1}{2}\left(\frac{N}{V}\right)^2 - 2\left(\frac{1}{V}\sum_{k<k_0} e^{i(\boldsymbol{k}[\boldsymbol{r}_1-\boldsymbol{r}_2])}\right)^2,$$

since $\sum\limits_{k<k_0} 1 = \frac{1}{2}N$. Changing from the sum to an integral we have (in polar coordinates)

$$F(r) = \frac{1}{V}\sum_{k<k_0} e^{i(\boldsymbol{k}\cdot\boldsymbol{r})} = \frac{2\pi}{(2\pi)^3}\int_0^{k_0} k^2\,dk \int_{-1}^{+1} e^{ikr\cos\vartheta}\, d(\cos\vartheta)$$

$$= \frac{1}{2\pi^2 r^3}(\sin k_0 r - k_0 r \cos k_0 r).$$

Therefore

$$G_{\mathrm{p}}^{\mathrm{HF}} = \frac{1}{2}\left(\frac{N}{V}\right)^2 - 2F^2(r),$$

where $r = |\boldsymbol{r}_1 - \boldsymbol{r}_2|$ is the distance of the two electrons. Since $F(0) = N/2V$, we have $G_{\mathrm{p}}^{\mathrm{HF}}(0) = 0$, that is, owing to the antisymmetry of the wave function, i.e. Pauli's principle, the probability of finding two electrons with parallel spin at the same point is equal to zero. In other words, each electron tends to remove a like electron (i.e. one with the same s in) from its neighbourhood. Therefore we also speak of a Fermi hole (Seitz, 1940; Mott and Jones, 1936). Together with the electron–electron interaction, this effect is fundamental for magnetism (cf. Section III.1).

In the Hartree–Fock approximation, the ground state of the electron gas consists of the contributions of free electrons, E_k, and the exchange energy E_{ex} describing the electron–electron interaction. Obviously we have (for $T = 0$)

$$E_k = \sum_{\substack{k,\, s \\ (k<k_0)}} \frac{\hbar^2 k^2}{2m} = \frac{3}{5} N\zeta_0.$$

As in electrostatics, the exchange energy is obtained from the two-particle density

$$E_{ex} = \frac{e^2}{2} \Sigma \int d^3\boldsymbol{r}_1 \, d^3\boldsymbol{r}_2 \frac{\varrho(\boldsymbol{r}_1, \boldsymbol{r}_2)}{|\boldsymbol{r}_1 - \boldsymbol{r}_2|},$$

where we have again to sum over the states occupied in the electron gas. It has to be noted that the first term in $\varrho(\boldsymbol{r}_1, \boldsymbol{r}_2)$ and $G_p^{HF}(r)$, the position-independent Hartree term, is exactly compensated in the total energy of the electron–ion system by the electron–ion and the ion–ion interactions. We can therefore neglect the Hartree term in $\varrho(\boldsymbol{r}_1, \boldsymbol{r}_2)$, and only the exchange density is left. Thus we have

$$E_{ex} = -e^2 \int d^3\boldsymbol{r}_1 \, d^3\boldsymbol{r}_2 \frac{F^2(\boldsymbol{r}_1 - \boldsymbol{r}_2)}{|\boldsymbol{r}_1 - \boldsymbol{r}_2|} = -e^2 V \int d^3\boldsymbol{r} \frac{F^2(r)}{r}.$$

It is easy to integrate over the exchange density, as the spatial integration may be extended to infinity because the integrand decreases rapidly. Using (1.17) we obtain

$$E_{ex} = -2 \frac{e^2 k_0^4}{(2\pi)^3} V = -\frac{3}{4\pi} e^2 k_0 N.$$

Expressing ζ_0 and k_0 in terms of N, we therefore obtain for the ground-state energy in the Hartree–Fock approximation

$$E_0^{HF} = E_k + E_{ex} = \frac{3h^2}{10m} \left(\frac{3}{8\pi V}\right)^{2/3} N^{5/3} - \frac{3}{2} e^2 \left(\frac{3}{8\pi V}\right)^{1/3} N^{4/3}.$$

In the case of high densities the exchange energy is only a correction to the kinetic energy. In contrast to the positive Hartree component of the electron–electron interaction it is negative, since the correlation between electrons with parallel spin reduces the weights of the energetically unfavourable close distances.

In a magnetic field the magnetic moments of the electrons will preferably lie in the direction of the field, i.e. the number of electrons N_+ whose moments are parallel to the field, is different from the number of electrons N_- whose moments are antiparallel. The Fermi distribution will thus be a little different for the two spin directions, and the ground-state energy in the magnetic field is now given by

$$E_0^{\mathrm{HF}}(H) = \frac{3h^2}{10m}\left(\frac{3}{4\pi V}\right)^{2/3}(N_+^{5/3}+N_-^{5/3}) \\ -\frac{3}{2}e^2\left(\frac{3}{4\pi V}\right)^{1/3}(N_+^{4/3}+N_-^{4/3})-\mu_{\mathrm{B}}H(N_+-N_-). \tag{2.9}$$

We put $N_\pm = N/2\pm p$ and expand $E_0^{\mathrm{HF}}(H)$ in a series breaking off after the terms quadratic in p; this is justified since p is proportional to the external field and therefore small. It is easy to obtain

$$E_0^{\mathrm{HF}}(H) = E_0^{\mathrm{HF}}+\frac{4}{3}\zeta_0\frac{p^2}{N}-\frac{2}{3\pi}e^2k_0\frac{p^2}{N}-2\mu_{\mathrm{B}}Hp.$$

The energy has a minimum if $\partial E_0^{\mathrm{HF}}/\partial p = 0$, i.e. if

$$p = \frac{3}{4}\mu_{\mathrm{B}}H\frac{1}{\zeta_0-\dfrac{e^2k_0}{2\pi}}.$$

With this formula and (2.3) we can calculate the magnetisation or the susceptibility $\chi_{\mathrm{p}}^{\mathrm{ex}}$ corrected for the exchange. Using (2.1) and the abbreviations defined by (1.38), and taking the effective masses (of the crystal) into account, we then have

$$\frac{\chi_{\mathrm{p}}}{\chi_{\mathrm{p}}^{\mathrm{ex}}} = \frac{m}{m^*}-\frac{e^2k_0}{2\pi\zeta_0} = \frac{m}{m^*}-\frac{\alpha r^s}{\pi}. \tag{2.10}$$

This is analogous to Landau's theory (2.10a) (see page 62) with

$$\bar{\varphi}_1 = -\frac{4}{3\pi}\frac{e^2k_0}{N}.$$

The susceptibilities χ_p^{ex} corrected for exchange interaction are compiled in Table 4. The influence of the exchange energy is very strong; in the case of Rb and Cs the exchange energy exceeds the kinetic energy ($p < 0$), so that $E(p)$ is not a minimum. Both metals will therefore become ferromagnetic. This drastic change in susceptibility, however, is essentially compensated by the correlation energy. The correlation energy accounts for the fact that electrons with antiparallel spins also must be correlated as a consequence of the Coulomb interaction. It is difficult to calculate the correlation energy; it can be expanded in a power series in r_s which probably converges only if $r_s < 1$ (cf. Pines, 1961). Since the metals have smaller electron densities ($1{\cdot}5 < r_s < 6$), these series expansions cannot be applied to metals. It is therefore necessary to have recourse to an approximate calculation of correlation energies in this range. The susceptibilities given in Table 4 for alkali metals, which account for exchange and correlation energies, are based upon such a calculation by Silverstein (1963; Rice, 1965). We see that the correlation energy compensates considerably the influence of the exchange energy; in particular for Rb and Cs, the electron gas susceptibilities do not correspond to ferromagnetism. The spin susceptibilities measured for lithium (Schumacher and Vehse, 1960) and sodium (Schumacher and Vehse, 1963) are in good agreement with the calculated values.

The total susceptibility of the alkali metals, $\chi_{tot} = \chi_p + \chi_L + \chi_{ion}$, is theoretically connected with considerable uncertainties, cf. Table 2. The calculated and experimental values agree none the less relatively well, both in order of magnitude and in their relative magnitudes.

Such a discussion is still less reasonable for other metals, as the deviations from the free-electron model increase more and more when we pass over to multivalent metals. Generally we can predict an anisotropy of the susceptibility, and the formulas given above for the susceptibilities can be at most in a qualitative agreement with experiment (cf., for example, Wilson, 1958).

2. Classical Theory of Paramagnetism

Unlike diamagnetism, atomic paramagnetism can be described according to Langevin by means of a classical formalism. Paramagnetic behaviour is observed for atoms possessing a permanent magnetic moment $\boldsymbol{\mu}$ which adjusts itself more or less freely in an external field $\boldsymbol{H}$. The moment $\boldsymbol{\mu}$ in the magnetic field has an energy

$$-(\mu \cdot \boldsymbol{H}) = -\mu H \cos \vartheta.$$

If the N atoms are identical, their energy will be given by

$$E = -N\mu H \overline{\cos \vartheta} = -MH. \tag{2.11}$$

M is the magnetisation and the average value of $\cos \vartheta$ is calculated using Boltzmann statistics

$$\overline{\cos \vartheta} = \frac{\int_{-1}^{+1} \cos \vartheta \, e^{\beta \mu H \cos \vartheta} \, d(\cos \vartheta)}{\int_{-1}^{+1} e^{\beta \mu H \cos \vartheta} \, d(\cos \vartheta)} = L(\alpha)$$

where $\alpha = \beta \mu H$. The Langevin function $L(\alpha)$ is defined by

$$L(\alpha) = -1/\alpha + \coth \alpha. \tag{2.12}$$

At ordinary temperatures and not too high field strengths we always have $\alpha \ll 1$ so that $L(\alpha) \cong \alpha/3$; we then obtain the Curie law

$$M = \chi H = N \frac{\mu^2}{3kT} H = \frac{C}{T} H. \tag{2.13}$$

The quantity C is called the Curie constant. Many paramagnetic substances satisfy a Curie law in their temperature dependence of the susceptibility; the range of validity of this law, however, is often quite different. Of course, the Curie constant can be reduced to the characteristic parameters of the atom only within the framework of a quantum-theoretical consideration.

3. The Zeeman Effect of Free Atoms

A quantum-mechanical treatment of the susceptibility of a system of free atoms requires the knowledge of the energy states of these atoms in an external magnetic field. The simplest case is that of an electron in the spherical-symmetric potential $e^{-1}V(r)$ of the nucleus and, if there exist other closed shells, also in the potential of the other electrons of the atom. The time-independent Schrödinger equation for such an electron reads

$$\left\{\frac{1}{2m}\left(\boldsymbol{p}-\frac{e}{c}\boldsymbol{A}\right)^2+V(r)+\mu_{\mathrm{B}}(\boldsymbol{\sigma}\cdot\boldsymbol{H})\right\}\psi = E\psi.$$

Here $\hbar \boldsymbol{s} = \hbar\boldsymbol{\sigma}/2$ is the spin operator of the electron and $\sigma_x, \sigma_y, \sigma_z$ are the Pauli matrices. The field is again assumed to be parallel to the z-axis and $\boldsymbol{A}$ shall again be given in the form $\boldsymbol{A} = \frac{1}{2}H(-y, x, 0)$. The Schrödinger equation can then be written explicitly (when we omit the small diamagnetic term) as:

$$\left\{\frac{p^2}{2m}+V(r)+\mu_{\mathrm{B}}H(l_z+2s_z)\right\}\psi = E\psi. \tag{2.14}$$

The unperturbed states ($H = 0$) are assumed to be known and given by

$$\mathscr{H}_0\psi_{n,l} \equiv \left(\frac{p^2}{2m}+V(r)\right)\psi_{n,l} = E^0_{nl}\psi_{nl}.$$

The unperturbed eigenvalues $E^0_{n,l}$ depend only on the principal quantum number n and the orbital quantum number l; this is due to the rotational symmetry of the problem considered. These eigenvalues are therefore $2(2l+1)$-fold degenerate. When we now choose basis functions in the $2(2l+1)$-dimensional space which are eigenfunctions of l_z and s_z, i.e.

$$l_z\psi_{n,l,m_l,m_s} = m_l\psi_{n,l,m_l,m_s}, \qquad -l \leq m_l \leq +l$$
$$s_z\psi_{n,l,m_l,m_s} = m_s\psi_{n,l,m_l,m_s}, \qquad m_s = \pm\tfrac{1}{2},$$

we obtain the eigenvalues of (2.14) in the form of

$$E_{n,l,m_l,m_s} = E^0_{n,l} + \mu_B H(m_l + 2m_s). \tag{2.15}$$

Thus the original degeneracy of the electron level is partly removed (see Fig. 8).

It is reasonable to define a total angular momentum of the electron, $\boldsymbol{j} = \boldsymbol{l}+\boldsymbol{s}$, which in any case is a conserved quantity. Accordingly we have a total angular momentum quantum number, j, which, for given

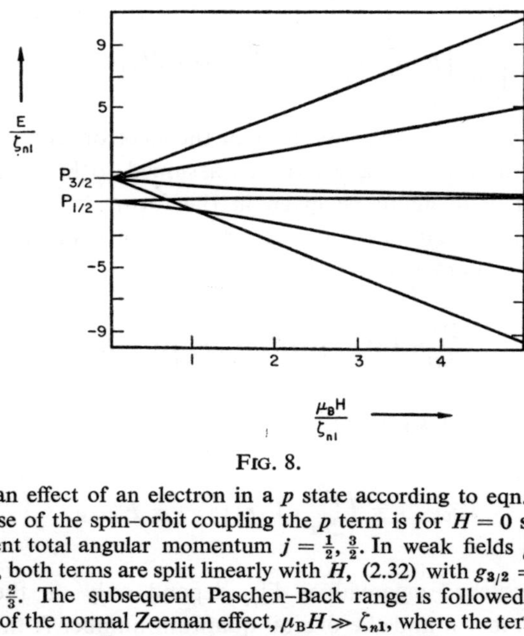

FIG. 8.

Zeeman effect of an electron in a p state according to eqn. (2.31). Because of the spin–orbit coupling the p term is for $H = 0$ split for different total angular momentum $j = \frac{1}{2}, \frac{3}{2}$. In weak fields $\mu_B H \ll \zeta_{nl}$, both terms are split linearly with H, (2.32) with $g_{3/2} = \frac{4}{3}$ and $g_{1/2} = \frac{2}{3}$. The subsequent Paschen–Back range is followed by the range of the normal Zeeman effect, $\mu_B H \gg \zeta_{nl}$, where the term energies are again linear functions of H.

l, can only take the values $j = l \pm 1/2$; the component of the total angular momentum (along the direction of the field) is given by the quantum number m_j whose values lie in the interval $(-j, +j)$. Since $m_j = m_l + m_s$, we can express one of the quantum numbers in (2.15) m_s or m_l, in terms of m_j.

The case here described is called the normal Zeeman effect for atoms which, apart from closed shells, possess but a single electron.

In the case of atoms with more electrons outside the closed shells, in addition to the spherical-symmetric potential each electron is subject to a Coulomb interaction with the other electrons (in a first approximation). For these N electrons the Hamiltonian reads

$$\mathcal{H} = \sum_{i=1}^{N} \left(\frac{p_i^2}{2m} + V(r_i) \right) + \mu_B H \sum_{i=1}^{N} (l_{iz} + 2s_{iz}) + \sum_{i<j=1} \frac{e^2}{|\boldsymbol{r}_i - \boldsymbol{r}_j|}. \quad (2.16)$$

The operator L of the total orbital angular momentum is defined by

$$\boldsymbol{L} = \sum_{i=1}^{N} \boldsymbol{l}_i$$

and the total spin by

$$\boldsymbol{S} = \sum_{i=1}^{N} \boldsymbol{s}_i .$$

The unperturbed Hamilton operator $\mathcal{H}_0$ (2.16) with $H = 0$ commutes with $\boldsymbol{S}$ since it does not contain $\boldsymbol{S}$; it also commutes with $\boldsymbol{L}$ since $\mathcal{H}_0$ is rotationally invariant, i.e. it remains unchanged when all coordinates are rotated. The eigenvalues of $\mathcal{H}_0$ will therefore depend on the eigenvalues of $\boldsymbol{L}^2$ and $\boldsymbol{S}^2$ in the case of a given electron configuration (equal distribution of electrons over the shells). The eigenvalues of $\boldsymbol{L}^2$ are $L(L+1)$ and those of $\boldsymbol{S}^2$ are $S(S+1)$. The energy terms of $\mathcal{H}_0$ are therefore denoted by E_{LS}^0, and the usual symbol ^{2S+1}L is applied to the pertinent levels (also called multiplets). For L the same letters are used as in the case of the electron levels, with the only exception that now capitals are used, i.e. S, P, D, F, etc. for a multiplet with orbital angular momentum $L = 0, 1, 2, 3$, etc. The total spin quantum number S gives the so-called multiplicity of the multiplet, $2S+1$. As a rule, atomic paramagnetism depends only on the energetically lowest multiplet, since the other multiplets are higher by about 1 eV or 10^4 cm^{-1}. (For comparison it should be mentioned that at room temperature kT is about $\frac{1}{40}$ eV.) The lowest multiplet of a configuration is determined by Hund's rule, which says that the multiplet of highest multiplicity and maximum possible quantum number of orbital angular momentum (L)

has the lowest energy. This rule is easy to understand. According to the Pauli principle two electrons with the same spin cannot be close to one another (Fermi hole, II.1). The Coulomb energy will therefore be relatively low for a state of maximum multiplicity, that is, for a state where as many electrons as possible have the same spin, and therefore the total energy of the state will be small. The other part of Hund's rule, maximum L, can be explained in a similar way. In a state of maximum orbital angular momentum the electrons can possess parallel and almost equal components of orbital angular momentum (m_l), i.e. roughly speaking, they can revolve in the same sense about the nucleus. This reduces the electrostatic repulsion and thus the total energy. The effects of the higher multiplets of a configuration are, as a rule, not so simple to explain (cf. Slater 1960; Condon and Shortley, 1959; Sobelman, 1972).

The multiplets have degeneracies $(2S+1)\cdot(2L+1)$. In a magnetic field this degeneracy is partly removed. Denoting by M_L and M_S the eigenvalues of L_z and S_z,

$$(-L \leqslant M_L \leqslant L; \quad -S \leqslant M_S \leqslant S),$$

we obtain from (2.16), in analogy with (2.15), for the terms of a multiplet in a magnetic field

$$E_{L,S,M_L,M_S} = E^0_{L,S} + \mu_B H(M_L + 2M_S). \tag{2.17}$$

This splitting of the multiplet, which is characteristic of the normal Zeeman effect, can be observed only in the case of extraordinarily high field strengths (because of the spin–orbit coupling). Generally such field strengths cannot be reached, and one remains in the range of the anomalous Zeeman effect (weak fields) or in the Paschen-Back range (medium fields). In these ranges the conditions are again simplest for hydrogenic atoms. In the rest system of the nucleus we have the electric field $e\boldsymbol{E} = -\nabla V(r)$. Since the electron moves, however, it is also under the action of a magnetic field of $(-1/c)[\boldsymbol{v} \wedge \boldsymbol{E}]$, $\boldsymbol{v}$ being the velocity of the electron. In the Hamilton operator this additional field gives rise to the term

$$\frac{2\mu_B}{c}([\boldsymbol{E} \wedge \boldsymbol{v}]\cdot\boldsymbol{s}) = \frac{\hbar^2}{m^2c^2}\,\frac{1}{r}\,\frac{dV}{dr}(\boldsymbol{l}\cdot\boldsymbol{s}),$$

describing the electron spin energy in the additional field, i.e. the spin–orbit coupling. Here $\hbar \boldsymbol{l} = -m[\boldsymbol{v} \wedge \boldsymbol{r}]$ is the orbital angular momentum of the electron. A more accurate calculation shows that this expression for the spin–orbit coupling is too large by a factor of two. This is due to the fact that the electron's motion about the nucleus is not uniform but accelerated. We see that the spin-orbit coupling, which is of the order of $1/c^2$, is a relativistic correction to the main terms of the Hamiltonian (c being the velocity of light). It is therefore more correct to start from the exact Dirac equation, and to expand it up to terms of the order of $1/c^2$ in order to obtain the required non-relativistic Schrödinger equation. It is easy (Becker and Sauter, 1959; Akhiezer and Berestetskii, 1962) to obtain for the Hamiltonian

$$\mathcal{H} = \frac{p^2}{2m} + V(r) + \mu_B H(l_z + 2s_z) - \frac{p^4}{8m^3c^2} + \frac{\hbar^2}{8m^2c^2}(-\nabla^2)V(r) + \frac{\hbar^2}{2mc^2}\frac{1}{r}\frac{dV(r)}{dr}(\boldsymbol{l}\cdot\boldsymbol{s}). \tag{2.18}$$

The first three terms agree with (2.14), the fourth term is simply due to the relativistic mass variation, the fifth one has no classical analogue, and the last term, finally, is the correct spin–orbit coupling with the Thomas factor 1/2. It should be mentioned that in this approximation the magnetic moment of the spin which, so far, was put equal to μ_B, must now be replaced by $\mu = \mu_B(1+\alpha/2\pi)$, where α is the fine structure constant, given by $\alpha = e_0^2/hc \cong 1/137$; $\mu_B\alpha/2\pi$ is called the anomalous magnetic moment of the electron. It results from the interaction of the electron with the vacuum (Akhiezer and Berestetskii, 1962). Because of its smallness we shall disregard this correction; we shall also neglect the fourth and fifth terms in (2.18), as they result only in a small shift of the electron doublet.

The unperturbed Hamiltonian ($H = 0$) will no longer commute with $\boldsymbol{l}$ and $\boldsymbol{s}$, but only with $\boldsymbol{l}^2$, $\boldsymbol{s}^2$, and, of course, with the total angular momentum $\boldsymbol{j} = \boldsymbol{l}+\boldsymbol{s}$ as is easy to see. The initially $2(2l+1)$-fold degeneracy of the doublets is therefore partly removed and the unperturbed energy eigenvalues are, with $2(\boldsymbol{l}\cdot\boldsymbol{s}) = \boldsymbol{j}^2 - \boldsymbol{l}^2 - \boldsymbol{s}^2$, now given by

$$E^0_{n,l,j} = E^0_{n,l} + \tfrac{1}{2}\zeta_{n,l}\big(j(j+1) - l(l+1) - s(s+1)\big) \tag{2.19}$$

where

$$\zeta_{n,l} = \frac{\hbar^2}{2m^2c^2} \langle n, l| \frac{1}{r}\frac{dV}{dr} |n, l\rangle \quad \text{and} \quad s = \frac{1}{2}. \tag{2.20}$$

For a Coulomb potential $V(r) = -Ze_0^2/r$ we obtain (Bethe and Salpeter, 1957):

$$\zeta_{n,l} = \frac{Z^4 e_0^2}{2m^2c^2} \cdot \frac{1}{n^3(l+1)(l+\frac{1}{2})l \cdot a_B^3}. \tag{2.21}$$

Because of the spin–orbit coupling the original level $E^0_{n,l}$ is split according to the eigenvalues of total angular momentum, $j = l \pm \frac{1}{2}$; both doublet terms have the degeneracy $2j+1$. It is easy to see that the mean splitting provided with the degeneracy as a weight is vanishing, so that the centre of gravity of the doublet terms remains unchanged, at $E^0_{n,l}$. Hence we can draw the conclusion that for $l = 0$ the indeterminate term arising with (2.21) in (2.19) can be set equal to zero. For a more exact discussion we refer to the literature (Bethe and Salpeter, 1957; Condon and Shortley, 1959). If $l = 0$, the doublet splitting is given by

$$E^0_{n,l,j+1/2} - E^0_{n,l,j-1/2} = \zeta_{n,l}(l+\tfrac{1}{2}).$$

For alkali metals values between 1 cm^{-1} and 500 cm^{-1} are obtained. Before entering into further details of the anomalous Zeeman effect, we shall discuss the Wigner–Eckart theorem which determines the matrix elements of irreducible tensor operators. In the following it will be used frequently. An irreducible tensor operator of order k is understood to be an operator whose components

$$T_q^{(k)}, \quad -k \leqslant q \leqslant +k,$$

have the following commutation relations with the angular momentum vector $\boldsymbol{j}$:

$$\begin{aligned} [j^{\pm}, T_q^{(k)}]_- &= \sqrt{k(k+1)-q(q\pm 1)}\, T_{q\pm 1}^{(k)} \\ [j_z, T_q^{(k)}] &= q T_q^{(k)} \end{aligned} \tag{2.22}$$

where $j^{\pm} = j_x \pm i\, j_y$.

An example of a tensor operator of the first rank ($k = 1$) is the angular-momentum operator $\boldsymbol{j}$, with $j_0 = j_z$ and

$$j_{\pm 1} = \mp\frac{1}{\sqrt{2}}(j_x \pm i j_y),$$

with which we obtain from (2.22) the usual commutation relations of the angular momentum components. Other examples of first-rank tensor operators, also called vector operators, are $\boldsymbol{s}$ or $\boldsymbol{l}$. The eigenfunctions of $\boldsymbol{l}^2$ to the eigenvalue $l(l+1)$ form a tensor operator of rank l when they are considered to be operators; in polar coordinates r, ϑ, φ, they are the spherical functions Y_l^m normalised to 1:

$$Y_l^m(\vartheta, \varphi) = (-)^m \sqrt{\frac{2l+1}{4\pi}\cdot\frac{(l-m)!}{(l+m)!}}\, P_l^m(\cos\vartheta)e^{im\varphi}$$

where (2.23)

$$P_l^m(u) = \frac{(1-u^2)^{m/2}}{2^l\cdot l!}\,\frac{d^{l+m}}{du^{l+m}}(u^2-1)^l.$$

It is easy to see that Y_l^m satisfies the relations (2.22), if we make use of the well-known relations

$$l^{\pm}Y_l^m = \sqrt{l(l+1)-m(m\pm 1)}\,Y_l^{m\pm 1},$$
$$l_z Y_l^m = mY_l^m$$

(cf., for example, Edmonds, 1957). Of course the spin operator commutes with Y_l^m.

The Wigner–Eckart theorem yields the following relation for the matrix elements of the tensor operator components:

$$\langle \tau j m_j |\, T_q^{(k)} \,| \tau' j' m_{j'}' \rangle = \frac{\langle \tau j \,||\, T^{(k)} \,||\, \tau' j' \rangle}{\sqrt{2j+1}} \langle j' k m_{j'}' q \,|\, j m_j \rangle. \qquad (2.24)$$

This theorem is valuable as it factorises the matrix element. Here we shall not consider its proof (for a proof, see, for example, Messiah, 1960 or Edmonds, 1957). The matrix element

$$\langle \tau j \,||\, T^{(k)} \,||\, \tau' j' \rangle$$

is called the reduced matrix element of $T^{(k)}$; τ, τ' stand for all other possible quantum numbers. The last factor in (2.24) is a Clebsch–Gordan coefficient. These coefficients are defined by

$$|j_1j_2jm\rangle = \sum_{m_1, m_2} |j_1m_1\rangle\,|j_2m_2\rangle\langle j_1j_2m_1m_2\,|\,jm\rangle. \qquad (2.25)$$

Thus the C.–G. coefficients determine the state vector of two particles with the angular momenta j_1 and j_2, whose combined state has the angular momentum j and the angular-momentum component m. From this interpretation it becomes evident that the C.–G. coefficients are subject to certain restrictions, that is, they vanish if certain conditions are not satisfied. Thus, for example, it is obvious that the angular-momentum components are additive quantities, i.e.

$$m = m_1+m_2, \qquad (2.26a)$$

so that the sum in (2.25) is limited. From a vector representation of the angular momenta we may also derive at once the triangle inequality

$$|j_1-j_2| \leqslant j \leqslant j_1+j_2. \qquad (2.26b)$$

In addition to this we have a series of symmetry and recursion relations for the C.–G. coefficients (Edmonds, 1957). One of the symmetry relations shall be given without proof:

$$\langle j_1j_2m_1m_2\,|\,jm\rangle = (-)^{j_1+j_2-j}\langle j_1j_2-m_1-m_2\,|\,j-m\rangle. \qquad (2.27)$$

The C.–G. coefficients are tabulated (Condon and Shortley, 1959; Edmonds, 1957); they can also be calculated with the help of the recursion formulas.

In the following we shall be interested in the matrix element of a vector in the space of the $|\tau jm_j\rangle$ for fixed τ and j. According to the Wigner–Eckart theorem, we have first

$$\langle \tau jm_j|\,T_q^{(1)}\,|\tau jm_j'\rangle = \frac{\langle \tau j|\,|T^{(1)}|\,|\tau j\rangle}{\sqrt{2j-1}}\langle j\,|m_j'q\,|\,jm_j\rangle. \qquad \text{(a)}$$

In addition to this, we have in the eigen-representation of the angular momentum vector

$$\langle \tau j m_j | j_q | \tau' j' m'_{j'} \rangle = \frac{\langle \tau j || \boldsymbol{j} || \tau' j' \rangle}{\sqrt{2j-1}} \langle j' | m'_{j'} q | j m_j \rangle \, \delta_{\tau,\tau'} \delta_{j,j'} . \tag{b}$$

so that we can write (a) in the form

$$\langle \tau j m_j | T^{(1)}_q | \tau j m'_j \rangle = \frac{\langle \tau j || T^{(1)} || \tau j \rangle}{\langle \tau j || \boldsymbol{j} || \tau j \rangle} \langle \tau j m_j | j_q | \tau j m'_j \rangle. \tag{c}$$

Hence we have the result that in this space the vector operator can be replaced by the angular-momentum vector $\boldsymbol{j}$ apart from a constant which is determined by calculating the expectation value of the scalar product of $T^{(1)}$ by $\boldsymbol{j}$; as it is a scalar quantity, the product cannot depend on m_j, so that we can take an arbitrary basis vector. Thus we obtain, with (2.24), (b) and (c):

$$\begin{aligned}
\langle (T^{(1)} \cdot j) \rangle &= \langle \tau j m_j | (T^{(1)} \cdot \boldsymbol{j}) | \tau j m_j \rangle \\
&= \sum_{q=-1}^{+1} \langle \tau j m_j | T^{(1)}_q j_q | \tau j m_j \rangle \\
&= \sum_{\tau', j', m'_{j'}} \sum_{q=-1}^{+1} \langle \tau j m_j | T^{(1)}_q | \tau' j' m'_{j'} \rangle \langle \tau' j' m'_{j'} | j_q | \tau j m_j \rangle \\
&= \frac{\langle \tau j || T^{(1)} || \tau j \rangle}{\langle \tau j || \boldsymbol{j} || \tau j \rangle} \sum_{m'_j, q} \langle \tau j m_j | j_q | \tau j m'_j \rangle \langle \tau j m'_j | j_q | \tau j m_j \rangle \\
&= \frac{\langle \tau j || T^{(1)} || \tau j \rangle}{\langle \tau j || \boldsymbol{j} || \tau j \rangle} \cdot j(j+1).
\end{aligned}$$

Using this we can write (c) in the form

$$\langle \tau j m_j | T^{(1)}_q | \tau j m'_j \rangle = \frac{\langle (T^{(1)} \cdot \boldsymbol{j}) \rangle}{j(j+1)} \langle \tau j m_j | j_q | \tau j m'_j \rangle. \tag{2.28}$$

Taking into account that the scalar product $(\boldsymbol{s} \cdot \boldsymbol{j}) = \boldsymbol{s}^2 + (\boldsymbol{s} \cdot \boldsymbol{l})$ in the space of the vectors $| sljm_j \rangle$ has the eigenvalue

$$\tfrac{1}{2} \big(j(j+1) - l(l+1) + s(s+1) \big)$$

we get as a special case for the z-component of the spin vector, s_z, the equation

$$\langle \tau sljm_j | s_z | \tau sljm_j \rangle = \frac{j(j+1)-l(l+1)+s(s+1)}{2j(j+1)} m_j . \quad (2.29)$$

The splitting of the electron doublet (2.19) in the magnetic field is more complicated than in the case of the normal Zeeman effect. The Hamiltonian (2.18)

$$\mathcal{H} = \mathcal{H}_0 + \mu_B H(j_z + s_z)$$

does not commute with $\boldsymbol{j}^2$ since $[\boldsymbol{j}^2, s_z] \neq 0$. The magnetic field induces transitions between doublet terms of different j. As j_z commutes with $\mathcal{H}$, it is reasonable to try the following solution for the wave function:

$$|\psi\rangle = c_+ |l+\tfrac{1}{2}, m\rangle + c_- |l-\tfrac{1}{2}, m\rangle,$$

m being the eigenvalues of j_z. From this we obtain with the help of the Schrödinger equation $\mathcal{H}|\psi\rangle = E|\psi\rangle$:

$$\begin{aligned} Ec_+ &= \varepsilon_+ c_+ + \mu_B H \langle l+\tfrac{1}{2}, m | s_z | l-\tfrac{1}{2}, m\rangle c_-, \\ Ec_- &= \varepsilon_- c_- + \mu_B H \langle l-\tfrac{1}{2}, m | s_z | l+\tfrac{1}{2}, m\rangle c_+. \end{aligned}$$

$\varepsilon_\pm$ are the two diagonal elements of $\mathcal{H}$:

$$\varepsilon_\pm = \langle \tau, l\pm\tfrac{1}{2}, m | \mathcal{H} | \tau, l\pm\tfrac{1}{2}, m\rangle = E^0_{n,\, l,\, l\pm 1/2} + g_{l\pm 1/2}\mu_B H m.$$

$g_{l\pm 1/2}$ is the Landé factor of the two unperturbed doublet terms, which is given by (2.29):

$$g_j = 1 + \frac{j(j+1)+s(s+1)-l(l+1)}{2j(j+1)} . \quad (2.30)$$

Here ($s = \frac{1}{2}$) it has the value $1 \pm 1/(2l+1)$, according to whether $j = l+\frac{1}{2}$ or $j = l-\frac{1}{2}$. $E^0_{n,\, l,\, l\pm 1/2}$ is given by (2.19). Taking into account that the determinant of the set of equations for $c_\pm$ vanishes, and that s_z is Hermitean, we obtain for the two eigenvalues

$$\begin{aligned} E_{1,\,2}(m) &= \tfrac{1}{2}(\varepsilon_+ + \varepsilon_-) \\ &\quad \pm \tfrac{1}{2}\sqrt{(\varepsilon_+ - \varepsilon_-)^2 + 4(\mu_B H)^2 |\langle l+\tfrac{1}{2}, m | s_z | l-\tfrac{1}{2}, m\rangle|^2} . \end{aligned}$$

The off-diagonal element of s_z is easy to calculate from

$$\tfrac{1}{4} = \langle jm|s_z^2|jm\rangle = \sum_{m'j'} \langle jm|s_z|j'm'\rangle\langle j'm'|s_z|jm\rangle.$$

Using (2.24), (2.29), and (2.30) we obtain

$$\begin{aligned}\tfrac{1}{4} &= \sum_{j'} \langle jm|s_z|j'm\rangle\langle j'm|s_z|jm\rangle \\ &= m^2(g_j-1)^2+|\langle jm|s_z|j\mp 1m\rangle|^2,\end{aligned}$$

or $j = l\pm\frac{1}{2}$. Assuming $j = l+\frac{1}{2}$ we obtain

$$\left|\left\langle l+\frac{1}{2},\ m|s_z|l-\frac{1}{2},\ m\right\rangle\right|^2 = \frac{1}{4}-\frac{m^2}{(2l+1)^2}.$$

Thus we obtain the $2(2l+1)$ levels in the magnetic field; with (2.19) we have

$$\begin{aligned}E_{1,2}(m) = E^0_{n,l}&-\tfrac{1}{4}\zeta_{n,l}+\mu_B Hm \\ &\pm\tfrac{1}{2}\sqrt{\zeta^2_{n,l}(l+\tfrac{1}{2})^2+2\zeta_{n,l}\mu_B H\cdot m+(\mu_B H)^2}, \quad (2.31)\end{aligned}$$

$|m| \leqslant l+\frac{1}{2}$. The splitting is more complicated than in the case of the normal Zeeman effect. The conditions are again simple if the fields are weak, $\zeta_{n,l} \gg \mu_B H$. We can then neglect the terms containing H^2 or, which is the same, omit the off-diagonal elements of s_z so that in a first approximation $\mathbf{j}^2$ commutes with $\mathcal{H}$. The eigenvalues are then the $\varepsilon_\pm$ or, simply

$$E_{n,l,j,m} = E^0_{nl,j}+g_j\mu_B m_j H. \quad (2.32)$$

This is the scheme of terms for the anomalous Zeeman effect (see Fig. 8). In the case of $\zeta_{n,l} \ll \mu_B H$, we obtain from (2.31) the scheme of terms for the normal Zeeman effect with $m = m_l\pm\frac{1}{2}$. Between these two extremes lies the transition region, the Paschen–Back range, which is described by (2.31) for $|m| \neq l+\frac{1}{2}$. In the case of $|m| = l+\frac{1}{2}$ eqn. (2.32) holds true as the off-diagonal element of s_z vanishes (see Fig. 8).

If several electrons are outside the closed shells, which cause only a spherical-symmetric potential, the problem becomes complicated because of the interaction between these electrons. In a non-relativistic approximation up to terms of the order of $1/c^2$ (see, for example, Bethe and Salpeter, 1957), the interaction energy of two electrons at $\boldsymbol{r}_1$ and $\boldsymbol{r}_2$, respectively, is given by

$$U(\boldsymbol{r}_1, \boldsymbol{r}_2) = \frac{e^2}{r} - \frac{e^2}{2m^2c^2}\left\{\frac{1}{r}(\boldsymbol{p}_1 \cdot \boldsymbol{p}_2) + \frac{1}{r^3}(\boldsymbol{r} \cdot (\boldsymbol{r} \cdot \boldsymbol{p}_1) \cdot \boldsymbol{p}_2)\right\}$$
$$-\frac{\pi\hbar^2 e^2}{m^2c^2}\delta(\boldsymbol{r}) + \frac{e^2\hbar^2}{m^2c^2}\left\{\frac{(\boldsymbol{s}_1 \cdot \boldsymbol{s}_2)}{r^3} - 3\frac{(\boldsymbol{s}_1 \cdot \boldsymbol{r})(\boldsymbol{s}_2 \cdot \boldsymbol{r})}{r^5}\right.$$
$$\left. -\frac{8\pi}{3}(\boldsymbol{s}_1 \cdot \boldsymbol{s}_2)\delta(\boldsymbol{r})\right\} - \frac{e^2\hbar}{2m^2c^2r^3}[(\{[\boldsymbol{r} \wedge \boldsymbol{p}_1] - 2[\boldsymbol{r} \wedge \boldsymbol{p}_2]\} \cdot \boldsymbol{s}_1)$$
$$+(\{2[\boldsymbol{r} \wedge \boldsymbol{p}_1] - [\boldsymbol{r} \wedge \boldsymbol{p}_2]\} \cdot \boldsymbol{s}_2)], \qquad (2.33)$$

where $\boldsymbol{r} = \boldsymbol{r}_1 - \boldsymbol{r}_2$. In this expression for the interaction, the position of the momentum and positron coordinates are important as they do not commute. The interpretation of the individual terms is relativelc simple. The first two terms represent the Coulomb interaction wity retardation; the following δ function is characteristic of a relativistih approximation of the Dirac equation, (2.18), with $\nabla^2 V = 4\pi Ze^2\delta(r)$. The following term describes the dipole interaction of the electron spins but with an additional term

$$-\frac{8\pi}{3}\frac{e^2\hbar^2}{m^2c^2}(\boldsymbol{s}_1 \cdot \boldsymbol{s}_2)\delta(\boldsymbol{r}),$$

which precisely corresponds to the Fermi contact potential, which is important in the case of hyperfine interactions. The terms of (2.33) mentioned so far are neglected in the following (with the exception of the main term), as they commute with the total spin $\boldsymbol{S} = \boldsymbol{s}_1 + \boldsymbol{s}_2$ and the total orbital angular momentum $\boldsymbol{L} = \boldsymbol{l}_1 + \boldsymbol{l}_2$ (with the exception of the middle term in the spin–spin interaction), so that their effect is but a small shift of the interesting multiplet and no splitting of it. The last terms of (2.33) are spin–orbit coupling terms which describe both the interaction between spin and orbital moment of the same electron, and that between spin and orbital moment of different electrons. Their effect on the multiplets is difficult to comprehend. Part of the expressions can be reduced to the form of the ordinary spin–orbit coupling,

$\sim(\boldsymbol{l}_i\cdot\boldsymbol{s}_i)$, so that the potential in (2.20) and (2.21) must be replaced by an effective potential. Although the other part cannot be neglected, particularly in the case of light atoms (Bethe and Salpeter, 1957), it plays only a minor role in most other atoms, compared with the ordinary spin–orbit coupling (Slater, 1960). We shall therefore disregard these terms in the following. The Hamiltonian of a system of N electrons not contained in closed shells can therefore be written in this approximation in the following form:

$$\mathcal{H} = \sum_{i=1}^{N}\left(\frac{p_i^2}{2m}+V(r_i)+\frac{\hbar^2}{2mc^2}\,\frac{1}{r_i}\,\frac{dV(r_i)}{dr_i}\,(\boldsymbol{l}_i\cdot\boldsymbol{s}_i)\right)+\sum_{i<j=1}^{N}\frac{e^2}{|\boldsymbol{r}_i-\boldsymbol{r}_j|}\,. \tag{2.34}$$

Here the potential in the spin–orbit coupling term might be replaced by an effective potential. We see that $\boldsymbol{L}^2$ and $\boldsymbol{S}^2$ no longer commute with $\mathcal{H}$; spin–orbit coupling will therefore induce transitions between the multiplets of a configuration. Since, however, for not too heavy atoms the multiplet distances exceed by one or two powers of ten the multiplet splittings caused by spin–orbit coupling, an admixture of states of other multiplets to those of a certain multiplet can be disregarded so that, in this approximation—the Russell–Saunders approximation—$\boldsymbol{L}$ and $\boldsymbol{S}$ remain "good" quantum numbers. It is well known that in the case of heavy atoms the Russell–Saunders approximation fails, since spin–orbit coupling exceeds the Coulomb interaction between the electrons; this case, however, shall not be considered (see Condon and Shortley, 1959).

When we replace for brevity the factor in front of $(\boldsymbol{l}_i\cdot\boldsymbol{s}_i)$ by g_i and choose the vectors $|\tau LSM_LM_S\rangle$ as a basis in the space of a multiplet (τ denoting the configuration), we obtain the following for the matrix elements of spin–orbit coupling in a multiplet of a configuration:

$$\begin{aligned}&\sum_{j=1}^{N}\langle\tau LSM_LM_S|\,g_i(\boldsymbol{l}_i\cdot\boldsymbol{s}_i)\,|\tau LSM_L'M_S'\rangle\\ &=\sum\langle\tau|\,g_i\,|\tau\rangle\,\frac{\langle(\boldsymbol{l}_i\cdot\boldsymbol{L})\rangle\langle(\boldsymbol{s}_i\boldsymbol{S})\rangle}{S(S+1)\,L(L+1)}\,\langle\tau LSM_LM_S|\,(\boldsymbol{L}\cdot\boldsymbol{S})\,|\tau LSM_L'M_S'\rangle\\ &=\zeta(\tau, L, S)\cdot\langle\tau LSM_LM_S|\,(\boldsymbol{L}\cdot\boldsymbol{S})\,|\tau LSM_L'M_S'\rangle;\end{aligned} \tag{2.35}$$

here we have applied (2.28) to both $\boldsymbol{l}_i$ and $\boldsymbol{s}_i$. In the Russell–Saunders approximation (assuming transitions to other configurations to be negligible), we have thus to replace the spin–orbit coupling in (2.34) by $\zeta(\tau, L, S)\cdot(\boldsymbol{L}\cdot\boldsymbol{S})$. A case interesting for the following is that of a configuration nl^N, i.e. a case where all electrons are in the same shell with the orbital angular momentum l (for example, the $3d$ shell, the $4f$ shell, ...). All $\langle\tau|\, g_i\, |\tau\rangle$ are then equal and identical with the $\zeta_{n,l}$ defined in (2.19) and (2.20). In the following we are interested only in the lowest multiplet, i.e. the multiplet of maximum multiplicity. It is then easy to obtain $\zeta(\tau, L, S)$. We calculate the diagonal matrix element of (2.35) with $M_S = M_S' = S$. Because of the relations

$$s^{\pm}\,|sm\rangle \equiv (s_x \pm is_y)\,|sm\rangle = \sqrt{s(s+1)-m(m\pm 1)}\,|sm\pm 1\rangle, \quad (2.36)$$
$$s_z\,|sm\rangle = m\,|sm\rangle,$$

which apply to every angular momentum, only the products of the z-components in the scalar products $(\boldsymbol{L}\cdot\boldsymbol{S})$ and $(\boldsymbol{s}_i\cdot\boldsymbol{l}_i)$ contribute to the matrix element, and we arrive at

$$\zeta(\tau, L, S)M_L S = \zeta_{n,l}\sum_{i=1}^{N} m_{si}m_{li} \quad \text{with} \quad \sum_{i=1}^{N} m_{li} = M_L \quad \text{and} \quad \sum_{i=1}^{N} m_{si} = S.$$

If the shell is less than half filled, i.e. $N < 2l+1$, all $m_{si} = \frac{1}{2}$ so that $2S\zeta(\tau, L, S) = \zeta_{n,l}$. If the shell is more than half filled, i.e. $n > 2l+1$, a number of $N_1 = 2l+1$ electrons will have parallel spins, $m_{si} = \frac{1}{2}$, and no resulting orbital angular momentum,

$$\sum_{i=1}^{2l+1} m_{li} = 0,$$

whereas the other $N-N_1$ electrons have antiparallel spins, $m_{si} = -\frac{1}{2}$, and the necessary resulting orbital moment

$$\sum_{i=N_1+1}^{N} m_{li} = M_L,$$

i.e. we have

$$\zeta(\tau, L, S)M_L S = \zeta_{n,l}\left\{\frac{1}{2}\sum_{i=1}^{N_1} m_{li} - \frac{1}{2}\sum_{i=N_1+1}^{N} m_{li}\right\} = -\frac{1}{2}M_L\zeta_{n,l}.$$

Hence it follows that

$$\zeta(\tau, L, S) = \pm\frac{1}{2S}\zeta_{n,l}. \tag{2.37}$$

The upper (lower) sign applies to a shell which is filled less (more) than half.

Since $(\boldsymbol{L}\cdot\boldsymbol{S})$ commutes with $\boldsymbol{J}^2$, $\boldsymbol{J} = \boldsymbol{L}+\boldsymbol{S}$, the multiplet splits its energy levels as in the one-electron system with respect to the quantum numbers of total angular momentum, $|L-S| \leqslant J \leqslant L+S$. It is thus more precise to use the symbol ${}^{2S+1}L_J$ for the multiplet terms. The energy of the term ${}^{2S+1}L_J$ is, because $2(\boldsymbol{L}\cdot\boldsymbol{J}) = \boldsymbol{J}^2-\boldsymbol{L}^2-\boldsymbol{S}^2$, given by

$$E^0_{L,S,J} = E^0_{L,S}+\tfrac{1}{2}\zeta(\tau, L, S)\cdot(J(J+1)-L(L+1)-S(S+1)) \tag{2.38}$$

in analogy to (2.19). Owing to the $2J+1$ possibilities of adjustment of the total angular momentum, $-J \leqslant M_J \leqslant J$, each term has a degeneracy of $2J+1$. The mean of the multiplet splitting with the degeneracy factor as weight is again equal to zero as is easy to verify.

In the presence of a magnetic field the atomic Hamiltonian reads

$$\mathscr{H} = \mathscr{H}_0+\mu_B H(L_z+2S_z) = \mathscr{H}_0+\mu_B H(J_z+S_z) \tag{2.39}$$

where $\mathscr{H}_0$ is the unperturbed Hamilton operator from (2.34). Since (2.39) does not commute with L_z and S_z (because of spin–orbit coupling), the eigenvalues of (2.39) are not classified according to the quantum numbers M_L and M_S as in (2.17). Only in the case of high field strengths, $\zeta(\tau, L, S) \ll \mu_B H$, can spin–orbit coupling be neglected, and the normal Zeeman effect is obtained (2.17).

The normal case of $\zeta(\tau, L, S) \gg \mu_B H$, results in the term splitting of the anomalous Zeeman effect. In a first approximation, i.e. when the off-diagonal elements of S_z with respect to the base $|JM_J\rangle$ are neglected, we obtain

$$\begin{aligned} E_{L,S,J,M_J} &= E^0_{L,S,J}+\mu_B H(M_J+\langle JM_J|S_z|JM_J\rangle) \\ &= E^0_{L,S,J}+g_J\mu_B HM_J \end{aligned} \tag{2.40}$$

with the Landé factor g_J

$$g_J = 1 + \frac{J(J+1)+S(S+1)-L(L+1)}{2J(J+1)} \tag{2.41}$$

according to (2.28) and (2.29), and in analogy to (2.30).

4. Paramagnetism of Free Atoms (Rare Earths)

The considerations in connection with the Zeeman effect can also be used to determine the energies and thus the magnetic moments to be expected for the atoms. In the classical derivation of the Curie law (2.13) this step was unnecessary, as fixed magnetic moments were assumed *a priori*, and we had only to average thermally over the orientation of these moments relative to the magnetic field. The quantum-mechanical derivation of the paramagnetic susceptibility therefore will always comprise two averaging processes: the quantum-mechanical expectation value of the magnetic moment with respect to the state of an atom, and the thermal mean over all states.

The statistical weight of a state ν is given by the Boltzmann factor

$$w_\nu = [\exp(-\beta E_\nu)]/Z \text{ where } \beta = 1/kT \text{ and } Z = \sum_\nu \exp(-\beta E_\nu),$$

E_ν being the energy eigenvalue of this state. Z is the partition function covering all states of the system considered. The thermal mean of an operator A, whose expectation value in the state ν is equal to A_ν, is given by

$$\bar{A} = \sum_\nu w_\nu A_\nu.$$

This formulation of the thermal mean requires the knowledge of the eigenvalues and eigenstates of the Hamiltonian. One can get rid of this restriction by introducing the density matrix ϱ, which is also called the statistical operator (ter Haar, 1961). It is defined by

$$\varrho = \frac{e^{-\beta\mathcal{H}}}{Z}, \qquad Z = \mathrm{Tr}\, e^{-\beta\mathcal{H}}. \tag{2.42}$$

We see that, when forming the trace with the eigenstates of $\mathcal{H}$, we rearrive at the old definition of the partition function. The eigenvalues of ϱ are the w_ν. Since the trace is invariant with respect to a unitary transformation of the basis vectors, we obtain for the thermal mean of A

$$\bar{A} = \text{Tr}\,\varrho A. \tag{2.43}$$

Thus in the energy representation we arrive back at the original form.

Here we are interested in the operator of the magnetic moment which, according to (2.39), is given by

$$m_z = -\frac{\partial \mathcal{H}}{\partial H} = -\mu_B(J_z+S_z),$$

if the magnetic field is parallel to the z-axis. The thermal mean of the magnetisation of a system of N independent atoms is given by

$$M = N\bar{m}_z = -N\mu_B \text{Tr}\,\varrho(J_z+S_z) = NkT\frac{\partial}{\partial H}\ln Z. \tag{2.44}$$

We therefore obtain for the susceptibility of such a system

$$\chi = -N\frac{\mu_B}{H}\,\text{Tr}\,\varrho(J_z+S_z) = N\frac{kT}{H}\frac{\partial}{\partial H}\ln Z. \tag{2.45}$$

Before discussing this formula in detail we want to consider the limit as $H \to 0$. In this case we can derive a general formula for the susceptibility which was first given by Kubo (1957, 1958). At the same time we shall generalise it for the case of fields which are inhomogeneous in space and time. A time-dependent field will cause transitions between the steady states of the system considered, i.e. the density matrix ϱ, which is used to calculate the expectation value of the magnetisation according to eqn. (2.43), will be time-dependent. Instead of (2.43) we therefore have

$$\bar{A}(t) = \text{Tr}\,\varrho(t)A. \tag{2.43a}$$

The time dependence of the density matrix is easy to obtain on the basis of the consideration that the density matrix ϱ describes the time dependence of the states (Landau and Lifshitz, 1958), while the operators used in our Schrödinger representation are independent of time. In the Heisenberg representation, however, the density matrix is time-independent, and the operators of the observable quantity A are time-dependent according to

$$A_H(t) = \mathrm{e}^{i\mathcal{H}t/\hbar} A\, \mathrm{e}^{-i\mathcal{H}t/\hbar},$$

i.e. they satisfy the equation of motion

$$i\hbar \dot{A}_H = [A_H, \mathcal{H}]_-.$$

The expectation value of an operator A is, of course, independent of the representation so that, according to eqn. (2.43a),

$$\bar{A}(t) = \mathrm{Tr}\, \varrho(t) A = \mathrm{Tr}\, \varrho_H A_H(t) = \mathrm{Tr}\, (\mathrm{e}^{-i\mathcal{H}t/\hbar} \varrho_H\, \mathrm{e}^{i\mathcal{H}t/\hbar} A)$$

will hold for every operator since cyclic permutation is possible within the trace. Comparison shows that

$$\varrho(t) = \mathrm{e}^{-i\mathcal{H}t/\hbar} \varrho_H\, \mathrm{e}^{i\mathcal{H}t/\hbar},$$

and the equation of motion for $\varrho(t)$ will read

$$i\hbar \dot{\varrho}(t) = [\mathcal{H}, \varrho(t)]_-.$$

This equation determines the dynamic behaviour of the system.

According to what has been said above, we are only interested in the susceptibility in the lowest order of the field strength. It is thus sufficient to solve the equation for ϱ only up to terms linear in the field strength. We write

$$\mathcal{H} = \mathcal{H}_0 + \mathcal{H}_1,$$

where $\mathcal{H}_0$ describes the unperturbed system and $\mathcal{H}_1$ contains the external field. Analogously we put

$$\varrho(t) = \varrho_0 + \varrho_1(t),$$

where ϱ_0 is the density matrix (2.42) of the unperturbed system (in it $\mathcal{H}$ is replaced by $\mathcal{H}_0$), and $\varrho_1(t)$ is a linear function of $\mathcal{H}_1$. After linearisation the equation of motion will then read

$$i\hbar\dot{\varrho}_1(t) = [\mathcal{H}_0, \varrho_1]_- + [\mathcal{H}_1, \varrho_0]_-.$$

As to its form, this is a linear inhomogeneous differential equation for $\varrho_1(t)$, the last term being the inhomogeneity. The solution is

$$\varrho_1(t) = \frac{1}{i\hbar} \int\limits_{-\infty}^{t} dt'\, e^{-i\mathcal{H}_0(t-t')/\hbar}[\mathcal{H}_1(t'), \varrho_0)]_- e^{i\mathcal{H}_0(t-t')/\hbar},$$

which is easy to verify by differentiation. Here we made use of the idea that the external field, and thus $\mathcal{H}_1$, was slowly switched on in the distant past, i.e. $\mathcal{H}_1(t)$ is assumed to contain an adiabatic switching-on factor $\exp \varepsilon t$ (ε being an infinitesimal positive quantity) so that at $t = -\infty$ it vanishes, and the system was then described by the density matrix ϱ_0.

Considering only the linear terms in the external perturbation, we have

$$\bar{A}(t) = \langle A \rangle + \mathrm{Tr}\, \varrho_1(t) A,$$

where $\langle \quad \rangle$ is the expectation value with respect to the unperturbed system. Here we are interested in a system of magnetic moments $\boldsymbol{m}_j$ at the points $\boldsymbol{r}_j$ which, at the time t, are under the influence of the external field $\boldsymbol{H}_j(t)$, i.e. $\mathcal{H}_1$ is given by

$$\mathcal{H}_1 = -\sum_{j,\nu} m_{j\nu} H_{j\nu}(t);$$

ν characterises the three vector components. If we had a continuous magnetisation density $\boldsymbol{m}(\boldsymbol{r})$ instead of the discrete moments, we would obtain

$$\mathcal{H}_1 = -\sum_{\nu} \int d^3\boldsymbol{r}\, m_\nu(\boldsymbol{r})\, H_\nu(\boldsymbol{r}, t).$$

Hence we have for $\varrho_1(t)$

$$\varrho_1(t) = -\frac{1}{i\hbar} \sum_{j,\nu} \int\limits_{-\infty}^{t} [m_{j\nu}(t'-t), \varrho_0]_-\, H_{j\nu}(t')\, dt',$$

where

$$m_{j\nu}(t) = e^{i\mathcal{H}_0 t/\hbar} m_{j\nu}\, e^{-i\mathcal{H}_0 t/\hbar}.$$

From this we obtain for the μth component of the magnetic moment at point $\boldsymbol{r}_i$, at time t:

$$\bar{m}_{i\mu}(t) = \langle m_{i\mu}\rangle - \frac{1}{i\hbar}\sum_{j,\nu}\int_{-\infty}^{t} dt'\ \mathrm{Tr}\ \{m_{i\mu}[m_{j\nu}(t'-t), \varrho_0]_-\}\, H_{j\nu}(t')$$

$$= \langle m_{i\mu}\rangle - \frac{1}{i\hbar}\sum_{j,\nu}\int_{-\infty}^{t} dt' \langle [m_{i\mu}(t-t'), m_{j\nu}]_-\rangle\, H_{j\nu}(t'),$$

which is again due to the cyclic invariance of the trace. The first term on the right-hand side is non-zero only if the unperturbed system contains a permanent magnetic moment as, for example, in the case of a ferromagnet; this case will be disregarded. When we define the tensor of the local time-dependent susceptibility (for $H = 0$) by the general linear relation

$$\bar{m}_{i\mu}(t) = \sum_{j,\nu}\int_{-\infty}^{+\infty} dt' \chi_{\mu\nu}(\boldsymbol{r}_i, \boldsymbol{r}_j; t-t)\, H_{j\nu}(t'),$$

a comparison with our previous considerations shows that

$$\chi_{\mu\nu}(\boldsymbol{r}_i, \boldsymbol{r}_j; t-t') = -\frac{1}{i\hbar}\,\theta(t-t')\,\langle [m_{i\mu}(t-t'), m_{j\nu}]_-\rangle.$$

In the following we shall consider only translationally invariant systems where the susceptibility can depend only on the coordinate difference. In this way we arrive at the Kubo formula for the isothermal susceptibility

$$\chi_{\mu\nu}(\boldsymbol{r}_i - \boldsymbol{r}_j; t-t') = \frac{i}{\hbar}\,\theta(t-t')\,\langle [m_{i\mu}(t-t'); m_{j\nu}]_-\rangle, \qquad (2.45a)$$

where $\theta(t)$ is the Heaviside step function

$$\theta(t) = \begin{cases} 1 & t > 0, \\ 0 & t < 0. \end{cases}$$

It is obvious in which way this is rewritten for continuous magnetisation. The susceptibility $\chi_{\mu\nu}$ is of the type of a retarded Green function (Zubarev, 1960).

When we change to Fourier transforms with respect to space and time, eqn. (2.45a) yields

$$\chi_{\mu\nu}(\boldsymbol{k}, \omega) = \frac{i}{\hbar} \sum_{i,j} \int_{-\infty}^{+\infty} dt\ e^{i\omega t - i(\boldsymbol{k}\cdot[\boldsymbol{r}_i - \boldsymbol{r}_j])} \langle [m_{i\mu}(t), m_{j\nu}]_- \rangle\ \theta(t). \quad (2.45b)$$

If ω is considered to be a complex quantity, the integral will certainly exist (because of the step function) in the upper half-plane of ω, i.e. $\chi_{\mu\nu}(k, \omega)$ is regular in this plane. In this property of regularity the principle of causality becomes apparent (cf., for example, Landau and Lifshitz, 1965). It is a remarkable feature of Kubo's formula that the susceptibility can be represented by two spacetime correlation functions of magnetisation, $\langle m_{i\mu}(t) m_{j\nu} \rangle$ and/or $\langle m_{j\nu} m_{i\mu}(t) \rangle$. Kubo's formula can be applied under quite general conditions; it is an essential assumption that the density matrix can be expanded into a power series of the external perturbation, i.e. that all quantities are analytical functions of the external field. This need not necessarily always be true; for example, at the Curie point the physical quantities can be non-analytical functions of H, so that a linear response theory will not exist for this case; cf Section III.4.

As regards the canonical distribution (2.42), we are interested in, Kubo's formula can be given another form which is often expedient.

It is first necessary to verify the validity of the relation

$$\langle [m_{i\mu}(t), m_{j\nu}]_- \rangle = -\frac{i}{Z} \operatorname{Tr} \left\{ \int_0^\beta d\lambda\ e^{-\beta \mathcal{H}_0} \frac{d}{d\lambda} \left(e^{\lambda \mathcal{H}_0} m_{i\mu}(t)\ e^{-\lambda \mathcal{H}_0} m_{j\nu} \right) \right\},$$

by integrating over λ. When we extend the definition of the Heisenberg representation to complex times, it is easy to see that

$$e^{\lambda \mathcal{H}_0} m_{i\mu}(t)\ e^{-\lambda \mathcal{H}_0} = m_{i\mu}(t - i\hbar\lambda).$$

With $d/d\lambda = -i\hbar d/dt$ we see that

$$\langle [m_{i\mu}(t), m_{j\nu}]_- \rangle = i\hbar \int_0^\beta \langle \dot{m}_{i\mu}(t - i\hbar\lambda) m_{j\nu} \rangle \, d\lambda,$$

so that now we obtain for the (isothermal) susceptibility

$$\chi_{\mu\nu}(\boldsymbol{k}, \omega) = -\sum_{i,j} \int_{-\infty}^{+\infty} dt \theta(t) \int_0^\beta d\lambda \langle \dot{m}_{i\mu}(t - i\hbar\lambda) m_{j\nu} \rangle \, e^{i\omega t - i(\boldsymbol{k} \cdot [\boldsymbol{r}_i - \boldsymbol{r}_j])}. \quad (2.45c)$$

This form easily yields the static homogeneous susceptibility $\chi\,(0, 0)$. We can immediately integrate with respect to time when we take into account that, because there is no long-range order in the system in time (which is not true for magnetically ordered systems), the mean value of the product at the upper limit can be written as the product of the mean values

$$\langle m_{i\mu}(\infty) m_{j\nu} \rangle = \langle m_{i\mu} \rangle \langle m_{j\nu} \rangle = 0,$$

so that, as there is no non-vanishing magnetisation in the unperturbed system, the upper limit makes no contribution to (2.45c).† Thus we obtain for the static susceptibility

$$(\chi_{\text{st}})_{\mu\nu} \equiv \chi_{\mu\nu}(0, 0) = \sum_{i,j} \int_0^\beta d\lambda \langle m_{i\mu}(-i\hbar\lambda) m_{j\nu} \rangle.$$

If we consider only cubic or isotropic solids or fluids, the tensor must be diagonal and, for reasons of symmetry, the three diagonal elements must be equal. Thus the formula for the static susceptibility can be simplified as follows:

$$\chi_{\text{st}} = \tfrac{1}{3} \sum_{i,j} \int_0^\beta d\lambda \langle (\boldsymbol{m}_i(-i\hbar\lambda) \cdot \boldsymbol{m}_j) \rangle; \quad (2.46)$$

this is due to the fact that the three components give identical contributions. Equation (2.46) represents a quantum-mechanical generalization of the Curie law (2.13) involving all correlations, i.e. taking all interactions into account. Under certain conditions we can integrate with respect to λ; this is certainly possible in the classical limiting

† In systems with long-range order we must replace the operators $m_{j\nu}$ by $m_{j\nu} - \langle m_{j\nu} \rangle$ in (2.45a–c), that is, the susceptibility is given by the fluctuation of the magnetisation.

case ($\hbar = 0$), and also if the moments $\boldsymbol{m}_i$ are proportional to the angular momenta $\boldsymbol{J}_i$ of the individual atoms, with the same proportionality factor for all atoms. In this case the total magnetic moment $\boldsymbol{M} = \sum_i m_i$ contained in (2.46) will be proportional to the total angular momentum of the system and thus to a conserved quantity, i.e. $\boldsymbol{M}$ commutes with the Hamilton operator $\mathcal{H}_0$. Thus we have

$$\boldsymbol{M}(-i\hbar\lambda) = \boldsymbol{M},$$

and therefore

$$\chi_{\text{st}} = \frac{N}{3kT} \sum_i \langle(\boldsymbol{m}_i \cdot \boldsymbol{m}_0)\rangle, \tag{2.46a}$$

where we made use of the fact that the correlation functions $\langle(\boldsymbol{m}_i \cdot \boldsymbol{m}_j)\rangle$ depend only on the distance of the two moments; N is the number of atoms. It the atoms are independent of one another we have

$$\langle(\boldsymbol{m}_i \cdot \boldsymbol{m}_0)\rangle = \mu^2 \delta_{i,0},$$

where $\delta_{i,0}$ is Kronecker's symbol, and we rearrive at Curie's law (2.13).

Equation (2.46a) is valid if, for example, the atoms have zero orbital momentum, (cf. Section II.5), or if only one of the multiplet states (2.38) contributes to the magnetic moment since the other terms are too high. If, however, several multiplet terms are to be taken into account, $\boldsymbol{M}$ will no longer be proportional to the total angular momentum, since the individual angular momentum states contributing to the magnetic moment have different Landé factors. In this case we might say that we have a system of different atoms with different g factors.

Let us now return to eqn. (2.45); since in general the summation cannot be carried out in (2.45), we have to distinguish between the following two cases:

(α) Multiplet splitting $\gg kT$ or $\beta\zeta(L, S) \ll 1$

(β) Multiplet splitting $\ll kT$ or $\beta\zeta(L, S) \gg 1$.

(α) In a first approximation the spin–orbit coupling can be neglected, as all terms of the multiplet are uniformly populated. The term splitting in the magnetic field will then correspond to that in the normal Zeeman effect, (2.17), and we have

$$Z = \exp[-\beta E^0_{L,S}] \sum_{M_L=-L}^{+L} \sum_{M_S=-S}^{+S} \exp[-\beta\mu_B H(M_L+2M_S)]$$

$$= \exp[-\beta E^0_{L,S}] \frac{\sinh \beta\mu_B H(L+\frac{1}{2})}{\sinh \frac{1}{2}\beta\mu_B H} \cdot \frac{\sinh \beta\mu_B H(2S+1)}{\sinh \beta\mu_B H};$$

this is easy to verify by summing the geometrical series. Hence we obtain for the susceptibility according to (2.45):

$$\chi = N\frac{\mu_B}{H}\{2SB_S(2\beta S\mu_B H)+LB_L(\beta L\mu_B H)\}. \tag{2.47}$$

The functions $B_K(x)$ are called Brillouin functions; they are defined by

$$B_K(x) = \frac{2K+1}{2K} \coth\left(\frac{2K+1}{2K}x\right) - \frac{1}{2K}\coth\frac{x}{2K}. \tag{2.48}$$

As $K \to \infty$ this corresponds to the classical case where the magnetic moment can have any arbitrary orientation with respect to the field; here the Brillouin function is identical with the classical Langevin function (2.12).

In the normal case, at not too low temperatures, we also have $\beta\mu_B H \ll 1$; it is then no longer reasonable to ignore spin–orbit coupling (described by a term $\mathcal{H}_{so}$), when the influence of the magnetic field is weak. With

$$\mathcal{H}_{so} = \zeta(L, S)(\boldsymbol{L}\cdot\boldsymbol{S}) \quad \text{and} \quad \mathcal{H}_H = \mu_B H(J_z+S_z)$$

χ can be obtained from (2.45), breaking off after the first-order terms in $\mathcal{H}_{so}$ and $\mathcal{H}_H$:

$$\chi = -N\frac{\mu_B}{H}\frac{\mathrm{Tr}\,(J_z+S_z)(1-\beta\mathcal{H}_{so}-\beta\mathcal{H}_H)}{\mathrm{Tr}\,(1-\beta\mathcal{H}_{so}-\beta\mathcal{H}_H)}.$$

In the eigen-representation of J^2 and J_z according to (2.29) we have

$$\mathrm{Tr}\,(J_z+S_z) \sim \sum_J \sum_{M_J=-J}^{+J} M_J,$$

i.e. the trace vanishes; the same is true for $\mathrm{Tr}(J_z+S_z)\mathcal{H}_{so}$. Since the mean of the multiplet splitting, Tr H_{so}, also vanishes (see above), we finally have

$$\chi = N\frac{\beta\mu_B^2}{(2S+1)(2L+1)} \cdot \sum_{M_S=-S}^{+S} \sum_{M_L=-L}^{+L} (M_L+2M_S)^2,$$

with Tr 1 = $(2S+1)(2L+1)$. Since

$$\sum_{x=-n}^{+n} x^2 = \tfrac{1}{3}\, n(n+1)(2n+1),$$

we obtain for the susceptibility

$$\chi = N\frac{\mu_B^2}{3kT}\{L(L+1)+4S(S+1)\}. \tag{2.49}$$

This formula applies regardless, whether we are in the range of the normal Zeeman effect or in that of the anomalous Zeeman effect; a difference will appear only in the higher powers of β, i.e. when the slightly different populations of the fine-structure terms are taken into account. According to (2.49), χ has a temperature dependence as given by the Curie law, with a Curie constant

$$C = N\frac{\mu_B^2}{3k}\{L(L+1)+4S(S+1)\}.$$

The atoms behave like classical magnetic moments with an effective moment

$$\mu_{\text{eff}} = \mu_B p_{\text{eff}} = \mu_B\sqrt{L(L+1)+4S(S+1)}, \tag{2.50}$$

as is shown by comparison with (2.13).

(β) If $\beta\zeta(L, S) \gg 1$, it will be mainly the lowest fine-structure term of the multiplet which is populated. In the lowest approximation the off-diagonal elements of S_z can be neglected and the eigenvalues of the anomalous Zeeman effect (2.40) can be used in the calculation.

The partition function will then be

$$Z = \exp\left[-\beta E^0_{L,S,J}\right] \cdot \sum_{M_J=-J}^{+J} \exp\left[-\beta g_J \mu_B H M_J\right]$$
$$= \exp\left[-\beta E^0_{L,S,J}\right] \frac{\sinh \beta g_J \mu_B H (J+\frac{1}{2})}{\sinh \frac{1}{2}\beta g_J \mu_B H},$$

and, according to (2.45) and (2.48), the susceptibility will be given by

$$\chi = N \frac{g_J \mu_B J}{H} B_J(\beta g_J \mu_B J H), \tag{2.51}$$

where J is the quantum number of total angular momentum of the lowest multiplet term. In the case of $\beta\mu_B H \ll 1$, which is important in practice, we hence obtain once more a Curie law:

$$\chi = N \frac{(g_J \mu_B)^2 J(J+1)}{3kT}, \tag{2.52}$$

with an effective magnetic moment

$$\mu_{\text{eff}} = \mu_B p_{\text{eff}} = g_J \mu_B \sqrt{J(J+1)}. \tag{2.53}$$

If the multiplet splitting is not very large compared with kT, both the off-diagonal terms of S_z and the higher fine-structure terms must be taken into account. Since here S_z is no longer diagonal, $\mathcal{H}_H$ will not commute with $\mathcal{H}_{so}$. It is therefore necessary to expand the exponential functions in the traces in (2.45). For the partition function we obtain

$$Z = \exp(-\beta E^0_{L,S}) \operatorname{Tr} \exp(-\beta \mathcal{H}_{so} + \mathcal{H}_H)$$
$$= \exp(-\beta E^0_{L,S}) \sum_{J,M_J} \sum_{n=0}^{\infty} \frac{(-\beta)^n}{n!} \langle JM_J | (\mathcal{H}_{so} + \mathcal{H}_H)^n | JM_J \rangle$$
$$\cong \exp(-\beta E^0_{L,S}) \sum_{n=0}^{\infty} \frac{(-\beta)^n}{n!} \sum_J \left\{ (2J+1) E_J^n \right.$$
$$\left. + \sum_{M_J=-J}^{+J} \langle JM_J | \sum_{\mu=0}^{n-1} \mathcal{H}_{so}^{n-1-\mu} \mathcal{H}_H \mathcal{H}_{so}^{\mu} | JM_J \rangle \right\}$$
$$= \exp(-\beta E^0_{L.S}) \sum_{n=0}^{\infty} \frac{(-\beta)^n}{n!} \sum_J \left\{ (2J+1) E_J^n \right.$$
$$\left. + n E_J^{n-1} \sum_{M_J=-J}^{+J} \langle JM_J | \mathcal{H}_H | JM_J \rangle \right\},$$

having broken off after the linear terms of the field strength. The symbol E_J denotes the eigenvalues of $\mathcal{H}_{\text{so}}$. The sum over M_J in the last line is obviously vanishing as, according to (2.29), the matrix element of $\mathcal{H}_H$ is proportional to M_J. In this approximation we obtain for Z simply

$$Z = \exp(-\beta E^0_{L,S}) \sum_{J=|L-S|}^{L+S} (2J+1) \exp(-\beta E_J) = Z(H=0).$$

Similarly we obtain for the other trace (again in the lowest approximation):

$$\begin{aligned}
&\operatorname{Tr} \mathcal{H}_H \exp[-\beta(\mathcal{H}_{\text{so}} + \mathcal{H}_H)] \\
&\quad \cong \sum_{n=0}^{\infty} \frac{(-\beta)^n}{n!} \sum_{J, M_J} \langle JM_J | \mathcal{H}_H \left(\sum_{\mu=0}^{n-1} \mathcal{H}_{\text{so}}^{n-1-\mu} \mathcal{H}_H \mathcal{H}_{\text{so}}^{\mu} \right) | JM_J \rangle \\
&\quad = \sum_{n=0}^{\infty} \frac{(-\beta)^n}{n!} \sum_{J, M_J} \sum_{J', M'_{J'}} \langle JM_J | \mathcal{H}_H | J'M'_{J'} \rangle \\
&\qquad\qquad \times \langle M'_{J'} J' | \mathcal{H}_H \backslash JM_J \rangle \left(\sum_{\mu=0}^{n-1} E_{J'}^{n-1-\mu} \cdot E_J^{\mu} \right) \\
&\quad = \sum_{n=0}^{\infty} \frac{(-\beta)^n}{n!} \sum_{J, M_J} \sum_{J', M'_{J'}} |\langle JM_J | \mathcal{H}_{\text{H}} | J'M'_{J'} \rangle|^2 \frac{E_{J'}^n - E_J^n}{E_{J'} - E_J},
\end{aligned}$$

since $\mathcal{H}_H$ is Hermitean. The matrix element of $\mathcal{H}_H$ is proportional to the Kronecker symbol $\delta_{M_J, M'_{J'}}$ since $\mathcal{H}_H$ commutes with J_z; this can also be seen from (2.24) and (2.26a). In addition to this, according to (2.24) and (2.26b), only those matrix elements of S_z are different from zero for which $J' = J$ or $J' = J \pm 1$; we therefore obtain for the sums

$$\begin{aligned}
&= -\sum_{J, M_J} \exp(-\beta E_J) \sum_{J'} \langle JM_J | \mathcal{H}_H | J'M_J \rangle |^2 \frac{1 - \exp[\beta(E_{J'} - E_J)]}{E_{J'} - E_J} \\
&= -\sum_{J, M_J} \exp(-\beta E_J) \Bigg\{ \beta |\langle JM_J | \mathcal{H}_H | JM_J \rangle|^2 \\
&\qquad + 2 \frac{|\langle JM_J | \mathcal{H}_H | (J+1) M_J \rangle|^2}{E_{J+1} - E_J} + 2 \frac{|\langle JM_J | \mathcal{H}_H | (J-1) M_J \rangle|^2}{E_{J-1} - E_J} \Bigg\}.
\end{aligned}$$

Again, using (2.29) and (2.30), we obtain for the diagonal element of $\mathcal{H}_H$:

$$|\langle JM_J|\,\mathcal{H}_H\,|JM_J\rangle|^2 = (g_J\mu_B)^2 M_J^2 H^2.$$

The off-diagonal elements of S_z are not so easy to calculate. We obtain (Condon and Shortley, 1959):

$$|\langle LSJM_J|\,S_z\,|LS(J-1)M_J\rangle|^2 = \frac{(J^2-(L-S)^2)\,((L+S+1)^2-J^2)}{4J^2(2J+1)\,(2J-1)}\,(J^2-M_J^2).$$

The sum over M_J can be obtained in the same way as in (2.49) and we finally obtain for the susceptibility

$$\chi = N\frac{\sum\limits_{J=|L-S|}^{L+S}(2J+1)\exp(-\beta E_J)\left(\frac{\beta}{3}(g_J\mu_B)^2\,J(J+1)+\alpha_J\right)}{\sum\limits_{J=|L-S|}^{L+S}(2J+1)\exp(-\beta E_J)} \tag{2.54}$$

where

$$\alpha_J = \mu_B^2\left\{\frac{(J^2-(L-S)^2)\,((L+S+1)^2-J^2)}{6J(2J+1)\,(E_{J-1}-E_J)} + \frac{((J+1)^2-(L-S)^2)\,((L+S+1)^2-(J+1)^2)}{6(J+1)\,(2J+1)\,(E_{J+1}-E_J)}\right\}.$$

If the multiplet splitting is large compared with kT, all other terms can be omitted with the exception of the lowest one, since their contribution is exponentially small. We then obtain from (2.54)

$$\chi = N\left(\frac{g_J^2\mu_B^2 J(J+1)}{3kT}+\alpha_J\right), \tag{2.55}$$

where J is the angular momentum of the lowest term. The difference between this formula and (2.52) is the temperature-independent so-called Van Vleck paramagnetism α_J. It is due to the admixture of other angular momentum states to the unperturbed states arising when a magnetic field is applied. The parameter α_J differs as to its order of magnitude from the first term in (2.55) by about a factor of $(\beta\zeta(L, S))^{-1}$ In general this factor is small compared with 1 so that Van Vleck paramagnetism plays a minor part. Only if the expectation value of the magnetic moment vanishes in the unperturbed ground state ($J = 0$), does it determine the temperature dependence of susceptibility.

In principle, the formulas derived above can be applied only to gases. Because of the low densities of gases their susceptibilities cannot be measured. It is therefore necessary to go to solid (liquid) substances. Owing to the strong binding forces, which mainly affect the outer shells of the atoms, only such solids are of interest (as regards the comparison with theory) for which the atomic magnetism is little influenced by the crystal fields. This is only true for rare-earth ions which have an exceptional structure: all states are populated up to the 4*d* shell and also the 5*s* and 5*p* shells are closed. This corresponds to the xenon configuration. These atoms also have three electrons in the 5*d* and 6*s* shells, which are lost in the triply-ionised state. Beginning with lanthanum with the configuration Xe $5d\,6s^2$, the 4*f* shell is filled successively up to lutetium. Comparing this with the configuration of a hydrogen-like atom, we see that the wave functions of the 4*f* electrons must be concentrated much more strongly around the nucleus than those of the other (5*s*, 5*p*, . . .) shells. The 4*f* electrons can thus be assumed to be only slightly influenced by the other atoms, since the outer shells screen the inner ones effectively against perturbing fields. Of course this is not true for the 5*d* and 6*s* electrons. The metals of rare earths are thus not suitable for a comparison with the theory; their magnetic behaviour is also very complicated as they can be ferromagnetic as well as antiferromagnetic. Both phases can be observed in the same metal in different ranges of temperature (Spedding *et al.*, 1957; Belov *et al.*, 1964). In principle, this by itself would not prevent a comparison in the paramagnetic range. It is, however, unknown to what extent the outer electrons are split off from the atoms bound in a metallic structure. It is therefore better to use rare-earth compounds, such as oxides, phosphates, sulphates, . . ., in which the triply charged ions appear. Apart from the xenon core, these ions have the simple configuration $4f^N$ with $N = 0, 1, 2, \ldots, 14$. The lowest multiplet is determined according to Hund's rule, i.e. the multiplicity increases up to gadolinium, $S = 7/2$, and decreases according to the Pauli principle in the lower half. The orbital quantum number, (L), is easy to obtain when we take into account that the electrons adjust their components m_l of the orbital angular momentum in parallel, as far as this is allowed by the exclusion principle. In the case of $N \leqslant 7$, if the shell is less than half populated, we therefore have

$$L = 3+(3-1)+\ldots(3-N) = 3N-\tfrac{1}{2}N(N-1).$$

The sequence of the multiplets from La^{3+} to Gd^{3+} is therefore the following:

$$^1S,\ ^2F,\ ^3H,\ ^4I,\ ^5I,\ ^6H,\ ^7F,\ ^8S.$$

In the other half with $N > 7$, the L appear in the same order, as all additional electrons have an m_s of $-\frac{1}{2}$ while the others had $m_s = +\frac{1}{2}$. The term structure of the multiplet is obtained from the constant ζ of the spin–orbit coupling. Since, because of their nature the $\zeta_{n,l}$ are positive, (2.20), ζ will be positive in the upper half with $N < 7$, according to (2.37), and negative in the lower half with $N > 7$. According to (2.38), for $N < 7$ the fine-structure terms with minimum J are the lowest ones, $J = |L-S|$, and for $N > 7$ those with maximum J are lowest, $J = L+S$. Table 5 illustrates the conditions in the case of rare-earth ions.

The distance between two neighbouring fine-structure terms is, according to (2.38), given by

$$E^0_{L,S,J}-E^0_{L,S,J-1} = \zeta(L, S)J.$$

The fact that the distance is proportional to J is called Landé's rule of intervals. We see from Table 5 that in the case of Eu^{3+} and Sm^{3+} the interval is particularly small. For rare earths it is usually only the lowest term of the multiplet which is populated, since 1 cm^{-1} corresponds to a temperature of about 1·439°K. Thus the conditions for a validity of eqn. (2.52) are satisfied. Accordingly, many rare-earth compounds behave in the paramagnetic range in such a way that the reciprocal sus-

TABLE 5

Z	Ion	Configu-ration	Ground term	ζ (cm^{-1})	g_J	$(p_{eff})_{theor}$	$(p_{eff})_{exp}$
57	La^{3+}	$4f^0$	$^1S_0$	—	0	0	0
58	Ce^{3+}	$4f^1$	$^2F_{5/2}$	640	6/7	2·54	2·5
59	Pr^{3+}	$4f^2$	$^3H_4$	470	4/5	3·58	3·6
60	Nd^{3+}	$4f^3$	$^4I_{9/2}$	270	8/11	3·62	3·6
61	Pm^{3+}	$4f^4$	$^5I_4$	—	3/5	2·68	—
62	Sm^{3+}	$4f^5$	$^6H_{5/2}$	310	2/7	0·84 (1·55)	1·6
63	Eu^{3+}	$4f^6$	$^7F_0$	300–340	0	0 (3·50)	3·6
64	Gd^{3+}	$4f^7$	$^8S_{7/2}$	—	2	7·94	7·9
65	Tb^{3+}	$4f^8$	$^7F_6$	—	3/2	9·72	9·7
66	Dy^{3+}	$4f^9$	$^6H_{15/2}$	−450	4/3	10·6	10·6
67	Ho^{3+}	$4f^{10}$	$^5I_8$	−630	5/4	10·6	10·6
68	Er^{3+}	$4f^{11}$	$^4I_{15/2}$	−1070	6/5	9·58	9·5
69	Tu^{3+}	$4f^{12}$	$^3H_6$	−1350	7/6	7·56	7·5
70	Yb^{3+}	$4f^{13}$	$^2F_{7/2}$	−2950	8/7	4·54	4·5
71	Lu^{3+}	$4f^{14}$	$^1S_0$	—	0	0	0

In the third and fourth columns the table gives the configurations and ground-state terms of the triply ionised rare-earth atoms. The spin–orbit coupling constant ζ is obtained from the spectroscopically observed excited levels by means of Landé's rule of intervals (Elliott and Stevens, 1952, 1953). The g_J are the Landé factors of the ground-state terms (2.41); the effective magnetic moments, p_{eff}, are obtained theoretically from (2.53). The parenthesised values for Eu^{3+} and Sm^{3+} result from (2.54), when the higher terms and Van Vleck paramagnetism are taken into account; they hold true for $T = 293°K$ (Van Vleck, 1932a, b.) The experimental values for p_{eff} (last column) are representative values, i.e. in the experiment deviations can be observed from these values, which can be explained by differences in the influence of the other constituents of the compound on the ions.

ceptibility is a linear function of the temperature (see Fig. 9); in addition to this, Table 5 shows that the theoretical effective magnetic moments (2.53) are in a good agreement with the moments measured.

So far the theoretical description is satisfactory. However, Fig. 9 also shows a typical deviation from (2.52), as it is often better to approximate the temperature dependence of χ^{-1} by means of a Curie–Weiss law

$$\chi = C/(T-\Theta);$$

Θ can be either positive or negative; it describes in a rough approximation the influence of the other atoms in the crystal. In the case of paramagnetic substances the constant Θ has no further physical significance as it has for ferromagnetic or antiferromagnetic substances.

The ions Eu^{3+} and, to a lesser extent, also Sm^{3+}, are exceptional cases. In both cases the higher multiplet terms will be populated also at

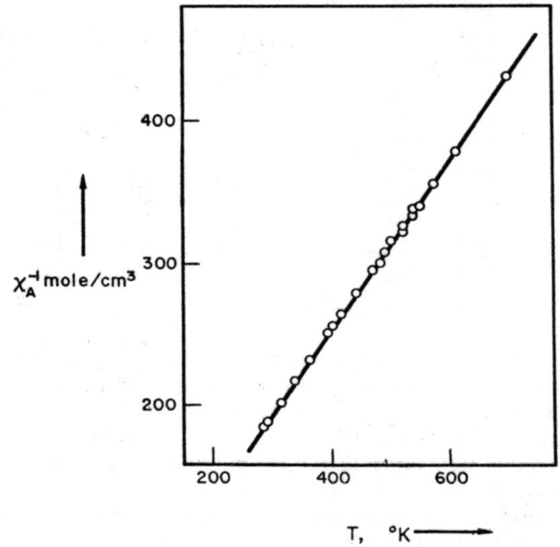

FIG. 9.

Reciprocal atomic susceptibility of $PrCl_3$ as a function of temperature. This temperature dependence can be approximated by a Curie–Weiss law.

room temperature. It is thus necessary to use the more general formula (2.54) for comparison and, according to Van Vleck (1932a) (see Table 5), good agreement is obtained with experiment. It is particularly interesting to consider the temperature dependence of susceptibility in the case of Eu^{3+}, as the expectation value of the magnetic vanishes in the unperturbed ground state, $J = 0$. The temperature-independent Van Vleck paramagnetism should therefore be observable particularly at low temperatures when the other fine-structure terms are not populated. Here, too, agreement with experiment is obtained as shown in Fig. 10.

Hitherto we discussed the case of an atomic magnetisation which is proportional to the external field strength. Deviations from linearity

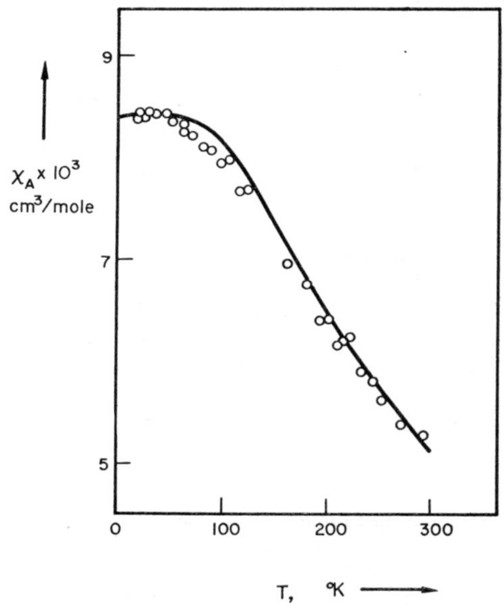

FIG. 10.

Temperature dependence of the susceptibility of Eu_2O_3 after Borovik–Romanov and Kreines (1956). The full line represents the theoretical curve (2.54) where the excited states are taken into account (Van Vleck, 1932a). In its ground state the Eu^{3+} ion has $J = 0$, so that at low temperatures the temperature-independent Van Vleck paramagnetism becomes essential.

must be obtained (saturation), if the temperatures are low enough or the fields strong enough. According to (2.51), the susceptibility as a function of H and T is given by the Brillouin function. Measurements by Henry (1952) at high field strengths and low temperatures prove the validity of eqn. (2.51) (see Fig. 11).

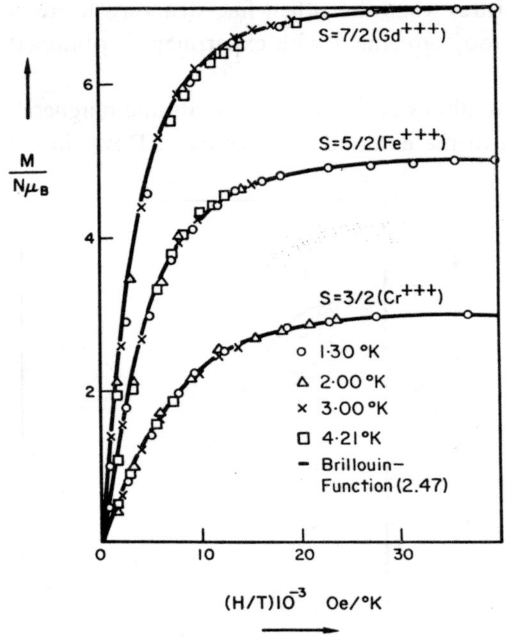

FIG. 11.

Dependence of the mean magnetic moment per atom in units μ_B on the ratio H/T for three paramagnetic ions after Henry (1952). The three ions, namely Gd^{3+} in $Gd_3SO_4 \cdot 8H_2O$, Fe^{3+} in $NH_4Fe(SO_4)_2 \cdot 12H_2O$, and Cr^{3+} in $KCr(SO)_4 \cdot 12H_2O$ are in S states with $S = \frac{7}{2}$ for Gd^{3+}, and $S = \frac{5}{2}$ for Fe^{3+} (see Tables 5 and 7). The orbital moment of Cr^{3+} is quenched (Section II.5); it has an effective spin of $S = \frac{3}{2}$.

5. Paramagnetism of Atoms in a Crystal

Two limiting cases of paramagnetism have been considered so far: the case of an atom in a crystal, whose outer electrons are separated and form a more or less free Fermi gas (Section II.1). This results in the appearance of the weak Pauli paramagnetism which, to a first approximation, is temperature-independent and superimposed on the likewise weak diamagnetism of the ionic core and the Landau diamagnetism. The other limiting case is realised satisfactorily for rare-earth compounds, in which the outer electrons are taken into the shell of the other constituent of the compound (which so becomes a diamagnetic anion), and whose magnetic properties are determined by the incomplete $4f$ shell. Unlike the wave functions of the conduction electrons in a metal, the wave functions are here strongly localised around the atoms, so that they virtually represent eigenfunctions of the free atom. This results in a strong Langevin paramagnetism discussed above. The localisation of the $4f$ electrons is favoured by the $5s$ and $5p$ electrons, which screen the $4f$ shell against the influences of the other lattice atoms.

This screening, however, is incomplete since at moderately low temperatures ($\leqslant 70°$K) the deviations from the Curie law are no longer negligible. This is due to the fact that the $(2J+1)$-fold degenerate ground term of the free ion in the crystal is split. Since the wave functions of the $4f$ electrons are concentrated around the nucleus, they will not overlap with the wave functions of the electrons of other atoms, i.e. under these assumptions the influence of the other atoms can be described by a field whose potential φ_{cr} satisfies the Laplace equation

$$\nabla^2 \varphi_{cr} = 0. \tag{2.56}$$

In the case of overlapping this equation must be replaced by the Poisson equation. Approximation (2.56) is called the crystal-field approximation. Since the polynomials $r^l Y_l^m(\vartheta, \varphi)$ with Y^m from (2.23) are solutions to (2.56), we can write a general solution to (2.56) in polar coordinates (r, ϑ, φ) in the form of

$$\varphi_{cr}(r, \vartheta, \varphi) = \sum_{l=1}^{\infty} \sum_{m=-l}^{+l} A_l^m r^l Y_l^m(\vartheta, \varphi). \tag{2.57}$$

The coefficients A_l^m are limited by several requirements. First, φ_{cr} must be real. From (2.23) it is easy to verify that $(Y_l^m)^* = (-)_l^m Y_l^{-m}$ so that φ_{cr} is real if $(A_l^m)^* = (-)^m A_l^{-m}$.† The other limitations on the coefficients are less general and differ from problem to problem. Again considering the case of the $4f$ electrons, we see that the crystal field will

† The asterisk labels the complex conjugate quantities.

induce transitions inside the multiplets, between the multiplets, and between the various configurations. The latter will be ignored, since their energies exceed that of the ground-state configuration of the rare earths by more than 10^4 cm^{-1}, and the influence of the lattice field is here relatively small, smaller than that of the spin–orbit coupling. All matrix elements of φ_{cr} are therefore formed with $4f$ wave functions. The product of two $4f$ functions, which are proportional to $Y_3^m(\vartheta, \varphi)$, remains unchanged in a reflection of space (inversion), since in an inversion, i.e. when $(\vartheta, \varphi) \to (\pi - \vartheta, \varphi + \pi)$, the Y_l^m are changed to $(-)^l Y_l^m(\vartheta, \varphi)$ so that only terms with even l must be taken into account in (2.57). The matrix elements of the other terms vanish. Another simplification results from the Wigner–Eckart theorem. According to (2.24) the matrix element of $Y_l^m(\vartheta, \varphi)$ formed with $4f$ functions is equal to

$$\langle l'm' | Y_l^m | l'm'' \rangle \sim \langle l'lm''m | l'm' \rangle$$

where $l' = 3$. By virtue of the properties of the C.–G. coefficients (2.26b) all matrix elements vanish unless $l \leqslant 2l'$; thus we can drop all terms with $l > 6$ in (2.56).

In this way the number of terms in (2.57) is appreciably reduced. It can still be lowered if the lattice field is required to be symmetric. By way of example, we consider an ion from the iron group with its incomplete $3d$ shell. The ions of this group in a solid are often bound in a lattice field with predominately cubic symmetry, i.e. there exist three pairwise conjugated four-fold axes of rotation. Thus the lattice potential must remain invariant if a rotation through $+\pi/2$ is carried out about one of these three axes. Assuming a frame where the coordinate axes are parallel to the symmetry axes, and carrying out such a rotation about the z-axis, we see that in (2.57) all A_l^m vanish unless their $m = 0$, ± 4, since $Y_l^m \sim e^{im\varphi}$ and the rotation corresponds to a transition $\varphi \to \varphi + \pi/2$. With these restrictions, the potential φ_{cr} is given by

$$\varphi_{cr} = A_2^0 r^2 Y_2^0 + A_4^0 r^4 Y_4^0 + A_4^4 r^4 Y_4^4 + (A_4^4)^* Y_4^{-4}.$$

Because of the $3d$ shell we have only terms with $l \leqslant 4$. It is easy to verify by means of (2.23) that φ_{cr} in Cartesian coordinates will be

$$\varphi_{cr} = \sqrt{\frac{5}{16\pi}} A_2^0(3z^2-r^2) + \frac{3}{16\sqrt{\pi}} A_4^0(35z^4-30r^2z^2+3r^4)$$
$$+\frac{3}{8}\sqrt{\frac{35}{2\pi}} A_4^{4(r)}(x^4-6x^2y^2+y^4)+\frac{3}{2}\sqrt{\frac{35}{2\pi}} A_4^{4(i)}(y^2-x^2),$$

$A_4^4 = A_4^{4(r)} + iA_4^{4(i)}$. Since the three coordinate axes are equivalent symmetry axes, φ_{cr} must remain invariant in a permutation of x, y, z. Hence we see immediately that $A_2^0 = 0$ and $A_4^{4(i)} = 0$, and also

$$A_4^{4(r)} = \sqrt{\frac{5}{14}} A_4^0 .$$

Having accounted for all symmetries, we finally arrive at

$$\varphi_{cr} = \frac{15}{4\sqrt{\pi}} A_4^0\left(x^4+y^4+z^4-\frac{3}{5}r^4\right). \tag{2.58}$$

As a rule the physical properties of a substance will essentially depend on the parameters A_l^m. It is therefore impossible to derive a general formula for the susceptibility. Let us therefore consider cerium ethyl sulphate, $Ce(C_2H_2SO_4)_3 \cdot 9H_2O$, which is a suitable example. This salt has been dealt with in detail in the literature (see for example, Elliott and Stevens, 1952, 1953; Meyer and Smith, 1959) and displays some general features. The structure of this complex salt is simple: the cerium ion is surrounded symmetrically by the three ethyl sulphate ions, i.e. the anions form an equilateral triangle with the cerium ion in its centre. In the same plane three molecules of crystal water are arranged symmetrically with respect to the cerium ion, so that in this plane a regular hexagon is formed by the three radicals and the three molecules of crystal water, which contains the cerium ion in its centre. The six vertices are occupied alternately by a radical and a crystal-water molecule. Other molecules of crystal water are arranged in planes above and below this principal plane, mirror-symmetrical with respect to it: three above the radicals, and three below them. Since all cerium ions have the same positions in the lattice, the problem of the lattice field can be reduced to a consideration of this structure. The point symmetry of the cerium ion thus consists of a trigonal axis perpendicular to the three planes through the cerium ion, and a horizontal plane of reflection which represents the principal plane. For brevity this point symmetry is denoted by $\bar{6}$ or C_{3h} (Schönflies symbols) (for details see, for example, Kittel, 1962; Sachs, 1963; or Heine, 1960). The lattice potential for the cerium ion can now be obtained by means of similar considerations as in the case of cubic symmetry. Owing to the trigonal symmetry with respect to the preferred hexagonal axis, which is taken as the polar axis, the series (2.57) will only contain terms with $m = 0, \pm 3$, and ± 6. For the $4f$ electron φ_{cr} must, of course, be expanded up to $l = 6$. The presence of a mirror plane requires the invariance of φ_{cr} with respect to the transformation $\vartheta \to \pi - \vartheta$, so that, since $Y_l^m(\pi - \vartheta, \varphi) = (-)^{l+m} Y_l^m(\vartheta, \varphi)$, all terms with $m = \pm 3$ will vanish. Thus the lattice potential is given by

$$\begin{aligned}\varphi_{\text{cr}} = {} & A_2^0 r^2 Y_2^0 + A_4^0 r^4 Y_4^0 + A_6^0 r^6 Y_6^0 + A_6^{6(r)} r^6 (Y_6^6 + Y_6^{-6}) \\ & + i A_6^{6(i)} r^6 (Y_6^6 - Y_6^{-6}),\end{aligned}$$

$A_6^6 = A_6^{6(r)} + i A_6^{6(i)}$. Applying the transformation $\varphi \to \varphi + \varphi_0$ makes it possible to remove the last term in the expression for φ_{cr}. It is easy to see that $\tan 6\varphi_0 = A_6^{6(i)}/A_6^{6(r)}$, so that $A_6^{6(r)}$ changes to $A_6^{6(r)} \cos 6\varphi_0 - A_6^{6(i)} \sin \varphi_0$. Formally the same potential (with real A_6^6) would be obtained if, in addition to the symmetries mentioned above, a vertical mirror plane would be assumed to pass through the radicals. When the azimuthal angle is referred to a radical in the principal plane, we have, because of the additional symmetry, $\varphi_{\text{cr}}(\vartheta, \varphi) = \varphi_{\text{cr}}(\vartheta, -\varphi)$ and therefore $A_6^{6(i)} = 0$, as $Y_6^6 - Y_6^{-6}$ is an odd function of the azimuthal angle. Thus the higher symmetry $\bar{6}m^2$, or D_{3h}, which should be based on a symmetry of the radical, yields the same potential for the $4f$ electron as the lower symmetry $\bar{6}$ or C_{3h}. In cartesian coordinates the potential will then read

$$\begin{aligned}\varphi_{\text{cr}}(x, y, z) = {} & B_2^0(3z^2 - r^2) + B_4^0(35z^2 - 30z^2r^2 + 3r^4) \\ & + B_6^0(231z^6 - 315r^2z^4 + 105z^2r^4 - 5r^6) \\ & + B_6^6(x^6 - 15x^4y^2 + 15x^2y^4 - y^6). \qquad (2.59)\end{aligned}$$

This is easy to prove by means of the definition (2.23) of the Y_l^m. The B_l^m are, of course, proportional to the A_l^m; we have

$$B_2^0 = \sqrt{\frac{5}{16\pi}}\,A_2^0, \quad B_4^0 = \frac{3}{16\sqrt{\pi}}\,A_4^0, \quad B_6^0 = \frac{1}{32}\sqrt{\frac{13}{\pi}}\,A_6^0,$$

$$B_6^6 = \frac{1}{32}\sqrt{\frac{3003}{\pi}}\,A_6^6,$$

if A_6^6 is considered *a priori* to be real. In its ground state the Ce^{3+} ion has $J = \frac{5}{2}$; the next higher level with $J = \frac{7}{2}$ is 2240 cm^{-1} above the ground state, see Table 5. In a first approximation we can neglect the $\frac{7}{2}$ level. We then see at once that the matrix elements of the last two terms of (2.59), formed with the basis vectors $|\frac{5}{2}\,M\rangle$, vanish by virtue of the Wigner–Eckart theorem (2.24) and the property (2.26b) of the C.–G. coefficients. The effective lattice potential for the $\frac{5}{2}$ level will therefore consist only of the first two terms of (2.59). These two terms commute with J_z, since they are rotationally symmetrical with respect to the z-axis. Thus the $\frac{5}{2}$ level in the potential (2.59) is split approximately with respect to states which are eigenstates of J_z. Since, because of the symmetry relation (2.27) of the C.–G. coefficients, the matrix elements of φ_{cr} depend only on the magnitude of M, the $\frac{5}{2}$ level is split into three doubly degenerate levels, the Kramers doublets. As φ_{cr} is diagonal, the splitting is determined by the diagonal elements. According to (2.24) we have

$$\begin{aligned}&\langle\tau\tfrac{5}{2}M\,|\,A_2^0r^3Y_2^0\,|\,\tau\tfrac{5}{2}M\rangle\\ &= (A_2^0\overline{r^2}1/\sqrt{6})\,\langle\tfrac{5}{2}\,||\,Y_2\,||\,\tfrac{5}{2}\rangle\,\langle\tfrac{5}{2}2M0\,|\,\tfrac{5}{2}M\rangle.\end{aligned} \tag{2.60}$$

Here again τ is used to symbolise the other quantum numbers of the cerium ion. The C.–G. coefficient is easily calculated from the recursion formulas or from tables (see, for example, Condon and Shortley, 1959; Edmonds, 1957). We have

$$\langle\tfrac{5}{2}2M0\,|\,\tfrac{5}{2}M\rangle = (1/\sqrt{280})\,(3M^2 - \tfrac{5}{2}(\tfrac{5}{2}+1)). \tag{2.61}$$

(From (2.59), (2.60) and (2.61) we see that we would have obtained the correct matrix elements of the lattice potential, if we had replaced the operator $3z^2-r^2$ by $3J_z^2-\boldsymbol{J}^2$, to within a proportionality constant; this correspondence is denoted operator equivalence.) The reduced matrix element in (2.60) must be calculated explicitly. For this purpose the state $|\frac{5}{2}\frac{5}{2}\rangle$ is resolved with respect to the orbital angular momentum and the spin; owing to the definition of the C.–G. coefficients (2.25), we have

$$|\tfrac{5}{2}\tfrac{5}{2}\rangle = |33\rangle\,|\tfrac{1}{2}-\tfrac{1}{2}\rangle\langle 3\tfrac{1}{2}3-\tfrac{1}{2}\,|\tfrac{5}{2}\tfrac{5}{2}\rangle + |32\rangle\,|\tfrac{1}{2}\tfrac{1}{2}\rangle\langle 3\tfrac{1}{2}2\tfrac{1}{2}\,|\tfrac{5}{2}\tfrac{5}{2}\rangle$$
$$= \sqrt{\tfrac{6}{7}}\,|33\rangle\,|\tfrac{1}{2}-\tfrac{1}{2}\rangle-(1/\sqrt{7})\,|32\rangle\,|\tfrac{1}{2}\tfrac{1}{2}\rangle.$$

Since φ_{cr} is independent of the spin, we have explicitly

$$\langle\tau\tfrac{5}{2}\tfrac{5}{2}\,|\,A_2^0r^2Y_2^0\,|\,\tau\tfrac{5}{2}\tfrac{5}{2}\rangle = A_2^0\overline{r^2}(\tfrac{6}{7}\langle 33\,|\,Y_2^0\,|\,33\rangle+\tfrac{1}{7}\langle 32\,|\,Y_2^0\,|\,32\rangle)$$
$$= A_2^0\overline{r^2}(\tfrac{6}{7}\int (Y_3^3)^*Y_2^0Y_3^3\,d^2\Omega+\tfrac{1}{7}\int (Y_3^2)^*Y_3^0Y_3^2\,d^2\Omega).$$

By $d^2\Omega = \sin\vartheta d\vartheta d\varphi$ we denote the solid-angle element. From the definition (2.23) of the Y_l^m it is easy to obtain the integrals. We obtain

$$\langle\tfrac{5}{2}\,||Y_2||\,\tfrac{5}{2}\rangle = -\sqrt{15/7\pi}.$$

The same is done with the second non-vanishing component of φ_{cr}. We have

$$\left\langle\frac{5}{2}4M0\,\middle|\,\frac{5}{2}M\right\rangle = (-)^{1/2-M}\cdot\frac{48}{\sqrt{42}}\,\frac{M}{(\frac{5}{2}-M)!\,(\frac{5}{2}+M)!}$$

and

$$\langle\tfrac{5}{2}\,||Y_4||\,\tfrac{5}{2}\rangle = 3/\sqrt{7\pi}.$$

In this way we obtain the eigenvalues of the doublets; they are compiled in Table 6. The parameters $B_2^0\overline{r^2}$ and $B_4^0\overline{r^4}$ must be chosen according to experimental results. From measurements of susceptibility and paramagnetic resonance we see that in cerium ethyl sulphate in concentrations so low that there is no mutual magnetic influence of the cerium ions (for this purpose cerium is added as an impurity to the chemically similar but non-magnetic lanthanum ethyl sulphate), the doublet with $M=\pm\frac{1}{2}$ is lowest and, about 3 cm^{-1} above it, there is the doublet with $M=\pm\frac{5}{2}$. The third doublet with $M=\pm\frac{3}{2}$ is at about 130 cm^{-1}

(Elliott and Stevens, 1952, 1953). With these data we obtain $eB_2^0\overline{r^2} = 24\ \text{cm}^{-1}$ and $eB_4^0\overline{r^4} = -73\ \text{cm}^{-1}$. If the cerium ions are concentrated, the two lower terms appear in inverse order (Bogle *et al.*, 1951; Meyer and Smith, 1959). Measurements of the Schottky anomaly of specific heat carried out between 1°K and 20°K with the concentrated salt (Meyer and Smith, 1959) showed that the level with $\pm\frac{1}{2}$ is $7{\cdot}6\pm0{\cdot}1\ \text{cm}^{-1}$ above the level with $\pm\frac{5}{2}$. With these data we obtain for the parameters $eB_2^0\overline{r^2} = 13{\cdot}6\ \text{cm}^{-1}$ and $eB_4^0\overline{r^4} = -82\ \text{cm}^{-1}$.

The Schottky anomaly consists of the following: the energy of a system of N independent paramagnetic atoms which, referred to the ground state, can be in the energy states $E_0 = 0, E_1, E_2, \ldots$ is given by

$$E = N\frac{\sum_{i=0} g_i E_i \exp(-\beta E_i)}{\sum_{i=0} g_i \exp(-\beta E_i)}. \tag{2.62}$$

The g_i are the weight factors of the energy states. The specific heat of the system, $C = \partial E/\partial T$, has peaks (the so-called Schottky anomalies) as long as other components of specific heat (e.g. that of the lattice) are smaller. For an atom with two levels we obtain from (2.62)

$$C = Nkg_0 g_1(\beta E_1)^2 \frac{\exp(-\beta E_1)}{(g_0 + g_1 \exp(-\beta E_1))^2} \tag{2.63}$$

and hence for the maximum

$$\frac{C(T_{\max})}{Nk} = \frac{\beta_{\max}^2 E_1^2}{4} - 1$$

where

$$-\beta_{\max}E_1 = \ln\frac{g_0}{g_1} + \ln\frac{\beta_{\max}E_1 - 2}{\beta_{\max}E_1 + 2}.$$

At low temperatures, when the Debye contribution of the lattice, which decreases proportional to T^3, is very small, the contribution of (2 63) to the specific heat will be overwhelming. Before the maximum is reached ($\beta E_1 \ll 1$), the specific heat increases proportional to T^{-2}, according to (2.63), as T is reduced. This increase with T^{-2} is not typical for Schottky anomalies; for example, it is easy to show that magnetic dipole interactions also give rise to a T^{-2} law, although this increase usually takes place at lower temperatures than that of the Schottky anomalies; a rough estimate yields $kT = \mu^2/a^3$ for them, where μ is the mean moment of the atoms and a is their mean mutual distance.

Owing to the lattice potential, the magnetic susceptibility of cerium ethyl sulphate becomes anisotropic, since the hexagonal symmetry axis is the natural axis of quantisation for the angular momentum and not, as hitherto, the magnetic field. The conditions are simple if the field is parallel to the axis of symmetry, and if the term with $J = \frac{7}{2}$ is neglected. Just as in the anomalous Zeeman effect the three Kramers doublets are split linearly in H. Using (2.29) we have

$$\begin{aligned} E_{L,\,S,\,5/2,\,M} &= E^0_{L,\,S,\,5/2} + E(M) + \langle \tfrac{5}{2} M \,|\, \mu_B H (J_z + S_z) \,|\, \tfrac{5}{2} M \rangle \\ &= E^0_{L,\,S,\,5/2} + E(M) + M g_{5/2} \mu_B H. \end{aligned} \tag{2.64}$$

$E(M)$ is the splitting of the level with $J = \frac{5}{2}$ calculated above and listed in Table 6a; it is due to the lattice potential. Each of the doubly degenerate Kramers doublets splits according to (2.64) by an amount of

$$\Delta(|M|) = 2\,|M|\,g_{5/2}\mu_B H = g_{||}(|M|)\mu_B H. \tag{2.65}$$

It is reasonable to call $g_{||}(|M|)$ the spectroscopic splitting factor; in particular, for the Kramers doublet $|M|$ and the case of a field $\boldsymbol{H}$ parallel to the axis of symmetry. According to (2.41) $g_{5/2} = \frac{6}{7}$.

By analogy with (2.51), with (2.64) and (2.65), and using (2.45) we obtain the susceptibility $\chi_{||}$ in the form

$$\chi_{||} = -N\frac{\mu_B}{2H}\,\frac{\sum\limits_{M=-5/2}^{+5/2} g_{||}(M)\, e^{-\beta(E(M) - g_{||}(M)\mu_B H/2)}}{\sum\limits_{M=-5/2}^{+5/2} e^{-\beta(E(M) - g_{||}(M)\mu_B H/2)}}$$

and, for weak fields,

$$\chi_{||} = N\frac{\mu_B^2}{4kT}\,\frac{\sum\limits_{M=1/2}^{5/2} g_{||}^2(M)\, e^{-\beta E(M)}}{\sum\limits_{M=1/2}^{5/2} e^{-\beta E(M)}}. \tag{2.66a}$$

Of course, the term corresponding to the Van Vleck paramagnetism does not appear in (2.66a), since in the whole calculation we have disregarded the other fine-structure term.

Let us now assume the magnetic field to be perpendicular to the symmetry axis, e.g. parallel to the x-axis. The Hamiltonian for the cerium atom is then given by

$$\mathcal{H} = \mathcal{H}_0 + e\varphi_{\mathrm{cr}} + \mathcal{H}_H$$

with $\mathcal{H}_0$ from (2.34) and $\mathcal{H}_H = \mu_{\mathrm{B}} H(J_x + S_x)$. The magnetic field induces transitions between the Kramers doublets, which are ordinary dipole transitions $\Delta M = \pm 1$, since, according to (2.28) and (2.36), $\mathcal{H}_H$ contains only the matrix elements

$$\langle JM|\,\mathcal{H}_H\,|J(M\pm 1)\rangle = \frac{g_J \mu_{\mathrm{B}} H}{2}\sqrt{J(J+1)-M(M\pm 1)}.$$

Owing to the selection rule $\Delta M = \pm 1$, the field will only give rise to a splitting linear in H in the doublet with $|M| = \frac{1}{2}$. Because of the definition (2.65), the splitting factors of the other two doublets must be zero. From the secular equation we obtain for the doublet with $|M| = \frac{1}{2}$ the splitting

$$\Delta(\tfrac{1}{2}) = g_{\perp}(\tfrac{1}{2})\mu_{\mathrm{B}} H = 3 g_{5/2} \mu_{\mathrm{B}} H,$$

so that $g_{\perp}(\frac{1}{2}) = \frac{18}{7}$. The susceptibility $\chi_{\perp}$ again follows from (2.45) if there the x-components of $\boldsymbol{J}$ and $\boldsymbol{S}$ are substituted appropriately. Carrying out a calculation similar to that which yielded (2.54), with the only exception that the trace is referred to the term with $J = \frac{5}{2}$ and the diagonal element of $\mathcal{H}_H$ vanishes, we obtain

$$\chi_{\perp} = -N(g_{5/2}\mu_{\mathrm{B}})^2 \times \frac{\sum\limits_{M, M'=-5/2}^{+5/2} |\langle M|\,J_x\,|M'\rangle|^2 \left(\mathrm{e}^{-\beta E(M)} - \mathrm{e}^{-\beta E(M')}\right)\left(E(M)-E(M')\right)^{-1}}{\sum\limits_{M=-5/2}^{+5/2} \mathrm{e}^{-\beta E(M)}}.$$

Substituting the matrix element, we obtain for $\chi_\perp$

$$\chi_\perp \cong N\frac{\mu_B^2}{4kT}\cdot\frac{g_\perp^2(\frac{1}{2})e^{-\delta/T}}{1+e^{-\delta/T}}+\frac{5}{2}N\frac{g_{5/2}^2\mu_B^2}{(E(\frac{3}{2})-E(\frac{5}{2}))(1+e^{-\delta/T})} \quad (2.66b)$$

since at low temperatures the $\frac{3}{2}$ doublet can be neglected; $k\delta = E(\frac{1}{2})-E(\frac{5}{2})$. From (2.66b) we see to what extent other states are admixed by the field to the unperturbed states. The strong susceptibility anisotropy is a characteristic feature, typical of many rare-earth salts. This is verified satisfactorily by experiments (see Table 6). In the range from 1°K to 20°K the following values are obtained for the molar susceptibilities of the concentrated salt (Bogle *et al.*, 1951):

$$\chi_{\parallel} = \frac{0{\cdot}094}{T}\,\frac{14{\cdot}44+e^{-7\cdot5/T}}{1+e^{-7\cdot5/T}}, \qquad \chi_\perp = \frac{0{\cdot}094}{T}\cdot\frac{5{\cdot}0+e^{-7\cdot5/T}}{1+e^{-7\cdot5/T}}+4{\cdot}7\times10^{-4}.$$

The splitting factors hence obtained are compiled in Table 6. A comparison with (2.66) shows that $\delta > 0$ so that the state with $|M| = \frac{5}{2}$ is the lowest state. Using the approximate formula (2.26b), we obtain from the temperature-independent component of $\chi_\perp$ a distance of the doublet with $|M| = \frac{3}{2}$ from the ground state which is too large. It may be assumed—and this assumption is verified by calculations (Elliott and Stevens, 1952, 1953)—that the results are improved if the term of the multiplet with $J = \frac{7}{2}$ is taken into account. In this case all the eigenstates are no longer eigenstates of J_z, since the perturbation potential, (2.59), does not commute with J_z. In particular the doublet with $|M| = \frac{5}{2}$ will be strongly influenced, since the parameters $B_6^0\overline{r^6}$ and $B_6^6\overline{r^6}$ are large. The secular-equation problem then arising can be solved. Even with optimum choice of the parameters in agreement with experimental data, however, it is impossible to reproduce the splitting factors measured. We must therefore assume slight deviations from the C_{3h} symmetry. Among all possible deviations we are particularly interested in those which mix the states of the two lowest levels, i.e. those which induce transitions with $\Delta M = \pm 3$, since they lie close to one another and can be strongly affected by small changes in φ_{cr}. Assuming this type of deviations, we can obtain agreement with experiment (Elliott and Stevens, 1952, 1953); see Table 6b.

TABLE 6a

M	$E(M)$	$g_{\parallel}$	$g_{\perp}$	$(g_{\parallel})_{exp}$	$(g_{\perp})_{exp}$
$\pm\frac{1}{2}$	$(16/35)eB_2^0\overline{r^2}+(16/21)eB_4^0\overline{r^4}$	6/7	18/7	0·955±0·005 (1·0±0·2)	2·185±0·01 (2·25±0·2)
$\pm\frac{3}{2}$	$(4/35)eB_2^0\overline{r^2}-(8/7)eB_4^0\overline{r^4}$	18/7	0	—	—
$\pm\frac{5}{2}$	$-(4/7)eB_2^0\overline{r^2}+(8/21)eB_4^0\overline{r^4}$	30/7	0	3·72±0·01 (3·80±0·04)	0·2±0·05 (0±0·4)

Table 6a. The second column shows the level splitting of the ground term of cerium ethyl sulphate with $J=\frac{5}{2}$ in the potential (2.59) in a first perturbation-theoretical approximation; in the same approximation the spectroscopic splitting factors $g_{\parallel}$ and $g_{\perp}$ are given. The experimental values of $g_{\parallel}$ and $g_{\perp}$ were obtained in paramagnetic resonance measurements on the dilute salt (Elliott and Stevens, 1952, 1953). The data in brackets were obtained in susceptibility measurements with single crystals of higher concentrations (Bogle *et al.*, 1951).

TABLE 6b

T (°K)	1	4	14	20	70	170	290
$(\chi_{\parallel}T)_{theor}$	1·38	1·21	0·935	0·89	0·81	0·77	0·745
$(\chi_{\parallel}T)_{exp}$	1·36	1·19	0·89	0·84	0·80	0·75	0·75
$(\chi_{\perp}T)_{theor}$			0·26	0·31	0·53	0·62	0·725
$(\chi_{\perp}T)_{exp}$	$5{\cdot}6\times10^{-4}$	$6{\cdot}6\times10^{-2}$	0·18	0·20	0·49	0·64	0·70

Table 6b. The theoretical and experimental principal susceptibilities of cerium ethyl sulphate. The theoretical values were obtained by Elliot and Stevens (1952, 1953), the experimental values between 1° and 20°K stem from Bogle *et al.* (1951); the values at higher temperatures were determined by Becquerel *et al.* (1938). The latter authors determined $\chi_{\parallel}\cdot T$ not directly but from the Faraday rotation, i.e. from the rotation of the plane of polarisation of light in a magnetised medium. Under certain conditions this rotation is proportional to the susceptibility (Van Vleck and Hebb, 1934; van den Handel, 1956). The data show that the susceptibility is strongly anisotropic at low temperatures and that the anisotropy vanishes gradually at high temperatures. We also see that towards higher temperatures the susceptibility tends to satisfy a Curie law.

The term splitting of paramagnetic ions in the lattice field is very important, e.g. in adiabatic demagnetisation. This is a particular use of the magneto-caloric effect in paramagnetic substances for reaching low temperatures.

The entropy $S(H, T)$, the internal energy E,† and the magnetisation M of a medium are linked by the relation

$$T\,dS = dE + M\,dH = C_H\,dT + T\left(\frac{\partial M}{\partial T}\right)_H dH$$

where

$$C_H = \left(\frac{\partial E}{\partial T}\right)_H = T\left(\frac{\partial S}{\partial T}\right)_H$$

is the specific heat at constant field.

The method of adiabatic demagnetisation is the following: as a first step the paramagnetic medium is magnetised isothermally at a temperature T; this results in an entropy change which, according to (2.67), is given by

$$S(H, T) - S(0, T) = \int_0^H \left(\frac{\partial M}{\partial T}\right)_H dH.$$

In the second step, the substance is separated from the heat bath (e.g. liquid helium) and the magnetic field is switched off, i.e. the substance is demagnetised adiabatically. The entropy remains unchanged; the temperature, however, changes from T to T_0, which can be determined from the implicit equation $S(H, T) = S(0, T_0)$. According to (2.67) we have for small adiabatic changes of the magnetic field

$$C_H\,dT = -T\left(\frac{\partial M}{\partial T}\right)_H dH.$$

Since as a rule $(\partial M/\partial T)_0 < 0$, the substance is cooled, $dT < 0$, if the magnetisation is reduced. The degree of this cooling depends on the susceptibility and the specific heat at constant magnetic field. In an

† E is connected with the usual thermodynamic internal energy U by $E = U - MH$ E corresponds to the Hamiltonian operator.

integrated form the equation determining T_0 can be written as follows:

$$S(H, T)-S(0, T) = S(0, T_0)-S(0, T) = \int_{T}^{T_0} \frac{C_{H=0}(T)}{T}\, dT$$

or

$$\int_{0}^{H} \left(\frac{\partial M}{\partial T}\right)_H dH = \int_{T}^{T_0} \frac{C_{H=0}(T)}{T}\, dT. \tag{2.68}$$

In order to obtain an estimate, assume the paramagnetic ion to possess two doubly degenerate states with an energy difference E_1. Equation (2.63) with $g_0 = g_1 = 2$ will then be applicable to $C_{H=0}(T)$. At sufficiently high temperatures, $\beta E_1 \ll 1$, we shall then have

$$\int_{T}^{T_0} \frac{C_{H=0}(T)}{T}\, dT = -\frac{NE_1^2}{8k}\left(\frac{1}{T_0^2}-\frac{1}{T^2}\right).$$

Let us further assume the field strength to be so low that M can be considered to be a linear function of H, $M = \chi \cdot H$, and that χ is described by a Curie law with the same isotropic splitting factor for both levels:

$$\chi = N\frac{g^2\mu_B^2}{4kT}.$$

We then obtain for the final temperature T_0:

$$\frac{1}{T_0^2}-\frac{1}{T^2} = \frac{1}{T^2}\left(\frac{g\mu_B H}{E_1}\right)^2.$$

For adiabatic cooling it is thus necessary to work with strong fields and small-term splittings. This condition is satisfied mainly for paramagnetic ions in S states, since the expectation value of their spin–orbit coupling vanishes so that the lattice field can affect the spin, and cause splitting, only in a higher approximation. Thus, for example, gadolinium salts are particularly suitable for adiabatic cooling. In order to estimate T_0, we assume $E_1/k = 0{\cdot}1°K$, $H = 10^4$ oe, and $g = 1$; with an initial temperature of $T = 5°K$ we obtain $T_0 = 0{\cdot}74°K$ for the final temperature. As to details of adiabatic demagnetisation we refer to the literature (de Klerk, 1956; Casimir, 1961).

The term splitting of the rare-earth elements generally results in a marked anisotropy of the susceptibility. It is typical for rare-earth compounds. Of course, many other paramagnetic ions will also behave anisotropically in the lattice field; this effect, however, is usually weaker than in the case of the rare-earth elements; we shall discuss this in the following.

This example also shows that, because of the field-induced level splitting of purely paramagnetic substances, it is impossible to derive a Curie–Weiss law, in contrast to the ferromagnetic substances, where the parameter Θ in the Curie–Weiss law represents a measure for the exchange interaction between the magnetic moments. The quantity Θ of paramagnetic substances can thus be considered only an empirical parameter approximating the temperature dependence of the susceptibility in a certain range.

Finally, it must be stressed that the two-fold degeneracy of the cerium ethyl sulphate levels is not a random effect and cannot be removed by an asymmetric lattice field. The important Kramers theorem (1930) applies, according to which the states of a system with an odd number of electrons in an arbitrary electric field must have an at least two-fold degeneracy. This degeneracy is called the Kramers degeneracy, and the corresponding degenerate states the Kramers doublets, quartets, etc. This law results from the invariance of the Hamiltonian under time reversal. Consider the Hamiltonian of an electron in an arbitrary potential. The time reversal transformation is a transition from $\boldsymbol{r}$ to $\boldsymbol{r}$ and from $\boldsymbol{p}$ to $-\boldsymbol{p}$, as momentum is a time derivative of the coordinates. As regards the orbital angular momentum, time reversal means a transition from

$$\hbar \boldsymbol{l} = m[\boldsymbol{r} \wedge \boldsymbol{p}] \rightarrow -\hbar \boldsymbol{l}.$$

Since the spin, as an angular momentum, must display the same transformation properties, time reversal will also cause the transition

$s \to -s$. Since the Hamiltonian contains the momentum in a quadratic form and the angular momenta in a bilinear form, cf. (2.18) with $H = 0$, it must be invariant under time reversal, if there is no external magnetic field. The same is true for a many-electron system as may be verified from (2.16) and (2.33). Assume the transformation to be described by an operator K. Since an arbitrary operator A is transformed to KAK^{-1}, on the basis of what has been said above, K must have the following properties:

$$KrK^{-1} = r, \quad KpK^{-1} = -p, \quad KlK^{-1} = -l, \quad KsK^{-1} = -s. \tag{2.69}$$

In addition to this we have $KK^{+} = 1$, in order that all physically important properties, such as hermiticity, norm of the vectors: $\langle u|u\rangle = \langle u|K^{+}K|u\rangle, \ldots$ are conserved in the transformation. In the following we shall consider a representation where r is a real and p an imaginary operator, e.g. the coordinate representation ($p = \hbar/i\nabla$). The operator K_0, which causes the transition to the conjugate complex form, will then satisfy the first three requirements of (2.69). K_0 is a so-called anti-linear operator, or, more exactly, an anti-unitary operator, as $K_0K_0^{+} = 1$. We must bear in mind that anti-linear operators do not commute with c numbers: $cK_0 = K_0c^*$ (Messiah, 1960). The last relation in (2.69) is not satisfied by the operator K_0, since in the usual redresentation of the spin matrices in which s_z is diagonal, s_x, s_z are real operators and s_y is an imaginary operator, i.e. we have $K_0s_{x,z}K_0^{-1} = s_{x,z}$ and $K_0s_yK_0^{-1} = -s_y$. The necessary sign reversal of s_x and s_z is obviously obtained when a rotation through an angle π is carried out about the y-axis in the spin space. This rotation is achieved by the operator $\exp(-i\pi s_y)$ (see, for example, Messiah, 1960; Heine, 1960). Thus we have

$$K = e^{-i\pi s_y}K_0 = -2is_yK_0,$$

since $\exp(-i\pi s_y) = -2is_y$ as can easily be verified. For a system with N electrons we must take the y-component of the total spin

$$K = e^{-i\pi S_y}K_0 = (-2i)^N \prod_{\nu=1}^{N} s_{y,\nu}K_0 \tag{2.70}$$

The symbol $\prod_{\nu}$ indicates that we have to form the product of the spins of all electrons. A state $|v\rangle$ is said to be Kramers-conjugate to $|u\rangle$ if $|v\rangle = K|u\rangle$. This does not permit the conclusion that also $|u\rangle$ must be Kramers-conjugate to $|v\rangle$, since this would imply $K^2 = 1$. However, as $s_y^2 = \frac{1}{4}$, $K^2 = (-)^N$, according to (2.70). This makes it necessary to distinguish the following cases:

(a) The number of electrons N is even, $|u\rangle$ and $|v\rangle$ are Kramers-conjugate with respect to one another. It is then possible in the space of the state vectors of the system to construct a complete orthonormalised basis, so that $K|w\rangle = |w\rangle$ for each basis vector. To show this, consider an arbitrary vector $|u_1\rangle$ and form $|w_1\rangle = c_1|u_1\rangle + c_1^*|v_1\rangle$. Since $K^2 = 1$, the vector $|w_1\rangle$ has the necessary property; c_1 is a normalisation constant. By means of a second vector, $|u_2\rangle$, which is orthogonal to $|w_1\rangle$, a vector $|w_2\rangle = c_2|u_2\rangle + c_2^*|v_2\rangle$ can be formed, for which again $K|w_2 = |w_2\rangle$, and which is orthogonal to $|w_1\rangle$:

$$\begin{aligned}\langle w_2 | w_1\rangle &= c_2^*\langle u_2 | w_1\rangle + c_2\langle v_2 | w_1\rangle = c_2(\langle u_2 | K^+)|w_1\rangle \\ &= c_2\{\langle u_2 | (K^+ | w_2\rangle)\}^* = c_2\{\langle u_2 | (K | w_1\rangle)\}^* = c_2\langle w_1 | u_2\rangle = 0\end{aligned}$$

since with $K^2 = 1$ we also have $K^+ = K$. (Unlike the linear operators, anti-linear operators make it necessary in the formation of the scalar product to give the vector upon which the operator acts, since we have now

$$(\langle s | A)|t\rangle = \{\langle s | (A | t\rangle)\}^*,$$

(Messiah, 1960). This is easy to verify for the operator K_0, when the scalar product is formed with arbitrary complex Schrödinger functions.) Continuing this procedure, we can construct an obviously complete and orthonormalised basis whose vectors are eigenvectors of K and, since K commutes with $\mathscr{H}$, they are also eigenvectors of $\mathscr{H}$. In a sufficiently asymmetric electric field the energy eigenvalues pertaining to the vectors will all be different from one another. Let $|w\rangle$ denote the lowest state which, at sufficiently low temperatures, will be the only populated one. The value expected for the magnetic moment in this state (for N electrons) is

$$-\mu_B\langle w_0|\boldsymbol{L}+2\boldsymbol{S}|w_0\rangle = -\mu_B(\langle w_0|K)(\boldsymbol{L}+2\boldsymbol{S})(K^{-1}|w_0\rangle$$
$$= -\mu_B\{\langle w_0|K(\boldsymbol{L}+2\boldsymbol{S})K^{-1}|w_0\rangle\}^* = \mu_B\langle w_0|\boldsymbol{L}+2\boldsymbol{S}|w_0\rangle,$$

according to (2.69), since $\boldsymbol{L}$ and $\boldsymbol{S}$ are Hermitean operators. The expectation value of the magnetic moment will thus vanish, which means that also the susceptibility of the system will be zero, if the temperature-independent Van Vleck paramagnetism is disregarded.

(*b*) The number N of electrons is odd; $|v\rangle$ will be Kramers-conjugate to $|u\rangle$ but not vice versa. We then have $K^2 = -1$ and thus $K^+ = K^{-1} = -K$. The two states $|u\rangle$ and $|v\rangle$ are now linearly independent and orthogonal onto one another:

$$\langle u|v\rangle = \langle u|(K|u\rangle) = (\langle u|K)|u\rangle^* = \langle u|(K^+|u\rangle) = -\langle u|(K|u\rangle)$$
$$= -\langle u|v\rangle$$

if we make use of the definition of a Hermitean-conjugate operator

$$\langle s|(A^+|t\rangle) = \{(\langle t|A)|s\rangle\}^*.$$

Since the Hamiltonian commutes with K, two linearly independent solutions pertain to each eigenvalue of $\mathcal{H}$, so that each level is at least two-fold degenerate. Thus Kramers' theorem is proven.

In a magnetic field the Hamiltonian will no longer be invariant under time reversal, as it now contains the operators in a linear form, as scalar products with the magnetic field. As to the paramagnetic susceptibility, it is a consequence of the Kramers theorem that the low-temperature susceptibility of an atom or ion, with an even number of electrons and a singlet state as the lowest state, is lower than the value resulting from the Curie law (see Fig. 12).

In the case of rare-earth ions, the influence of the lattice is as a rule restricted to a relatively narrow temperature range; at higher temperatures, ($T > 100°$K), the magnetic behaviour becomes more and more similar to that of free ions. In the case of other paramagnetic substances this behaviour is different; here the elements of the iron group are known best. In the free state their atoms have an argon configuration, with one or two electrons in the 4*s* shell and an incomplete 3*d* shell which is filled when we go from scandium to copper. In an ionised state the 3*d* electrons of these atoms are thus exposed to the ambient fields without any screening. The elements of the palladium group with their incomplete 4*d* shell, those of the platinum group which follows upon the rare earths (the 5*d* shell is filled in it), and finally the elements of the uranium group, in which the 5*f* and 6*d* shells are filled, have a similar structure. These four groups are some-

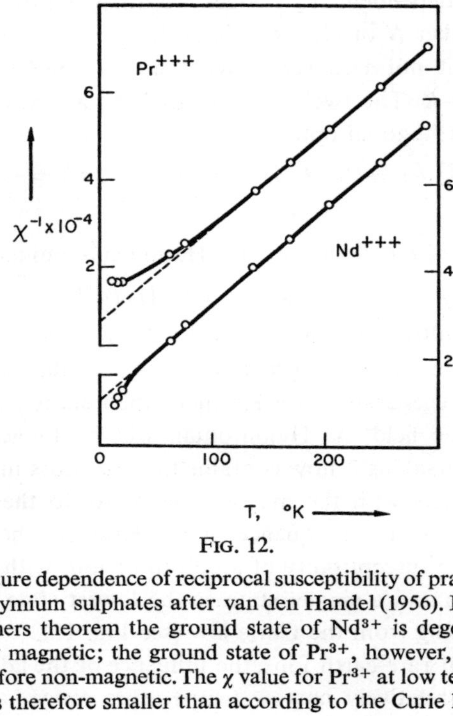

FIG. 12.

Temperature dependence of reciprocal susceptibility of praseodymium and neodymium sulphates after van den Handel (1956). By virtue of the Kramers theorem the ground state of Nd^{3+} is degenerate and obviously magnetic; the ground state of Pr^{3+}, however, is a singlet and therefore non-magnetic. The χ value for Pr^{3+} at low temperatures is therefore smaller than according to the Curie law.

how in contrast to the rare-earth elements, since their magnetic behaviour is decisively influenced by the lattice field; we shall discuss them in the following by way of example of the iron group.

The intensity of the lattice field is characterised by the term splitting in this field compared with the spin–orbit coupling constant. Typical

TABLE 7

Z	Ion	Configuration	Ground term	ζ (cm^{-1})	$p_{eff} = g_J\sqrt{J(J+1)}$	$p_{eff} = 2\sqrt{S(S+1)}$	$(p_{eff})_{exp}$
22	Ti^{3+}	$3d^1$	$^2D_{3/2}$	154	1·35	1·73	1·8
23	V^{3+}	$3d^2$	$^3F_2$	104	1·63	2·83	2·8
23	V^{2+}	$3d^3$	$^4F_{3/2}$	55	0·77	3·87	3·9
24	Cr^{3+}	$3d^3$	$^4F_{3/2}$	87	0·77	3·87	3·9
24	Cr^{2+}	$3d^4$	$^5D_0$	57	0	4·90	4·7
25	Mn^{3+}	$3d^4$	$^5D_0$	85	0	4·90	4·7
25	Mn^{2+}	$3d^5$	$^6S_{5/2}$	—	5·92	5·92	5·8
26	Fe^{3+}	$3d^5$	$^6S_{5/2}$	—	5·92	5·92	5·8
26	Fe^{2+}	$3d^6$	$^5D_4$	−100	6·70	4·90	5·5
27	Co^{2+}	$3d^7$	$^4F_{9/2}$	−180	6·63	3·87	4·8
28	Ni^{2+}	$3d^8$	$^3F_4$	−335	5·59	2·83	3·2
29	Cu^{2+}	$3d^9$	$^2D_{5/2}$	−828	3·55	1·73	1·9

The third and fourth columns of this table give the configuration and the ground terms of the doubly and triply ionised atoms of the iron group. The following columns contain the values for the spin–orbit coupling constant ζ after Bleaney and Stevens (1953), the effective magnetic moments obtained for free ions according to (2.53), and the values obtained when the orbital moment is neglected. In the last column representative values are compiled of the experimental effective magnetic moments of the ions.

values for this field-induced term splitting of iron-group ions are of the order of 10^4 cm^{-1}, whereas ζ is about 10^2 cm^{-1} (see Table 7). Hund's rule is fulfilled, i.e. the influence of the lattice field is smaller than that of electrostatic interaction of the electrons of an atom. Because of the relative weakness of the spin–orbit coupling, it can be neglected in the Schrödinger equation considered in a first approximation. Since the Hamiltonian, (2.16) with $H = 0$, does not contain the spin, the wave function is a product of a coordinate component and a spin component which can be split off. The Hamiltonian operator will now commute with K_0, the operator which corresponds to the transition to the complex conjugate form. As in the case of Kramers' theorem, we can construct a basis so that for each eigenfunction $K_0\psi = \psi$, i.e. ψ is real. For a sufficiently asymmetric lattice potential the eigenfunctions will not be degenerate, and the expectation value for the orbital moment will vanish for each state. Thus, if the lattice field is able to remove degeneracies, the magnetic behaviour of the ions must be such as if they were in S states, i.e. as if the resulting orbital angular momentum were zero. This property is called the quenching or "freezing in" of orbital angular momentum. From Table 7 we see that for the salts of the iron-group elements the orbital angular momentum is quenched to a substantial degree, since the effective magnetic moments are almost entirely determined by the spin. Deviations can be explained by spin–orbit coupling, which remixes the real states, albeit to a small extent. The g factors will therefore deviate from the pure spin value 2 by only about ζ/Δ, Δ being a measure for lattice-field splitting. On the other hand, the deviations can be partly attributed to the fact that the ions of the iron-group elements are frequently bound in a potential with predominantly cubic symmetry, cf. (2.58). In such a high-symmetry potential the terms are not completely split, so that in certain cases only partial quenching can be observed. The d^1 configuration is a good example of these conditions. A basic system for the coordinate constituent of the wave function is given by

$$\frac{xy}{r^2}, \quad \frac{yz}{r^2}, \quad \frac{zx}{r^2} \quad \text{and} \quad \frac{x^2-y^2}{r^2}, \quad \frac{y^2-z^2}{r^2}.$$

We see that in the cubic potential (2.58) the first three terms are degenerate, since they are obtained by a coordinate permutation. The same is true for the other two functions. The triplet is often denoted by Γ_5 and the doublet by Γ_3. The expectation value of the orbital angular momentum formed with an arbitrary linear combination of Γ_3 vanishes; this is easy to verify. The influence of the orbital angular momentum is suppressed if Γ_3 is the ground state of the ion, and the energetic distance between the term groups Γ_3 and Γ_5 is sufficiently large. On the other hand, in the case of the triplet, it is possible to pass from the basis given above to another basis,

$$\frac{z(x\pm iy)}{r^2}, \quad \frac{xy}{r^2},$$

so that the first two functions are eigenfunctions of l_z with the eigenvalues ± 1, and the expectation value of l_z vanishes for the third function. Considered in this way, the triplet behaves like a p state, and the orbital angular momentum of an ion in this state is partially quenched. It is therefore important to know which of the two groups of terms is lowest. For this purpose we consider a relatively realistic model, in which the cubic lattice field is produced by six groups of crystal water at equivalent points on the coordinate axes. These points are the face centres of a cube which contains the paramagnetic ion in its body centre. The positions of the H_2O groups are assumed to be such that the (diamagnetic) O^{2-} ion is oriented towards the paramagnetic ion. The triplet will then clearly lie below the doublet, since the triplet functions are maximum along the diagonals $x = y$, $y = z$, $z = x$, while the doublet functions are maximum along the coordinate axes. The electrons in Γ_3 states are closer to the oxygen ions and are thus subject to a stronger coulomb repulsion. It is obvious that we obtain the same sequence of terms if we assume positive charges at the corners of the cube, instead of negative charges at the centres of its faces. We also obtain this sequence of terms with a d^6 configuration, since it is easy to imagine that, because of Hund's rule, or, more exactly, because of the electron exchange interaction, the first five electrons will have parallel spins and occupy all five states of orbital angular momentum, so that their charge distribution is spherically symmetric, while the sixth electron has its spin in an antiparallel orientation and can be in any of the five states of orbital angular momentum. As to the orbital angular momentum, it thus behaves like an electron in a d^1 configuration. The d^4 and the d^9 configurations, however, can be imagined to arise when an electron is removed from the spherically symmetric d^5 and d^{10} configurations, respectively, i.e. when these charge distributions are provided with a "positive hole". It can thus be demonstrated that an atom or ion with, e.g. a configuration l^N has the same multiplet structure as an atom or ion with the configuration $l^{2(2l+1)-N}$, since for the electrostatic interaction of the holes determining the multiplet structure (see p. 80), the sign of the charge is inessential (Slater, 1960; Condon and Shortley, 1959).

Tables 5 and 7 illustrate this symmetry for the ground terms. The positive sign of the hole charges, of course, becomes apparent in the interaction with external fields. Thus, for example, the spin–orbit coupling constant ζ, which chiefly describes the interaction of electron and hole spins with the fields of the nucleus and the other closed shells, changes its sign in a transition from the upper to the lower half

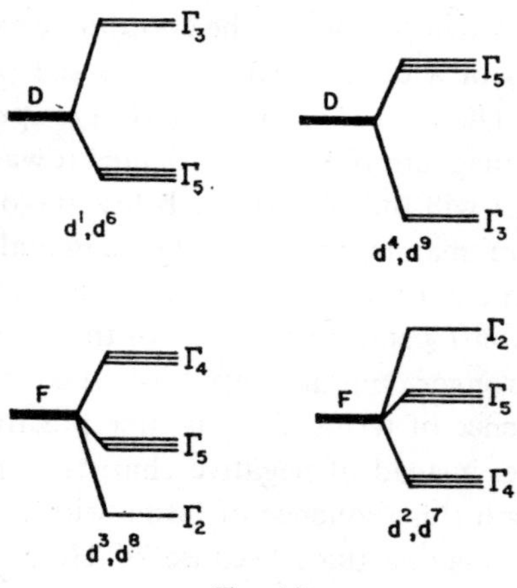

FIG. 13.

Splitting of the D and F terms of an atom with the configuration d^N in a lattice field of cubic symmetry, which is produced by negative charges on the coordinate axes, if the paramagnetic ion is placed in the origin of the coordinate frame. The terms are denoted according to Bethe (1929).

of a shell (2.37), which results in an inversion of the term order within the multiplets. The same consequence is obtained for the term order in a lattice field which is inverted in the d^4 and d^9 configurations, compared with the d^1 and d^6 configurations, respectively; see Fig. 13. A similar consideration for the F terms shows that they are split in a crystal field into two triplets, Γ_4 and Γ_5, and one singlet state, Γ_2 (Fig. 13). The Cr^{3+} ion is an example, in the singlet state Γ_2 its orbital angular momentum is entirely suppressed (Fig. 11).

As a rule, weaker components of lower symmetry (tetragonal, trigonal) are superimposed on the strong cubic component of the lattice potential which, together with spin–orbit coupling, give rise to a stronger splitting of the terms. It is easiest to determine the possible field-induced splittings by means of group-theoretical methods, since the eigenfunctions of the terms are irreducible representations of the symmetry group of the lattice potential. Here, however, we shall not enter into details of this method (see, for example, Heine, 1960; Bethe, 1929; Fick and Joos, 1957). Deviations from cubic symmetry are of course of decisive importance for the anisotropy of the susceptibility. In general, we can say that the anisotropy will be small, if in the cubic lattice field an (orbital) singlet state represents the lowest state, as such a state cannot be split up in an asymmetrical field. In this way we can explain the isotropy of the salts of Mn^{2+} or Fe^{2+}, and also the isotropy of the Ni^{2+} salts compared with the Co^{2+} salts,

TABLE 8

$CoSeO_4 \cdot 6H_2O$				$NiSeO_4 \cdot 6H_2O$		
T (°K)	p_1^2	p_2^2	p_3^2	T (°K)	$p_{\parallel}^2$	$p_{\perp}^2$
298·6	27·54	17·71	21·64	298·6	10·45	10·67
255·0	27·36	16·98	21·15	211·8	10·39	10·61
157·1	26·97	12·93	18·69	168·4	10·34	10·58
121·0	26·81	10·61	17·27	110·7	10·13	10·40
100·1	26·06	8·23	15·60	82·1	9·95	10·30
79·8	25·11	6·12	13·92			

In this table is compiled the squares of the effective magnetic moments in units of μ_B for $CoSeO_4 \cdot 6H_2O$ and $NiSeO_4 \cdot 6H_2O$ after Guha (1951); $p^2 = 3kT \chi(N\mu_B^2)^{-1}$. The cobalt salt crystallises in a monoclinic structure; we therefore give the three principal susceptibilities. The nickel salt crystallises in a tetragonal structure, so we give the susceptibilities parallel and perpendicular to the symmetry axis. In the cubic potential of the crystal water molecules, the Co^{2+} ion is in a triplet state and Ni^{2+} in an (orbital) singlet state. Co^{2+} deviates strongly from the pure spin value $p_{\text{eff}}^2 = 15$; it displays a clear magnetic anisotropy. In addition to this, the salt does not obey a Curie law, as the p values are not independent of temperature. Ni^{2+} deviates from the pure spin value $p_{\text{eff}}^2 = 8$ owing to the strong spin–orbit coupling; it is, however, isotropic and obeys a Curie law.

whose term order is inverted as compared to that of nickel (Table 8). Apart from this, the susceptibility is calculated similarly as in the example considered above; we shall therefore not discuss it again (Van Vleck, 1932a; Schlapp and Penney, 1932).

The compounds of other paramagnetic groups differ from those of the iron group as regards the nature of their binding, which is changed more and more; for the heavy elements ion binding plays a smaller part than covalent binding. Owing to this, the electron wave functions of the paramagnetic ions are extended more and more from the ion into the crystal, and thus lose their localised character. It is, however, precisely the localisation which makes the exchange forces effective (Hund's rule) and thus causes the strong paramagnetism. It is therefore easy to understand that, as a result of the spreading of the wave functions into the crystal, the effective magnetic moment becomes smaller. This can be observed for all transition elements. In this case the crystal-field approximation is no longer applicable.

As far as they are not ferromagnetic or antiferromagnetic, the metals of the transition elements are strongly paramagnetic (Table 9). However, they have no Langevin paramagnetism ($\chi \sim 1/T$), as this is typical for localised electrons, but rather a Pauli-type paramagnetism, since the susceptiblity depends only a little on the temperature (with the exception of palladium). This shows that the d electrons must be mobile. It is therefore more correct to use a description within the framework of the band model. The mobility of the electrons in a solid is plausible since, when a gas condenses into a solid, the outer electrons, which are less strongly bonded, may tunnel through the potential hills between the atoms and migrate through the crystal. This must result in a spreading of the initially distinct levels over a certain energy range and thus in the appearance of bands. The extent of this range, i.e. the band width, is a measure for the influence the transition to the solid phase exerts on the electrons. A narrow band therefore means slight deviations from the conditions for free atoms and, particularly in this connection, it indicates a localisation of the electrons. A narrow band also requires a high density of states, $Z'(E) = dZ/dE$ and, possibly, a strong Pauli paramagnetism; see (2.4)

TABLE 9

	Sc	Ti	V	Cr	Mn	Fe	Co	Ni
$\chi_p \times 10^6$	315	153	230	165	533	ferrom.	ferrom.	ferrom.
$d\chi_p/dT \times 10^8$	−24	+4	−1	+1	−14			
$\gamma_0 \times 10^4$ (cal/mole·deg²)	27	8	22	4	33–43	12	12	17·4
$\chi_p^0 \times 10^6$	155	46	126	25	190–247			
	Y	**Zr**	**Nb**	**Mo**	**Tc**	**Ru**	**Rh**	**Pd**
$\chi_p \times 10^6$	191	120	212	89	270	44	102	558
$d\chi_p/dT \times 10^8$	−7	+2·5	−2	+0·2		0	+3	−140
$\gamma_0 \times 10^4$ (cal/mole·deg²)	24	7	28	5		8	10–12	22
$\chi_p^0 \times 10^6$	138	40	161	29		46	57–69	126
	La	**Hf**	**Ta**	**W**	**Re**	**Os**	**Ir**	**Pt**
$\chi_p \times 10^6$	122	81	154	58	69	10	26	190
$d\chi_p/dT \times 10^6$	−3	+1	−1	0	0	+1	+2	−12
$\gamma_0 \times 10^4$ (cal/mole·deg²)	24	7	13–14	2	6	6	8	16
$\chi_p^0 \times 10^6$	138	40	75–80	11	34	34	46	92

This table shows the atomic susceptibility χ_p and its temperature derivative, $d\chi_p/dT$, as obtained experimentally at room temperature for the metals of the iron group, the palladium group, and the platinum group; γ_0 is the constant of specific heat of electrons at low temperatures; $c_{el} = \gamma_0 T$ with $\gamma_0 = \frac{2}{3}\pi^2 k^2 Z'(\zeta_0)$ (see Shimizu *et al.* (1962). χ_p^0 is the paramagnetic susceptibility calculated according to (2.4) from the experimental values γ_0. This approximation corresponds to the use of an isotropic effective mass for the electrons of the transition metals (cf. Section II.1).

or (2.7). The density of states can be determined experimentally from the electron contribution to the specific heat which, at low temperatures, exceeds the lattice contribution. (As one can easily verify from (1.14), we have $c_{el} = \gamma_0 T$ where $\gamma_0 = \frac{2}{3}\pi^2 k^2 Z'(\zeta_0)$; cf., for example, Section II.1.) Figure 14 shows a typical density of states behaviour in the

3*d* band. Substituting the values obtained for $Z'(E)$ in eqns. (2.4) and (2.7) for the susceptibility, we do not obtain quantitative agreement with the susceptibilities measured experimentally (Table 9). This may be explained from two reasons. On the one hand, as in the case of the alkali metals, exchange and correlation energies must be taken into account in a correct calculation. This is still more difficult than in the

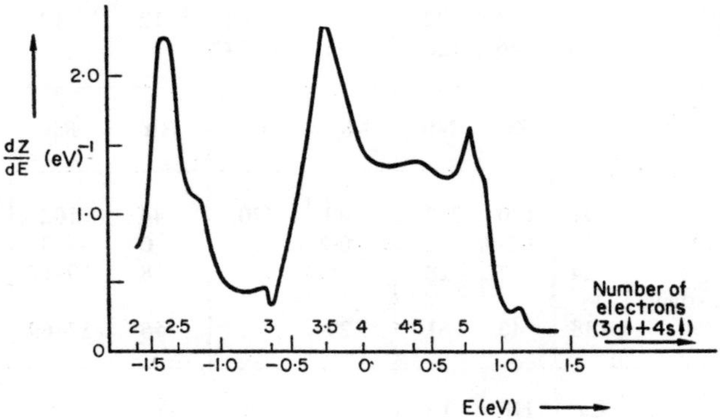

FIG. 14.

Possible density of states of the 3*d* band per atom, for one spin direction, depending on the number of electrons with a fixed direction, of spin, and as a function of energy (Mott, 1964). This curve is characterised by the typical minimum whose position is at about the point of the Cr Fermi edge. Because of its overlapping with the 4*s* band, this band comprises a little more than five electrons per spin.

case of the alkali metals, since the transition metals have a complex band structure. So far only roughly approximate calculations have been possible (Sections II.1 and III.5). On the other hand, in contrast to the alkali metals which have only *s* electrons, in transition metals the influence of the orbital angular momentum must be taken into account. It may result in a deviation of the effective magnetic moment from the pure spin value (if quenching is incomplete), as well as in the appearance of an induced Van Vleck paramagnetism (Denbigh and Lomer, 1963). Thus, in the case of the transition metals, the theoretical situation is unsatisfactory and we have to restrict ourselves to qualitative results.

Thus, for example, the Fermi edge of chromium seems to lie in a minimum of the density of states (Fig. 14). According to (2.7), this means that $d\chi_p/dT > 0$, since at the minimum $Z''(E) = 0$ and $Z'''(E) > 0$, in agreement with Table 9. In this connection the behaviour of alloys is interesting, since the population of the bands can be varied continuously by means of alloying; here, however, we shall not consider these possibilities but refer to review articles (Friedel, 1958; Vonsovskij and Izjumov, 1964; Mott, 1964).

6 Paramagnetic Resonance

Cyclotron (diamagnetic) resonance was shown to be a result of the fact that transitions between the Landau levels of electrons in a constant magnetic field were induced by a high-frequency electric field; this gave rise to a resonance absorption (Section I.6). In paramagnetic resonance, a high-frequency magnetic field induces transitions between the discrete levels which a particle possesses because of its permanent magnetic moment in a constant magnetic field. In principle this method is identical with the well-known Rabi atomic beam method used to determine the magnetic moments of free atoms (Kopfermann, 1956). The simplest case of such a resonance is that of a particle with spin $\frac{1}{2}$ which does not interact with other spins. The Hamiltonian of such a particle in a magnetic field $\boldsymbol{H}$ is given by

$$\mathcal{H} = -\mu(\boldsymbol{\sigma}\cdot\boldsymbol{H}).$$

μ is the magnetic moment of the electron or nucleon, σ denotes the three Pauli matrices. The magnetic field is assumed to possess a constant z-component H_0, and a circularly polarised transverse component:

$$\boldsymbol{H} = (H_1 \cos \omega t,\, H_1 \sin \omega t,\, H_0).$$

The time-dependent Schrödinger equation $\mathcal{H}|\psi\rangle = i\hbar|\dot{\psi}\rangle$ can be solved by means of the Ansatz

$$|\psi\rangle = c_+(t)e^{i\mu H_0 t/\hbar}|+\rangle + c_-(t)e^{-i\mu H t/\hbar}|-\rangle.$$

$|\pm\rangle$ are the eigenvectors of the z component of the angular momentum so that, according to (2.36),

$$\sigma_z|\pm\rangle = \pm|\pm\rangle, \quad \sigma_x|\pm\rangle = |\mp\rangle, \quad \sigma_y|\pm\rangle = \pm i|\mp\rangle. \tag{2.71}$$

Assuming the spin to be parallel to the z-axis at $t = 0$ a simple calculation leads us to the probability densities for the spin in the two possible states:

$$\begin{aligned} |c_+(t)|^2 &= \frac{(\omega-\omega_0)^2+\omega_1^2\cos^2\left((t/2)\sqrt{(\omega-\omega_0)^2+\omega_1^2}\right)}{(\omega-\omega_0)^2+\omega_1^2}, \\ |c_-(t)|^2 &= \frac{\omega_1^2\sin^2\left((t/2)\sqrt{(\omega-\omega_0)^2+\omega_1^2}\right)}{(\omega-\omega_0)^2+\omega_1^2} \end{aligned} \tag{2.72}$$

where

$$\omega_0 = -\gamma H_0 = -2\frac{\mu}{\hbar}H_0 \quad \text{and} \quad \omega_1 = -\gamma H_1 = -2\frac{\mu}{\hbar}H_1.$$

The parameter γ is called the gyromagnetic factor. For electron spins $\gamma = -g\mu_B/\hbar$ with $g = 2{\cdot}0023$. This example shows that for $\omega_1 \ll \omega_0$, i.e. for $H_0 \gg H_1$, a sharp resonance will arise at the point $\omega = \omega_0$, which corresponds to the ordinary magnetic dipole transitions, $\Delta m_s = \pm 1$. Measurements of the resonance frequency yield the magnetic moment of the particle. This method was first applied in the determination of the atomic moments in the atomic-beam method. It cannot be applied without restrictions to a system of magnetic dipoles (nuclear spins, paramagnetic ions, electrons), as we are then faced with the problem of the various interactions arising in such a system. It is, however, possible to give phenomenological equations for the system's magnetisation, which in many cases yield a suitable description, and were first given by Bloch (1946) in connection with problems of nuclear resonance. The equation of motion for the magnetic moment $\boldsymbol{M}$ of a substance consisting of free magnetic dipoles reads, in both classical and quantum-mechanical representations,

$$d\boldsymbol{M}/dt = \gamma[\boldsymbol{M}\wedge\boldsymbol{H}],$$

if $\boldsymbol{H}$ is a spatially homogeneous magnetic field. The equation says that the variation with time of the angular momentum $\hbar\boldsymbol{J}$ is equal to the torque since $\boldsymbol{M} = \gamma\hbar\boldsymbol{J}$. When we now assume theoretically the possible interactions to become active, the following will be changed: interaction between the magnetic moments and the lattice will result in a magnetisation of the system in the direction of the field; the magnetisation will mainly be parallel to the z-direction if $\boldsymbol{H} = (H_1 \cos \omega t, H_1 \sin \omega t, H_0)$ and $H_0 \gg H_1$. The magnetisation corresponding to the static field H_0 will be $M_0 = \chi H_0$. It will then be plausible to assume for the equation of motion of the z-component of magnetisation

$$\frac{dM_z}{dt} = \gamma[\boldsymbol{M} \wedge \boldsymbol{H}]_z - \frac{M_z - M_0}{T_1}. \tag{2.73a}$$

This means that M_z will go over into its equilibrium value M_0 if, for example, at the time $t = 0$ the transverse field is switched off, and if at this instant of time M_z was equal to zero; equilibrium is assumed to be established as an exponential function of time:

$$M_z(t) = M_0(1 - e^{-t/T_1}).$$

The time T_1, which characterises the time dependence of the magnetisation, is called the spin–lattice relaxation time. It is a measure of the interaction of the moments with the lattice. Similar equations of motion apply to the transverse components with the exception that their equilibrium magnetisation must vanish:

$$\left.\begin{aligned}\frac{dM_x}{dt} &= \gamma[\boldsymbol{M} \wedge \boldsymbol{H}]_x - \frac{M_x}{T_2}\\ \frac{dM_y}{dt} &= \gamma[\boldsymbol{M} \wedge \boldsymbol{H}]_y - \frac{M_y}{T_2}.\end{aligned}\right\} \tag{2.73b}$$

In general T_2 is essentially smaller than T_1. This shows that the transverse magnetisation is influenced not by interactions with the lattice but by a stronger mutual interaction of the moments. Their influence can be illustrated by means of the following model: assume the magnetisation to possess a non-zero transverse component. In the case of free, mutually independent magnetic moments it would be conserved as to its magnitude, since all moments would precess uniformly around the field H_0. Owing to the interactions (dipole–dipole, exchange interaction, ...), however, the magnetic field will be different at different points so that the moments precess at different rates around the preferred direction. The result is that after a certain time T_2, which is called the spin–spin relaxation time, a transverse component of magnetisation will be reduced to zero.

Equations (2.73) are called Bloch's equations. Of course they are not exact, since the relaxation processes can only be described approximately by means of constants in the equation of motion. Moreover, the preferred status of the z-component depends only on the strength of the magnetic field H_0, so that in the case of smaller field strengths Bloch's equations must be modified (see Abragam, 1962).

Experimentally, resonance is achieved by varying either the frequency ω of the transverse field or the intensity of field H_0; at the same time the voltage, which is induced in a coil by the variation with time of the magnetisation, is measured; this voltage will be highest if resonance is reached (nuclear induction). It is also possible to measure the energy absorbed in the flipping-over of the moments (nuclear resonance absorption). It is generally cumbersome to deal with the Bloch equations; in practice, however, only limiting cases are of interest; the most important case among them is that where ω or H_0 are varied so slowly that the system remains in thermal equilibrium, i.e. $dM_z/dt = 0$ is then always satisfied. Passing over temporarily to a complex formulation, we put $m = M_x + iM_y$. The Bloch equations (2.73) will then read in this limiting case

$$\dot{m} = i\omega_0 m - i\omega_1 M_z \, e^{i\omega t} - \frac{m}{T_2},$$

$$0 = -i\frac{\omega_0}{2}(m \, e^{-i\omega t} - m^* \, e^{i\omega t}) - \frac{M_z - M_0}{T_1}.$$

These two coupled equations can be solved by means of the Ansatz $m = m_0 e^{i\omega t}$ and we obtain

$$m_0 = -\frac{i\omega_1 T_2 M_z}{1 + i(\omega - \omega_0)T_2}, \qquad M_z = M_0 \frac{1 + (\omega - \omega_0)^2 T_2^2}{1 + (\omega - \omega_0)^2 T_2^2 + T_1 T_2 \omega_1^2}.$$

The energy withdrawn per unit time from the external field is given by $(\boldsymbol{H}\cdot d\boldsymbol{M}/dt)$, when the magnetisation varies by $d\boldsymbol{M}/dt$ per unit time. Averaging this expression for the energy over a time interval $2\pi/\omega$, we obtain the work A performed by the field:

$$A(\omega) = \frac{\omega}{2\pi}\int_0^{2\pi/\omega}\left(\boldsymbol{H}\cdot\frac{d\boldsymbol{M}}{dt}\right)dt = \frac{\omega H_1}{2\pi}\,\mathrm{Re}\int_0^{2\pi/\omega}\dot{m}\,\mathrm{e}^{-i\omega t}\,dt$$
$$= \omega H_1\,\mathrm{Re}\,im_0 = \frac{\omega\omega_0 H_1^2 T_2 \chi}{1+(\omega-\omega_0)^2 T_2^2 + T_1 T_2 \omega_1^2}\,. \qquad (2.74)$$

We obtain a resonance line which, as in (2.72), has its maximum at ω_0; it must be stressed, however, that A is proportional to the equilibrium magnetisation so that here, unlike in the atomic-beam method where every spin flip is recorded, the system must be polarised, i.e. the level population must be different in order to obtain a resonance effect. The shape of the absorption line depends on the relaxation times. For $H_0 \gg H_1$ and $\omega_0 T_2 \gg 1$ a sharp resonance line is obtained whose amplitude is given by

$$A_{\max} = \frac{\omega_0^2 H_1^2 T_2 \chi}{1+T_1 T_2 \omega_1^2}$$

and whose half-width is

$$|\Delta\omega| = \frac{\sqrt{1+\omega_1^2 T_1 T_2}}{T_2}.$$

The relaxation times are determined from the line shape. The values of T_1 and T_2 are strongly different; generally the relaxation times of the nuclei exceed those of the electrons, since electron–electron interactions (owing to the higher moments) and the electron–lattice interactions are stronger than the corresponding interactions with the nuclear moments. The nuclear relaxation times are between several seconds and 10^{-4} sec; the electron relaxation times can be as small as 10^{-9} sec. According to (2.74), the small relaxation times give rise to large line widths in electron resonance. The relaxation times can be raised if the concentration of the paramagnetic ions is reduced (i.e. T_2 becomes longer since, e.g. in dipole–dipole interaction, T_2 will be of the order of $\hbar a^3/\mu^2$ because of the uncertainty relation; a is the mean dipole distance) and if the temperatures are low (larger T_1). Here too the ions in S states are particularly important since their spin–lattice interaction is weak. For a detailed discussion of the various relaxation mechanisms and relaxation times we refer the reader to the literature (Abragam, 1962; Pake, 1962; Slichter, 1963).

Nuclear resonance makes it possible to determine the magnetic moments of nuclei with high accuracy, which is limited only by the deviations of the field from the external field H_0 at the position of the nucleus. Since nuclear-resonance experiments are essentially carried out with diamagnetic atoms, the currents induced in the electron shell by the external field and the corresponding additional field are small. For the latter we obtain about $\delta H = 10^{-4} H_0$ (Abragam, 1962; Pake, 1962).

Resonance experiments can also be carried out in metals if they are available in a colloidal solution, so that the h.f. magnetic field can penetrate into the metal. Relatively large resonance frequency shifts (strong additional fields) are obtained, which are caused by a polarisation of the metal electrons (Knight shift). This effect is important for a comprehension of the electron structure of metals and shall therefore be considered in detail.

A nuclear magnetic moment $\boldsymbol{\mu}$ produces a magnetic field $\boldsymbol{H}$ outside the nucleus, which can be derived from a vector potential $\boldsymbol{A}$. With div $\boldsymbol{A} = 0$ and $\boldsymbol{H} =$ curl $\boldsymbol{A}$ we obtain for $\boldsymbol{A}$

$$\boldsymbol{A} = \frac{[\boldsymbol{\mu} \wedge \boldsymbol{r}]}{r^3} = -\operatorname{curl} \frac{\boldsymbol{\mu}}{r},$$

r being the distance between nucleus and reference point. The interaction between the nucleus and the orbital and spin moments of an electron at point $\boldsymbol{r}$ is described by the operator

$$\mathcal{H}_{\text{int}} = \mathcal{H}_{\text{int}}^{(1)} + \mathcal{H}_{\text{int}}^{(2)} = -\frac{e}{mc}(\boldsymbol{p} \cdot \boldsymbol{A}) + \mu_{\text{B}}(\boldsymbol{\sigma} \cdot \boldsymbol{H}).$$

Introducing the orbital angular momentum operator, $\hbar \boldsymbol{l} = [\boldsymbol{r} \wedge \boldsymbol{p}]$, we can write $\mathscr{H}_{\text{int}}^{(1)}$ in the form of

$$\mathscr{H}_{\text{int}}^{(1)} = -\frac{e}{mcr^3}(\boldsymbol{p}\cdot[\boldsymbol{\mu}\wedge\boldsymbol{r}]) = 2\frac{\mu_B}{r^3}(\boldsymbol{l}\cdot\boldsymbol{\mu}).$$

For $\mathscr{H}_{\text{int}}^{(2)}$ we obtain with $\boldsymbol{p} = (\hbar/i)\nabla$, and curl $= [\nabla\wedge$:

$$\mathscr{H}_{\text{int}}^{(2)} = \mu_B\left(\boldsymbol{\sigma}\cdot\left[\nabla\wedge\left[\nabla\wedge\frac{\boldsymbol{\mu}}{r}\right]\right]\right) = \mu_B\{(\boldsymbol{\sigma}\cdot\nabla)(\boldsymbol{\mu}\cdot\nabla)-(\boldsymbol{\mu}\cdot\boldsymbol{\sigma})\nabla^2\}\frac{1}{r}.$$

(For the vector product of three vectors we have: $[\boldsymbol{a}\wedge[\boldsymbol{b}\wedge\boldsymbol{c}]] = (\boldsymbol{c}\cdot\boldsymbol{a})\boldsymbol{b}-(\boldsymbol{b}\cdot\boldsymbol{a})\boldsymbol{c}$.) Since $1/r$ is a solution to the differential equation

$$\nabla^2\varphi(r) = -4\pi\,\delta(\boldsymbol{r}),$$

$\mathscr{H}_{\text{int}}^{(2)}$ will be singular at $r = 0$. For $r \neq 0$ it is easy to see that

$$\mathscr{H}_{\text{int}}^{(2)} = \mu_B\left(\frac{3(\boldsymbol{\sigma}\cdot\boldsymbol{r})(\boldsymbol{\mu}\cdot\boldsymbol{r})}{r^5} - \frac{(\boldsymbol{\sigma}\cdot\boldsymbol{\mu})}{r^3}\right),$$

i.e. the usual dipole–dipole interaction. In the case of $r \to 0$ the operator must be treated with caution. First $1/r$ is expanded in a Fourier integral

$$\frac{1}{r} = \frac{1}{2\pi^2}\int\frac{e^{i(\boldsymbol{k}\cdot\boldsymbol{r})}}{k^2}\,d^3\boldsymbol{k}.$$

Then we have

$$\begin{aligned}\lim_{r\to 0}\{(\boldsymbol{\sigma}\cdot\nabla)(\boldsymbol{\mu}\cdot\nabla)\}\frac{1}{r} &= -\lim_{r\to 0}\frac{1}{2\pi^2}\int\frac{d^3\boldsymbol{k}}{k^2}\,e^{i(\boldsymbol{k}\cdot\boldsymbol{r})}(\boldsymbol{\sigma}\cdot\boldsymbol{k})(\boldsymbol{\mu}\cdot\boldsymbol{k})\\ &= -\lim_{r\to 0}\frac{1}{2\pi^2}\int\frac{d^3\boldsymbol{k}}{k^2}\,e^{i(\boldsymbol{k}\cdot\boldsymbol{r})}\sum_{\lambda,\nu=1}^{3}\sigma_\lambda k_\lambda\mu_\nu k_\nu\\ &= -\lim_{r\to 0}\frac{1}{2\pi^2}\int\frac{d^3\boldsymbol{k}}{k^2}\,e^{i(\boldsymbol{k}\cdot\boldsymbol{r})}\sum_{\lambda=1}^{3}\sigma_\lambda\mu_\lambda k_\lambda^2,\end{aligned}$$

since, because of symmetry relations, in the $\lim_{r\to 0}$ the integrals over odd powers of k_λ vanish. For the same reasons we have

$$\lim_{r\to 0}\int \frac{k_\lambda^2}{k^2}\,e^{i(k\cdot r)}\,d^3k = \lim_{r\to 0}\int \frac{k_\mu^2}{k^2}\,e^{i(k\cdot r)}\,d^3k$$

$$= \frac{1}{3}\lim_{r\to 0}\int e^{i(k\cdot r)}\,d^3k = \frac{(2\pi)^3}{3}\,\delta(r),$$

so that

$$\lim_{r\to 0}\{(\sigma\cdot\nabla)(\mu\cdot\nabla)\}\frac{1}{r} = -\frac{4\pi}{3}\,\delta(r)\,(\sigma\cdot\mu).$$

We therefore obtain for $\mathcal{H}^{(2)}$

$$\mathcal{H}^{(2)}_{\text{int}} = \mu_B\left(\frac{3(\sigma r\cdot\mu r)}{r^5} - \frac{(\sigma\cdot\mu)}{r^3}\right) + \frac{8\pi}{3}\,\mu_B(\mu\cdot\sigma)\,\delta(r).$$

The last term is the Fermi contact term, cf. (2.33). $\mathcal{H}_{\text{int}} = \mathcal{H}^{(1)}_{\text{int}} + \mathcal{H}^{(2)}_{\text{int}}$ is called the hyperfine interaction; it is responsible for the hyperfine structure of the electron terms (Kopfermann, 1956). It is reasonable to consider $\mathcal{H}_{\text{int}}$ the energy of the nuclear moment in the magnetic field of the orbiting electron, $\mathcal{H}_{\text{int}} = -(\mu\cdot H_{\text{el}})$, with

$$H_{\text{el}} = -\mu_B\left(2\frac{l}{r^3} + 3\frac{(\sigma\cdot r)r}{r^5} - \frac{\sigma}{r^3} + \frac{8\pi}{3}\,\sigma\,\delta(r)\right). \qquad (2.75)$$

Because of the smallness of the nuclear magnetic moments ($\mu \cong 10^{-3}\mu_B$) the hyperfine interaction is insignificant for static magnetism. In the case of magnetic resonance, however, it gives rise to frequency shifts in both nuclear resonance and electron resonance, and describes an important relaxation mechanism (nuclear spin–electron relaxation).

The metal electrons are frequently s electrons, or their wave functions are mixtures of s functions and other functions. The Fermi contact terms in $\mathcal{H}_{\text{int}}$ will then contribute most to the interaction and we can restrict ourselves to this term. The rest of the interaction will generally result in a slight anisotropy of the Knight shift (Abragam, 1962). In a metal the nuclear spin interacts with all conduction electrons. It is therefore necessary to average eqn. (2.75) over all electron states. The electron states are described by a product of Bloch and spin functions. In such a state the expectation value of (2.75) is given by

$$\langle\psi_k|\,H_{el}\,|\psi_k\rangle = -(\pm)\frac{8\pi}{3}\,\mu_B\,|\psi_k(\boldsymbol{r}=0)|^2.$$

The double sign indicates that the electron spin may be parallel or antiparallel to the z-direction. It is convenient to assume the z-axis parallel to the external field. We then obtain the thermal average of H_{el} when we multiply the expectation value by the population probability of the state, i.e. by $f(E_k \pm \mu_B H_0)$ according to the spin orientation, where $f(E)$ is the Fermi distribution and E_k is the energy of the Bloch state, and sum the whole expression over all electron states. We then obtain

$$\begin{aligned}\overline{H}_{el} &= -\frac{8\pi}{3}\,\mu_B\,\frac{V}{\hbar^3}\int d^3\boldsymbol{k}\,|\psi_k(0)|^2\,\{f(E_k+\mu_B H_0)-f(E_k-\mu_B H_0)\}\\ &= -\frac{16\pi}{3}\,\mu_B^2\,\frac{V}{\hbar^3}\int |\psi_k(0)|^2\frac{\partial f}{\partial E_k}\,d^3\boldsymbol{k}\\ &= \frac{16\pi}{3}\,\frac{\mu_B^2 H_0 V}{\hbar^3}\int |\psi_k(0)|^2\delta(E_k-\zeta)\,d^3\boldsymbol{k},\end{aligned}$$

if the Fermi function is again expanded in terms of H_0 and the derivative is replaced by the δ-function. Rearranging the expression for $\overline{H}_{el}$ we obtain

$$\begin{aligned}\overline{H}_{el} &= \frac{16\pi}{3}\,\frac{\mu_B^2 H_0 V}{\hbar^3}\int \delta(E_k-\zeta)\,d^3\boldsymbol{k}\,\frac{\int|\psi(_k0)|^2\,\delta(E_k-\zeta)\,d^3\boldsymbol{k}}{\int\delta(E_k-\zeta)\,d^3\boldsymbol{k}}\\ &= -\frac{16\pi}{3}\,\frac{\mu_B^2 H_0 V}{\hbar^3}\int\frac{\partial f}{\partial E_k}\,d^3\boldsymbol{k}\cdot\langle|\psi_k(0)|^2\rangle_{k=k_0},\end{aligned}$$

where $\langle|\psi_k(0)|^2\rangle_{k=k_0}$ denotes the mean of $|\psi_k(0)|^2$ on the Fermi surface. Since according to (2.2) and (2.3) we also have

$$-\mu_B\frac{V}{\hbar^3}\,2\mu_B H_0\int\frac{\partial f}{\partial E_k} = \mu_B(N_+-N_-) = \chi_p H_0,$$

we finally obtain for the magnetic field of the electrons at the nucleus

$$\overline{H}_{el} = \frac{8\pi}{3} \chi_p \langle |\psi_k(0)|^2 \rangle_{k=k_0} H_0 .$$

The magnetic field is proportional to the spin susceptibility and the averaged probability density of the electrons at the nucleus. The Knight shift is defined as the relative frequency shift with respect to the resonance frequency $\omega_0 = -\gamma H_0$:

$$K = \frac{\Delta\omega}{\omega_0} = \frac{\overline{H}_{el}}{H_0} = \frac{8\pi}{3} \chi_p \langle |\psi_k(0)|^2 \rangle_{k=k_0} . \qquad (2.76)$$

Table 10 shows the K values of a few alkali metals. K grows as the nuclear charge Z is increased, since with increasing Z the electron wave functions are drawn more and more into the atom. Knowing the spin susceptibility χ_p, we are able, from the K values measured, to draw conclusions about the wave functions of the electrons. In a few cases it is possible to determine χ_p from the electron resonance by measuring the surface area below the absorption line; according to (2.74) it is proportional to χ_p. The susceptibilities for Li and Na listed in Table 4 were measured in this way (Schumacher and Vehse, 1960, 1963). It can be seen that the electron density at the nucleus is considerably higher in the free atom than in an atom of the metal (see Table 10).

TABLE 10

	Li^7	Na^{23}	Rb^{87}	Cs^{133}
$K \cdot 10^4$	2·49	11·3	65·3	149
ξ_{exp}	0·44	0·70		
ξ_{theor}	0·49	0·80		

The table shows the Knight shifts of a few isotopes of alkali metals after Knight (1956). The lower lines show the ratio ξ of electron densities at a nucleus in the metal (from (2.76)) and in a free atom (measured spectroscopically). We have $\Omega\langle|\Phi(0)|^2\rangle\xi = V\langle|\psi_k(0)|^2\rangle_{k=k_0}$ since the wave function Φ of the electron in the free atom (or ψ_k) is normalised with respect to the atomic volume Ω (or the volume of the system V). The theoretical values were taken from papers by Kohn (1954) and Kjeldaas and Kohn (1956).

In connection with our discussion of the Knight shift it should be mentioned finally that the hyperfine interaction also results in a frequency shift of the electron resonance. The frequency shift of the electrons is given by (2.76) with the only exception that the electron spin susceptibility χ_p must be replaced by the nuclear spin susceptibility χ_{nucl}. Since the nuclear moments are small and the nucleus has no orbital momentum, eqn. (2.52) will apply with $g_J = g_I$ and $J = I$, I denoting the nuclear spin quantum number:

$$\chi_{nucl} = N \frac{(g_I \mu_{nucl})^2 I(I+1)}{3kT};$$

μ_{nucl} is the nuclear magneton $= 5{\cdot}05 \times 10^{-24}$ erg·gauss^{-1}.

The true paramagnetic or electron resonance differs from nuclear resonance only in quantitative respects. Thus, for example, at usual field strengths of 10^4 oe the resonance frequency is in the cm band instead of the metre band. It is of greater importance that the relaxation times are essentially shorter, on the one hand because of the higher magnetic moments of the electrons, and on the other hand, since the electrons are strongly coupled with the thermal motion of the lattice; this is due to their orbital moment and spin–orbit coupling. The absorption lines are therefore broadened, (2.74). This requires low temperatures, when we disregard the paramagnetic ions in S states (Mn^{2+}, Gd^{3+}) (with which spin resonance can also be observed at room temperature). However, of most importance is the fact that the orbital moment together with spin–orbit coupling yields spectroscopic splitting factors, which deviate considerably from the pure spin value 2. Electron resonance is therefore mainly used to determine the g factors. As already mentioned, conclusions can be drawn as to the lattice field and the field-induced splitting for ions, if the g factors are known as functions of the angle between the magnetic field and the crystal axes. Compared with the susceptibility, which is a mean over all ion levels, electron resonance has the advantage that only one of the many levels is investigated, and this with a higher accuracy (cf. Table 6a).

This is based on the supposition that this level splits in a magnetic field in such a way that magnetic dipole transitions ($\Delta M = \pm 1$) are possible. This restriction limits the applicability of electron resonance in solids. Another restriction is due to the fact that, as it is necessary to work at low temperatures, only the lowest states of an ion are occupied, so that there are only a few levels which can be in resonance. Thus usually atoms or ions with an even number of electrons are uninteresting as they, according to a theorem by Jahn and Teller (1937), always have a single state as the lowest state. This theorem says that a non-linear molecular complex (such as the paramagnetic ion with its neighbourhood), which because of the arrangement of atoms has a degenerate ground state, deforms itself in such a way that the ground state is not degenerate. As regards the magnitude of the Jahn–Teller deformation we have to have recourse to hypotheses. If it is small, resonance may arise none the less, when other terms lie a little above the ground state (a few cm^{-1}). In the case of ions with an odd number of electrons, Kramers degeneracy will always appear so that the ground state (without magnetic field) is at least a Kramers doublet.

A theoretical calculation of the g factors requires a knowledge of the lattice field. It has been dealt with for cerium ethyl sulphate so that it need not be repeated. Generally we do not proceed in this way because the lattice field is known only insufficiently. One tries instead to derive a so-called effective spin Hamiltonian (Pryce 1950; Abragam and Pryce 1951) for the lower terms, the only ones that are important for resonance, which can describe the term splitting in the magnetic field in a correct way. The factors appearing in it (the g factors, . . .) must be determined experimentally. Let the Hamiltonian of the electrons of a paramagnetic ion be given by

$$\mathcal{H} = \mathcal{H}_0 + \mathcal{H}_1,$$

$\mathcal{H}_1$ being an arbitrary perturbation operator describing, for example, the interaction of the ion with the magnetic field. The effective Hamiltonian, which represents correctly the eigenvalues of the term group denoted by 0, up to the second power of the perturbation operator, will then read

$$\mathcal{H}_{\mathrm{eff}} = E_0 P_0 + P_0 \mathcal{H}_1 P_0 + P_0 \mathcal{H}_1 Q_0 \mathcal{H}_1 P_0, \tag{2.77}$$

where

$$Q_0 = (1-P_0)\frac{1}{E_0-\mathcal{H}_0}(1-P_0).$$

E_0 is the energy of the non-perturbed terms, which are assumed to be either degenerate or quasi-degenerate so that all of them have the same energy. P_0 is the projection operator to the space R_0 formed by the non-perturbed eigenvectors $|E_0\alpha_0\rangle$ of the lower term group; α_0 denotes the various degenerate states pertaining to E_0.

To obtain a proof for (2.77), consider the subspace R formed by the exact eigenvalues of $\mathcal{H}$, which, if $\mathcal{H}_1 = 0$, is identical with R_0. The dimension of R, i.e. the number of the linearly independent vectors, will be equal to that of R_0, if the perturbation $\mathcal{H}_1$ is sufficiently small, since the dimension can be changed only discontinuously. It is obvious that we obtain a complete basis in R and thus all eigenvectors of $\mathcal{H}$ in R, if the projection operator P pertaining to R is used to map the unperturbed vectors from R_0 on R. If this were not the case, i.e. if the basis thus obtained in R were incomplete, there must exist one (or several) vectors outside R_0 whose projections on R would complete the basis. Then, however, the dimension of R would be different from that of R_0, which contradicts our assumption. It is thus possible to represent each eigenvector of $\mathcal{H}$ in R by $P\,|E_0\alpha_0\rangle$ if a suitable basis is chosen. Thus we have

$$HP\,|E_0\alpha_0\rangle = E_{\alpha_0}P\,|E_0\alpha_0\rangle.$$

Multiplication by P_0 yields

$$\tilde{\mathcal{H}}\,|E_0\alpha_0\rangle \equiv P_0\mathcal{H}PP_0\,|E_0\alpha_0\rangle = E_{\alpha_0}P_0PP_0\,|E_0\alpha_0\rangle. \tag{2.78}$$

Here we made use of the definition of the projection operator P_0:

$$P_0 = \sum_{\alpha_0}|E_0\alpha_0\rangle\langle E_0\alpha_0|$$

and the resulting relation $P_0\,|E_0\alpha_0\rangle = |E_0\alpha_0\rangle$. With $\mathcal{H} = \mathcal{H}_0+\mathcal{H}_1$ eqn. (2.78) can be written in the form

$$(E_0-E_{\alpha_0})P_0PP_0\,|E_0\alpha_0\rangle + P_0\mathcal{H}_1PP_0\,|E_0\alpha_0\rangle = 0. \tag{2.79}$$

With (2.78) and (2.79) the problem has been reduced to an eigenvalue problem in R_0. We now have to expand P in a power series in $\mathcal{H}_1$. According to (2.79), it is sufficient if the series is expanded up to terms linear in $\mathcal{H}_1$ to obtain the perturbed eigenvalues E_{α_0} up to quadratic terms in $\mathcal{H}_1$ For this purpose the Hamiltonian operator is represented by the projection operators pertaining to the eigenvalues. Obviously

$$\mathcal{H}_0 = \sum_{i=0}^{\cdots} E_iP_i \quad \text{where} \quad P_i = \sum_{\alpha_i}|E_i\alpha_i\rangle\langle E_i\alpha_i|$$

and we have the completeness relation $\sum_i P_i = 1$. The sum over i is to be extended over all eigenvalues of $\mathcal{H}_0$. We now define the Green function of $\mathcal{H}_0$, $G_0(z)$:

$$G_0(z) = \frac{1}{z-\mathcal{H}_0}.$$

z is an arbitrary complex variable. As $P_iP_j = P_i\delta_{ij}$ and $\mathcal{H}_0P_i = E_iP_i$ we have

$$G_0(z) = \sum_i \frac{P_i}{z-E_i}, \tag{2.80}$$

so that the projection operator P_0 can be represented by means of an integral over $G_0(z)$ in the complex z-plane:

$$P_0 = \frac{1}{2\pi i}\int_{C_0} G_0(z)\,dz.$$

C_0 is a closed contour around $z = E_0$; C_0 must be chosen in such a way that no other eigenvalue of $\mathcal{H}_0$ lies inside or on C_0. In the same way we define the Green function of $\mathcal{H}$:

$$G(z) = \frac{1}{z-\mathcal{H}}.$$

For a sufficiently small perturbation the projection operator can thus be represented in the form

$$P = \frac{1}{2\pi i}\int_{C_0} G(z)\,dz$$

where C_0 denotes the same contour as for P_0. With $\mathcal{H} = \mathcal{H}_0+\mathcal{H}_1$ we obtain the following equation for $G(z)$:

$$G(z) = G_0(z)\,(1+\mathcal{H}_1G(z));$$

this is easy to verify by a simple substitution. The equation can be solved by iteration. We have

$$G(z) = G_0(z)+\sum_{\nu=1}^{\infty} G_0(z)\,(H_1G_0(z))^{\nu}.$$

Substituting this in the equation for P we obtain (breaking off after terms linear in $\mathcal{H}_1$):

$$P = P_0+\frac{1}{2\pi i}\int_{C_0} G_0(z)\mathcal{H}_1G_0(z)\,dz = P_0+P_0\mathcal{H}_1Q_0+Q_0\mathcal{H}_1P_0$$

where Q_0 is obtained from (2.77), (2.80); here Cauchy's integral theorem was used. Substituting this in (2.78) or (2.79) we obtain the following eigenvalue equation ($P_0Q_0 = 0$):

$$(E_0P_0+P_0\mathcal{H}_1P_0+P_0\mathcal{H}_1Q_0\mathcal{H}_1P_0)\,|E_0\alpha_0\rangle = E_{\alpha_0}\,|E_0\alpha_0\rangle$$

and thus the equation (2.77) given above for the effective Hamiltonian.

In order to proceed further, we have to make assumptions about the perturbation operator $\mathscr{H}_1$, which are based on the nature of the paramagnetic ion. In the case of rare-earth elements, for example, we would introduce a spin–orbit interaction in $\mathscr{H}_0$ and consider both the lattice field and the magnetic field to be perturbations. On the other hand, in the case of the iron group, the cubic part of the lattice field is taken into $\mathscr{H}_0$ and all other influences must be comprised by $\mathscr{H}_1$. We shall consider such a case in the following, namely the case of an ion possessing an orbital singlet state as the lowest state (Cr^{3+}, Ni^{2+}, Fe^{3+}, Gd^{3+}). The degeneracy of the ground term is then given by the multiplicity S. The perturbation operator is equal to

$$\mathscr{H}_1 = \mu_B(\{\boldsymbol{L}+2\boldsymbol{S}\}\cdot\boldsymbol{H})+\zeta(\boldsymbol{L}\cdot\boldsymbol{S}).$$

Deviations from cubic symmetry are assumed not to exist. The expectation value of $\boldsymbol{L}$ vanishes in $\boldsymbol{R}_0$; moreover, $\boldsymbol{S}$ commutes with $\mathscr{H}_0$ and thus with $\boldsymbol{P}_0$; therefore

$$P_0\mathscr{H}_1P_0 = 2\mu_B(\boldsymbol{S}\cdot\boldsymbol{H})P_0\,.$$

Accordingly we obtain the second-order term of $\mathscr{H}_1$ by means of a simple calculation:

$$P_0\mathscr{H}_1Q_0\mathscr{H}_1P_0 = -\sum_{\mu,\,\nu=1}^{3} \Lambda_{\mu\nu}(\mu_B H_\mu+\zeta S_\mu)\,(\mu_B H_\nu+\zeta S_\nu)P_0$$

where

$$\Lambda_{\mu\nu} = \sum_{i\neq 0}\frac{\langle E_0|\,L_\mu\,|E_i\rangle\langle E_i|\,L_\nu\,|E_0\rangle}{E_i-E_0}\,.$$

Here we made use of the fact that the quantum numbers α_i in the vector $|\alpha_i, E_i\rangle = |\alpha_i\rangle|E_i\rangle$ refer to the spin. The vectors $|E_i\rangle$ are real† so that $\Gamma_{\mu\nu}$ is a positive, real, and symmetric tensor since $E_i > E_0$. The letters μ and ν denote the components of the vectors $\boldsymbol{L}$, $\boldsymbol{S}$, and $\boldsymbol{H}$, and the sum over i is taken over all higher terms of the multiplet. The projection operator P_0 can be omitted if we taken into account that the

† $\mathscr{H}_0$ is real so that the $|E_i\rangle$ can be chosen real; cf. the Kramers theorem, pp. 117 and 122.

effective Hamiltonian acts only in the space R_0. A rearrangement yields

$$\mathcal{H}_{\text{eff}} = E_0 + \mu_B \boldsymbol{H}\cdot g\cdot \boldsymbol{S} - \mu_B^2 \boldsymbol{H}\cdot \Lambda\cdot \boldsymbol{H} + \zeta^2 \boldsymbol{S}\cdot \Lambda\cdot \boldsymbol{S} \tag{2.81}$$

where

$$g = 2(1-\zeta\Lambda). \tag{2.82}$$

The products in (2.81) must be considered to be tensor products; g is the splitting factor. It deviates from the isotropic g factor 2 (characteristic

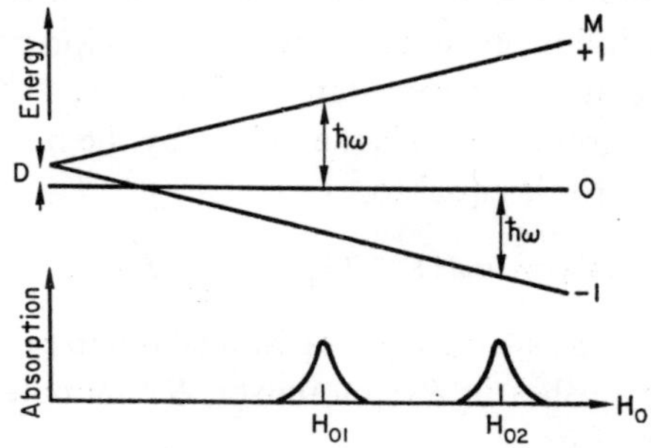

FIG. 15.

Term splitting of Ni^{2+} in a magnetic field, after Bowers and Owen (1955). Ni^{2+} has an effective spin $S' = 1$ (Table 11). In a lattice field with predominating cubic symmetry and a small axisymmetric component, the triple degeneracy is partly removed; fine structure, $D = 1\ \text{cm}^{-1}$. In a magnetic field H_0, which is parallel to the symmetry axis, the terms are split according to the projection of the effective spin. In the case of constant high frequency and variable magnetic field H_0 we obtain two allowed resonance transitions with $\Delta M_{S'} = \pm 1$.

of the spin) by about ζ/Δ, where Δ is the energy difference to the next higher level above the orbital singlet state. In the case of ions with a singlet state as the ground state the anisotropies of the g factor should not amount to much more than 10%, and should be strongest for Ni^{2+} because of the relatively strong spin–orbit coupling (Table 11). The term $\sim H^2$ in (2.81) is insignificant for paramagnetic resonance as it shifts all terms by the same amount. The last term in (2.81) shows that owing to spin–orbit coupling, the degenerate terms are also split for $H = 0$. As in the case of the free atom this splitting is called fine structure. It is determined experimentally by studying the resonance at various magnetic fields and extrapolating to $H = 0$. Figure 15 shows the conditions for a simple example. The nomenclature "effective spin Hamiltonian" for (2.81) is reasonable, since (2.81) contains only spin operators, and the form of the effective Hamiltonian (2.81) is not restricted to cases where the lowest state is an orbital singlet state. This can be seen at once when we consider the simple but practically important case where the ground state of the paramagnetic ion is a Kramers doublet. The effective Hamiltonian describing this Kramers doublet must be a matrix of rank 2 which, as the magnetic field can always be considered to be a small perturbation, will contain the field only in a linear form (quadratic terms in H would only shift the doublet as a whole). Such a matrix can be represented as a linear combination of the unit matrix (which, as it yields only a shift of the doublet, is again omitted) and the three Pauli matrices or the three angular momentum matrices for spin $S' = \frac{1}{2}$; the effective Hamiltonian of the doublet will therefore read in its most general form

$$\mathcal{H}_{\text{eff}} = \mu_{\text{B}} \boldsymbol{S}' \cdot g \cdot \boldsymbol{H}. \tag{2.83}$$

The product is again a tensor product. The effective spin has nothing to do with the actual spin; it may be considered a combination of spin and orbital angular momentum (cf. cerium ethyl suldhate), although in special cases, such as S states, or orbital singlet states, S' can coincide with the spin quantum number. If the degeneracy of the lowest state is higher than 2, the effective spin will be correspondingly higher. We then have in the effective Hamiltonian, in addition to the Zeeman term, also terms of second and higher order of S', which yield the fine structure. (In the case of $S' = \frac{1}{2}$ these terms only shift the doublet.) Neglecting terms higher than quadratic in S', we obtain

$$\mathcal{H}_{\text{eff}} = \mu_{\text{B}} \boldsymbol{S}' \cdot g \cdot \boldsymbol{H} + \boldsymbol{S}' \cdot D \cdot \boldsymbol{S}', \tag{2.84}$$

where g and D are again symmetric tensors. D is not identical with Λ, since in the general case considered here, we have to take into account both spin–orbit coupling and spin–spin interactions, which gives rise to an expression quadratic in S'. It is in general sufficient to take these quadratic terms into account when resonance experiments are described: however, for ions in S states, ($Gd^{3+}: S = \frac{7}{2}$), higher powers are necessary (Bleaney and Stevens, 1953; Bowers and Owen, 1955; Low, 1960).

The tensor formalism in (2.81), (2.83), and (2.84) is unnecessarily complicated. It is more expedient to use a coordinate system in which both tensors g and D are simultaneously on principal axes; this must be possible since g and D contain the lattice field symmetry in the same way, cf. (2.81). We then can rewrite (2.84) as follows:

$$\mathcal{H}_{\text{eff}} = \mu_B(g_x H_x S'_x + g_y H_y S'_y + g_z H_z S'_z) + DS'^2_z + E(S'^2_x - S'^2_y). \quad (2.85)$$

A term $S'^2_x + S'^2_y$ will not appear since it could be replaced by S'^2 (=unit matrix) and S'^2_z. The parameters g_x, g_y, g_z, S', D, and E are chosen in agreement with experiment. In the case of axial symmetry of the lattice field, (2.85) becomes still simpler since, if the z-axis coincides with the symmetry axis, $E = 0$ and $g_x = g_y$. Table 11 shows the g factors and the effective spin quantum numbers of several ions of the iron group whose lattice field has a strong cubic component (therefore $D \cong E \cong 0$).

The effective spin Hamiltonian is a very suitable description of paramagnetic resonance and of the ionic term structure in crystals. However, it must not be overlooked that this is a phenomenological description as, for example, it does not require detailed data on the lattice field.

TABLE 11

Ion	Configuration	Multiplicity of the free ion S	S'	g
Ti^{3+}	d^1	1/2	$\frac{1}{2}$	1·1–1·3
$V^{2+}Cr^{3+}$	d^3	3/2	$\frac{3}{2}$	2·0
Cr^{2+}	d^4	2	2	1·9–2·0
$Mn^{2+}Fe^{3+}$	d^5	5/2	$\frac{5}{2}$	2·0
Co^{2+}	d^7	3/2	$\frac{1}{2}$	1·4–7
Ni^{2+}	d^8	1	1	2·2–2·3
Cu^{2+}	d^9	1/2	$\frac{1}{2}$	2·0–2·5

Table 11 contains the effective spin quantum numbers and the g factors of several iron-group ions which have an essentially cubic lattice field (Bowers and Owen, 1955). The last column gives the variation of the g values in the various salts.

Considered from this point of view, the formalism is unsatisfactory as long as we cannot deduce the lattice field or other features from the empirically chosen parameters.

In particularly favourable cases spin–lattice relaxation can be made so small that the resonance lines obtained are very sharp. Of course, ions in S states are best suited for this purpose. If the nucleus of the ion possesses a magnetic moment, it is then possible to observe the hyperfine splitting of the electron terms. Besides the terms of (2.84) or (2.85), the effective Hamiltonian will then contain the hyperfine interaction and the Zeeman term of the nuclear spin:

$$\boldsymbol{I}\cdot\boldsymbol{A}\cdot\boldsymbol{S}' - g_I\mu_{\text{nucl}}(\boldsymbol{I}\cdot\boldsymbol{H}). \tag{2.86}$$

$\boldsymbol{I}$ is the angular momentum and $g_I\mu_{\text{nucl}}$ is the magnetic moment of the

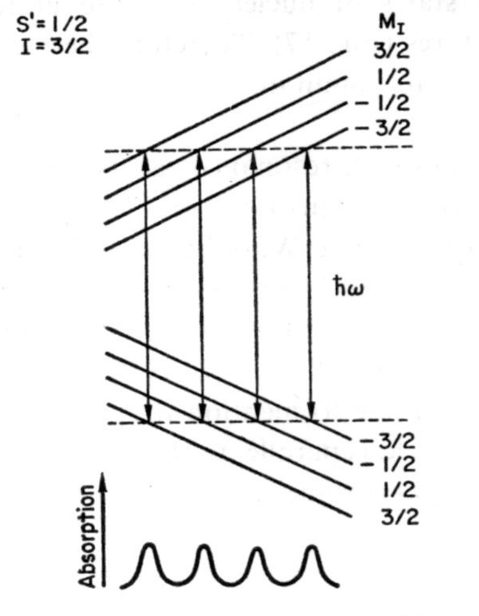

FIG. 16.

Hyperfine splitting of an electron term with an effective spin $S' = \frac{1}{2}$ and a nuclear spin $I = \frac{3}{2}$ in medium fields, after Bleaney and Stevens (1953), $\mu_B H_0 \gg A \gg \mu_{\text{nucl}} H_0$; According to (2.83) and (2.86), the splitting is given by $E(M_{S'}, M_I) = g\mu_B M_S H_0 + A M_{S'} M_I$. With constant high frequency and variable field H_0 four resonance lines are obtained which correspond to dipole transitions with $\Delta M_{S'} = \pm 1$.

nucleus. The tensor A can usually be diagonalised simultaneously with D and g (Abragam, 1962). The Fermi contact term in the hyperfine interaction would, of course, yield an isotropic coupling $A(\boldsymbol{I}\cdot\boldsymbol{S}')$ in the effective Hamiltonian; here, however, its influence must be very weak, since the paramagnetic ions have d^N or f^N configurations with, at most, small admixtures of other configurations with possible s character. Typical values of A are between 10 cm^{-1} (rare-earth elements) and 10^{-3} cm^{-1} (iron group). In stronger magnetic fields, $g_I \mu_{nucl} H_0 \gg A$, the Zeeman term exceeds the hyperfine interaction, and the electron terms are split in the magnetic field in the same way as in the normal Zeeman effect for electrons, according to the $2I+1$ projections of nuclear spin on the magnetic field direction (Fig. 16). The $2I+1$ resonance lines are at a constant distance from one another and have the same intensity, since the $2I+1$ states of nuclear spin are uniformly populated at normal temperatures (Fig. 17). Therefore there exist only a few cases where it is possible to obtain information on the nuclear spin of paramagnetic ions.

As already mentioned, resonance can be induced also for the free electrons of metals and semiconductors. This yields information on the band structure (Ludwig and Woodbury, 1963; Yafet, 1963).

For more detailed considerations of paramagnetic resonance, we refer the reader to the literature (Pake, 1962; Bleaney and Stevens, 1953; Bowers and Owen, 1955; Low, 1960).

Finally we should like to mention that no resonance absorption is obtained if the h.f. field is parallel to the constant field. The absorption observed is called paramagnetic relaxation (Gorter, 1947, 1957).

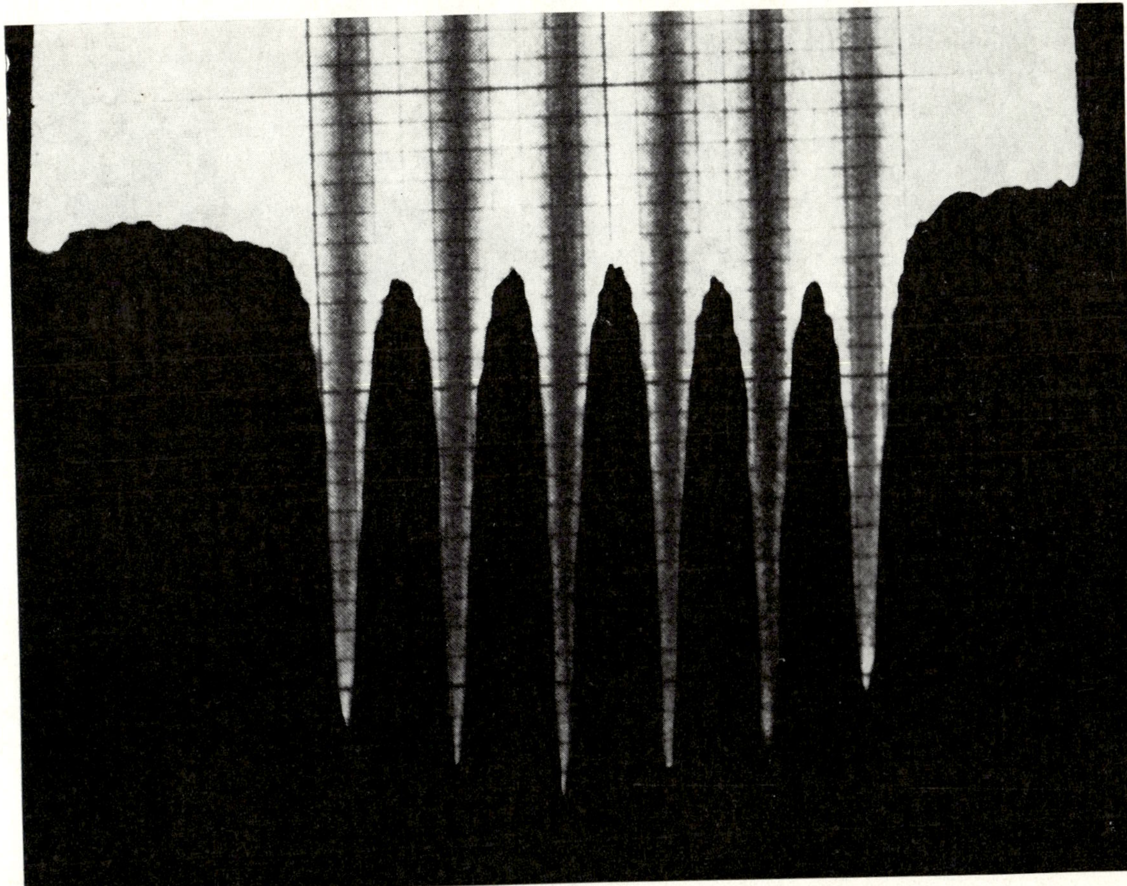

FIG. 17.

Hyperfine structure of the Mn^{2+} resonance line after Tinkham *et al.* (1951). The Mn isotope has a nuclear spin $I = \frac{5}{2}$.

III. Ferromagnetism

1. Interactions

Ferromagnetism, antiferromagnetism, and ferrimagnetism are magnetic phenomena, which differ essentially from the diamagnetism and paragnetism considered so far, since they are based on the interactions between the magnetic moments. These effects belong therefore to the class of co-operative phenomena. The theory of ferromagnetism (antiferromagnetism, ferrimagnetism) can be subdivided into two parts. On the one hand, we have the subject of spontaneous magnetisation (saturation magnetisation), where the very nature of ferromagnetism is the object of consideration. The second part comprises all effects caused by the magnetic anisotropy of the crystals: for example, the theory of the technical magnetisation curve, which deals with the adjustment of the individual uniformly magnetised domains in an external field and thus the problems of hysteresis, the theory of domain structure, the Bloch walls representing the transition region between two domains, and magnetostriction. This part, which is important for applications, will not be treated in this book; we refer the reader to standard books (Bates, 1961; Becker and Döring, 1939; Kneller, 1962; Bozorth, 1951; Chikazumi, 1964) and review articles (Kittel and Galt, 1957; Kanamori, 1963).

First we are confronted with the problem of interactions between the atomic magnetic moments. We see at once that it cannot be described by an ordinary dipole interaction, which is of the order of $\mu^2/a^3 \cong 10^{-16}$ erg, if we assume a mean distance a of about 1 Å between the atoms. The Curie temperatures T_c, however, can amount to several hundred degrees Kelvin so that a thermal energy of $kT_c \cong 10^{-14}$ erg is necessary to destroy the magnetically ordered state. Dipole interactions can therefore be at most a correction (anisotropy effect) to the real cause. Indeed, the interaction is an electrostatic one, as was first shown by Heisenberg (1928). To illustrate this, let us consider two atoms, A and B, at $\boldsymbol{r}_A$ and $\boldsymbol{r}_B$, each with one electron; to start with we assume them to be far from one another. The electron of atom A is assumed to be in the orbital state $\varphi_A(\boldsymbol{r}_1)$, that of atom B in the state $\varphi_B(\boldsymbol{r}_2)$. Both φ_A and φ_B are normalised eigenfunctions of the Hamiltonians of the atoms A and B, respectively:

$$\mathcal{H}_A\varphi_A(\boldsymbol{r}_1) = \varepsilon_A\varphi_A(\boldsymbol{r}_1), \qquad \mathcal{H}_B\varphi_B(\boldsymbol{r}_2) = \varepsilon_B\varphi_B(\boldsymbol{r}_2).$$

The total system of the two atoms has the energy $E_0 = \varepsilon_A+\varepsilon_B$. This energy state is degenerate, as both

$$\Phi(1,2) = \varphi_A(\boldsymbol{r}_1)\varphi_B(\boldsymbol{r}_2) \qquad \text{and} \qquad P\Phi(1,2) = \varphi_A(\boldsymbol{r}_2)\varphi_B(\boldsymbol{r}_1)$$

are eigenfunctions of $\mathcal{H}_A+\mathcal{H}_B$. P is the permutation operator of the two electron positions; $P^2 = 1$. The functions $\Phi(1,2)$ and $P\Phi(1,2)$ are orthogonal to one another, since the wave functions of the two atoms are assumed not to be overlapping: $(\varphi_A, \varphi_B) = 0$. It is reasonable to consider, instead of the functions Φ and $P\Phi$, the functions

$$\psi_+^0 = (1/\sqrt{2})(1+P)\Phi, \qquad \psi_-^0 = (1/\sqrt{2})(1-P)\Phi$$

to be the basis functions, as they are eigenfunctions of the permutation operator with eigenvalues ± 1. The function ψ_+^0 is symmetrical with respect to the electron coordinates, ψ_-^0 is antisymmetrical. They are normalised and orthogonal:

$$(\psi_\pm^0, \psi_\pm^0) = \tfrac{1}{2}(\Phi, (1\pm P)^2\Phi) = (\Phi, \Phi) = 1,$$
$$(\psi_+^0, \psi_-^0) = \tfrac{1}{2}(\Phi, (1-P^2)\Phi) = 0.$$

If the two atoms are now assumed to approach one another, an interaction $W(\boldsymbol{r}_1, \boldsymbol{r}_2, \boldsymbol{r}_A, \boldsymbol{r}_B)$ will arise, which results from the Coulomb interaction between the electrons, between the ionic cores, and between each electron with respect to the other and the ionic core of the other atom:

$$W = \frac{e_0^2}{|\boldsymbol{r}_1-\boldsymbol{r}_2|}+\frac{e_0^2}{|\boldsymbol{r}_A-\boldsymbol{r}_B|}-\frac{e_0^2}{|\boldsymbol{r}_A-\boldsymbol{r}_2|}-\frac{e_0^2}{|\boldsymbol{r}_B-\boldsymbol{r}_1|}.$$

The ion cores are assumed to carry one positive charge. The total Hamiltonian $\mathcal{H}_A+\mathcal{H}_B+W$ is symmetrical in the electron coordinates, i.e. it commutes with P. The new eigenfunctions can therefore again be classified according to the eigenfunctions of P. Denoting them by $\psi_{\pm}$ we can represent them in a first approximation (if the correlation owing to the Coulomb interaction between the electrons can be neglected as a small perturbation) as the product of the eigenfunctions u and v:

$$\psi_+ = (1/\sqrt{N_+})\,(1+P)\,u(\boldsymbol{r}_1)\,v(\boldsymbol{r}_2), \quad \psi_- = (1/\sqrt{N_-})\,(1-P)\,u(\boldsymbol{r}_1)\,v(\boldsymbol{r}_2),$$

where $N_{\pm}$ are normalisation factors. If, in addition to this, W is considered a perturbation, the functions u and v can be approximated by the old eigenfunctions φ_A and φ_B, so that

$$\psi_{\pm} = (1/\sqrt{N_{\pm}})\,(1\pm P)\Phi(1, 2) \tag{3.1}$$

to a first approximation will be eigenfunctions of the whole problem. In principle there exist another two functions of state besides the functions (3.1), namely

$$\varphi_A(\boldsymbol{r}_1)\,\varphi_A(\boldsymbol{r}_2)\pm\varphi_B(\boldsymbol{r}_1)\,\varphi_B(\boldsymbol{r}_2);$$

they are also eigenfunctions of P. However, since they describe states where two electrons are bound to the same atom, they are neglected as energetically unfavourable states; this is due to the Coulomb interaction between the electrons. As far as the ground state is concerned they play only a minor role; they are called polar states. This neglect of the polar states is the Heitler–London approximation in the problem of chemical binding.

The normalisation factors $N_{\pm}$ deviate from the value 2 since the wave functions now overlap with one another. We have

$$N_{\pm} = (\Phi, (1\pm P)^2\Phi) = 2(1\pm\tilde{S})$$

where

$$\tilde{S} = (\Phi, P\Phi) = (\varphi_A, \varphi_B)\,(\varphi_B, \varphi_A) = |(\varphi_A, \varphi_B)|^2. \tag{3.2}$$

The expectation value of the energy in both states ($\psi_{\pm}$) is obtained by means of (3.1) and (3.2):

$$\begin{aligned} E_{\pm} &= (\psi_{\pm}, (\mathcal{H}_A+\mathcal{H}_B+W)\psi_{\pm}) = E_0+(\psi_{\pm}, W\psi_{\pm}) \\ &= E_0+[1/(1\pm\tilde{S})]\,(\Phi, W(1\pm P)\Phi). \end{aligned} \tag{3.3}$$

The energy difference between the symmetrical and the antisymmetrical states is

$$E_{+}-E_{-} = 2J_{AB}$$

where

$$J_{AB} = [1/(1-\tilde{S}^2)]\{(\Phi, WP\Phi)-\tilde{S}(\Phi, W\Phi)\}. \tag{3.4}$$

J_{AB} is called the exchange integral. It can be both positive or negative.

As we ignore the spin–orbit interaction, the total wave function of the system is the product of a position coordinate component and a spin component. According to the Pauli principle it must be antisymmetric, i.e. it must change its sign if the coordinates and the spins of the electrons are exchanged. It is thus necessary to multiply the symmetrical function ψ_+ by an antisymmetrical spin function, and the function ψ_- by a symmetrical spin function. If α and β are the orthonormalised eigenfunctions of the z-component of the electron spin, which were introduced in (2.71), so that

$$s_z\alpha = \tfrac{1}{2}\alpha, \quad s_y\beta = -\tfrac{1}{2}\beta, \quad s^+\beta \equiv (s_x+is_y)\beta = \alpha,$$
$$s^-\alpha = (s_x-is_y)\alpha = \beta, \quad s^+\alpha = s^-\beta = 0,$$

then the only possible antisymmetric spin function will be

$$\chi_0 = (1/\sqrt{2})(\alpha(1)\,\beta(2)-\alpha(2)\,\beta(1)). \tag{3.5a}$$

The arguments of the spin function refer to the electrons. χ_0 is the eigenfunction with total spin $S = 0$, since

$$\boldsymbol{S}^2\chi_0 = (\boldsymbol{s}_1+\boldsymbol{s}_2)^2\chi_0 = 0,$$

as is easily verified by means of the definitions of α and β. Three functions, χ_1, χ_2, χ_3, pertain to a total spin $S = 1$; they are eigenfunctions of $S_z = s_{1z}+s_{2z}$ with eigenvalues $+1$, 0, -1. We have

$$\chi_1 = \alpha(1)\,\alpha(2), \quad \chi_3 = \beta(1)\,\beta(2),$$

$$\chi_2 = (1/\sqrt{2})\,(\alpha(1)\,\beta(2)+\alpha(2)\,\beta(1)). \tag{3.5b}$$

These four functions are mutually orthogonal and normalised to 1.

The energy splitting of the originally degenerate levels calculated in (3.3) can now be interpreted in such a way that the difference between the singlet state ($S = 0$) and the triplet amounts to $2J_{AB}$. As in (2.83), this can be expressed by means of an effective spin Hamiltonian:

$$\mathscr{H}_{\text{eff}} = E_0 + K - 2J_{AB}(\boldsymbol{s}_A \cdot \boldsymbol{s}_B), \tag{3.6}$$

since $2(\boldsymbol{s}_A \cdot \boldsymbol{s}_B) = \boldsymbol{S}^2 - \boldsymbol{s}_A^2 - \boldsymbol{s}_B^2$ has the eigenvalues $-\frac{3}{2}$ for $S = 0$ and $\frac{1}{2}$ for $S = 1$. Here $\boldsymbol{s}_A$ and $\boldsymbol{s}_B$ are the spins of the electrons of the atoms A and B, respectively. From a comparison with (3.3) we obtain for K

$$K = [1/(1-\mathcal{S}^2)]\{(\Phi, W(1-\mathcal{S}P)\Phi) - \tfrac{1}{2}(\Phi, W(P-\mathcal{S})\Phi)\}.$$

The spin-independent expressions E_0 and K can now be omitted in (3.6), as they shift all terms by the same amount.

Extending this concept to atoms with several electrons and thus arbitrary spin (Van Vleck, 1945), we can write for the Hamiltonian of a system of N atoms

$$\mathscr{H} = -\sum_{i=1}^{N} \sum_{j=1}^{N} J_{ij}(\boldsymbol{s}_i \cdot \boldsymbol{s}_j). \tag{3.7}$$

J_{ij} is a function of the atomic distance $|\boldsymbol{r}_i - \boldsymbol{r}_j|$; it will vary exponentially with the distance since the wave functions drop exponentially at large distances from the nucleus. J_{ii} is put equal to zero; $\boldsymbol{s}$ is the total spin of the ith atom.

In this way we have shown that the Pauli principle together with the Coulomb interaction makes the levels of an atomic spin system dependent on the relative orientations of the spins; they will be parallel in the ground state if $J_{ij} > 0$, and antiparallel if $J_{ij} < 0$.

The derivation of (3.6) and (3.7), however, is not unobjectionable. On the one hand, the initial function (3.1) was chosen arbitrarily, with the only restriction that the single-electron functions u and v must change into the atomic eigenfunctions φ_A and φ_B, if the distance between the atoms is large enough. It is thus conceivable, instead of putting $u = \varphi_A$ and $v = \varphi_B$ as in (3.1), to start from two **orthogonal** functions:

$$u = c_+\varphi_A + c_-\varphi_B, \quad v = c_+\varphi_B + c_-\varphi_A$$

where

$$c_{\pm} = \frac{1}{\sqrt{1+(\varphi_A, \varphi_B)}} \pm \frac{1}{\sqrt{1-(\varphi_A, \varphi_B)}}.$$

For the sake of simplicity φ_A and φ_B are assumed to be real. Although the functions u and v obviously have the correct asymptotic behaviour, the exchange integral (3.4) is now positive. It is easy to show that it is unimportant that u and v are not eigenfunctions of $\mathscr{H}_A$ and $\mathscr{H}_B$, respectively. Because of the orthogonality of u and v the overlap integral $\tilde{S}$ is equal to zero. At sufficiently large atomic distances, the Coulomb repulsion of the electrons will always dominate in the perturbation energy W; according to (3.4) we therefore obtain for the exchange integral J_{AB}

$$J_{AB} = e_0^2 \iint u^*(\boldsymbol{r}_1)\, v^*(\boldsymbol{r}_2) \frac{1}{|\boldsymbol{r}_1 - \boldsymbol{r}_2|} u(\boldsymbol{r}_2)\, v(\boldsymbol{r}_1)\, d^3\boldsymbol{r}_1\, d^3\boldsymbol{r}_2 . \tag{3.8}$$

This integral is positive. On the other hand, it can be proven that in a diatomic molecule and also in a linear arrangement of atoms the non-magnetic state, i.e. the state with the lowest total spin, will always be the lowest state (Lieb and Mattis, 1962). This requires $J_{AB} < 0$. The arbitrariness in the choice of the initial functions can be removed, however, and a Hamiltonian in the form of (3.7) can be justified in a modified Heitler–London approximation according to Herring (1962).

As already mentioned, the Heitler–London approximation is mainly based upon a neglect of the polar states. This means that the electrons are always assumed to be localised at the ions. Such an electron distribution will certainly hold true for insulators but not for metals. Thus eqn. (3.7) should only yield a correct description of the conditions in the case of non-conductors. In insulators the magnetic order is chiefly antiferromagnetic ($J_{ij} < 0$), however, in such a way that a diamagnetic ion (oxygen, fluorine) lies between two antiferromagnetically coupled spins. According to the above model there can be no exchange interaction over such large distances, since the exchange integral decreases rapidly as the distance between the atoms increases. Kramers called the presence of an exchange interaction over larger distances "superexchange". As an example we may mention MnO as a substance with superexchange; in it the manganese ions are antiferromagnetically coupled across the oxygen ion. If the spins of the paramagnetic ions are antiparallel, the electrons of the one ion can move over to those of the other ion via the oxygen bridge, while in the case of parallel spins the Pauli principle would partly forbid such a change-over. According to Anderson (1963), this exchange of electrons reduces the system's total energy so that the antiferromagnetic state is most favourable. As a result we obtain a Hamiltonian of the form (3.7) with $J_{ij} < 0$. Thus the mechanism of superexchange differs essentially from that of direct exchange.

Ferromagnetism is mostly encountered in metals, which are only rarely ferrimagnetic or antiferromagnetic. In the case of metals it is certainly not justified to neglect the the polar states, as the magnetic electrons are not localised. This can be seen immediately from the temperature-independent paramagnetism of the transition metals (cf. Table 9). Thus the correctness of the derivation of the Hamiltonian (3.7) is questionable. In addition to this, accurate calculations of the exchange integral (3.4) yield too small or negative values, so that the ferromagnetism of the metals cannot be explained by means of the Heitler–London approximation.

Finally, the fact that many rare-earth metals are ferromagnetic suggests the assumption that, besides the mechanism described above, other mechanisms must also play a part in the exchange interaction. Since the wave functions of the 4f electrons of neighbouring atoms can overlap only very little, a direct exchange interaction is impossible in the case of rare-earth elements. It is, however, possible that the spins are coupled indirectly via the conduction electrons. According to this idea, the conduction electrons (mostly *s* electrons) are polarised by the direct exchange interaction with the magnetic *d* or *f* electrons, i.e. they are partially oriented, so that the *d* electrons are coupled in this way. This mechanism is called the Ruderman–Kittel mechanism (Ruderman and Kittel, 1954; Kasuya, 1956; Van Vleck, 1962). It was first dealt with in detail in connection with the indirect coupling of nuclear spins via a hyperfine interaction with the conduction electrons.

According to (3.7), the direct exchange interaction between the s electrons and the d or f electrons is given by

$$\mathcal{H}_{sd} = -2 \sum_{i=1}^{N'} \sum_{m=1}^{N} J(|\boldsymbol{r}_i - \boldsymbol{r}_m|)(\boldsymbol{s}_i \cdot \boldsymbol{s}_m).$$

i and m denote the atoms with their d electrons and conduction electrons, respectively. The perturbation operator $\mathcal{H}_{sd}$ will not influence the s electrons essentially, since their energy amounts to about 10 eV, and the exchange integrals are of the order of 10^{-2} eV. Thus $\mathcal{H}_{sd}$ can be considered a small perturbation. If we assume that in the unperturbed state the electron spins are equally distributed over both directions, the expectation value of $\mathcal{H}_{sd}$ will be zero. The perturbation operator changes the electron energy only in second order. According to (2.77), the effective Hamiltonian,† which describes the perturbation of the ground-state energy of the s electrons in a second-order approximation, is

$$\mathcal{H}_{eff} = E_0 P_0 + P_0 \mathcal{H}_{sd} Q_0 \mathcal{H}_{sd} P_0$$

where

$$Q_0 = (1-P_0)\frac{1}{E_0 - \mathcal{H}_0}(1-P_0) = \sum_{\lambda \neq 0} \frac{P_\lambda}{E_0 - E_\lambda}.$$

Here E_0 is the energy of the ground state, E_λ is the energy of the excited states, the P_λ are the projection operators for these states. Substituting Q_0 in $\mathcal{H}_{eff}$, we obtain

$$\mathcal{H}_{eff} = \left(E_0 + \sum_{\lambda \neq 0} \frac{|\langle 0 | H_{sd} | \lambda \rangle|^2}{E_0 - E_\lambda}\right) P_0 .$$

The second term in the parentheses represents the change in energy of the unperturbed state of the conduction electrons,

$$\delta E_0 =$$

$$4 \sum_{\lambda \neq 0} \sum_{i,j} \sum_{n,m} \cdot \frac{\langle 0 | J(|\boldsymbol{r}_i - \boldsymbol{r}_m|)(\boldsymbol{s}_i \cdot \boldsymbol{s}_m) | \lambda \rangle \langle \lambda | J(|\boldsymbol{r}_j - \boldsymbol{r}_m|)(\boldsymbol{s}_j \cdot \boldsymbol{s}_n) | 0 \rangle}{E_0 - E_\lambda} .$$

† Here the atomic spins act upon the conduction electrons like external fields.

The sum over i and j contains two components: the terms with $i = j$ and those with $i \neq j$. The first terms may be considered a self-energy of the atoms which is due to the interaction with the s electrons. The other terms represent an interaction between different atomic spins; we shall consider only the latter. We therefore have

$$\delta E_0 = 4 \sum_{\lambda \neq 0} \sum_{i \neq j} \sum_{n,m} \sum_{\nu,\mu=1}^{3} s_i^\nu s_j^\mu \times \frac{\langle 0 | s_m^\nu J(|\boldsymbol{r}_i - \boldsymbol{r}_m|) | \lambda \rangle \langle \lambda | s_n^\mu J(|\boldsymbol{r}_j - \boldsymbol{r}_n|) | 0 \rangle}{E_0 - E_\lambda}. \tag{3.9}$$

where ν and μ label the components of the spin vectors. The vectors $|\lambda\rangle$ are state vectors for the total system of s electrons. Neglecting electron–electron interactions, we can construct them from the single-electron states $|k\rangle$ by means of an antisymmetrised product (Slater determinant) of N electron states; we must take this product because of the Pauli principle. Assume in the ground state the single-electron states $k_1 \ldots k_N$ to be occupied. We then have

$$|0\rangle = \frac{1}{\sqrt{N!}} \sum_P (-)^P P |k_1 \ldots k_N\rangle.$$

Summation is carried out over all permutations of the electrons; $(-)^P$ is positive, if P is an even permutation (i.e. if it contains an even number of transpositions) and negative, if P is odd. Accordingly the excited states can be described by

$$|\lambda\rangle = \frac{1}{\sqrt{N!}} \sum_P (-)^P P |k_1' \ldots k_N'\rangle.$$

In (3.9) we are interested only in the matrix elements of operators depending on a single electron. Because of the orthonormalisation of the vectors, such an operator will satisfy the relation

$$\langle 0 | O_1 | \lambda \rangle = \langle k_1 | O_1 | k_1' \rangle$$

if the set of single-electron states contained in $|0\rangle$ differs from the set of states contained in $|\lambda\rangle$ only because $|k_1\rangle$ has been replaced by $|k_1'\rangle$.

All other matrix elements vanish. Therefore

$$\delta E_0 = 4 \sum_{i \neq j} \sum_{k} \sum_{k' \neq k} \sum_{\nu,\mu=1}^{3} s_i^\nu s_j^\mu \times \frac{\langle k | s^\nu J(|\boldsymbol{r}_i - \boldsymbol{r}|) | k' \rangle \langle | k' | s^\mu J(|\boldsymbol{r}_j - \boldsymbol{r}|) | k \rangle}{E_k - E_{k'}}.$$

Since the s electrons have no spin–orbit coupling, the electron states can be described as a product between a Bloch state and a spin state:

$$|k\rangle = |\boldsymbol{k}\rangle |m_s\rangle.$$

Then

$$\delta E_0 = 4 \sum_{i \neq j} \sum_{k, m_s} \sum_{k', m_s'} \sum_{\mu,\nu} s_i^\nu s_j^\mu \langle m_s | s^\nu | m_s' \rangle \langle m_s' | s^\mu | m_s \rangle \times \frac{\langle \boldsymbol{k} | J(|\boldsymbol{r}_i - \boldsymbol{r}|) | \boldsymbol{k}' \rangle \langle \boldsymbol{k}' | J(|\boldsymbol{r}_j - \boldsymbol{r}|) | \boldsymbol{k} \rangle}{E_k - E_{k'}}.$$

$\boldsymbol{k}$ denotes all states occupied in the ground state and $\boldsymbol{k}'$ denotes the free states. The spin sum can be written explicitly in the form

$$\sum_{m_s, m_s'} \langle m_s | s^\nu | m_s' \rangle \langle m_s' | s^\mu | m_s \rangle = \mathrm{Tr}\,(s^\nu s^\mu) = \tfrac{1}{2}\, \delta_{\mu,\nu},$$

so that

$$\delta E_0 = -2 \sum_{i \neq j} f(\boldsymbol{r}_i, \boldsymbol{r}_j)(\boldsymbol{s}_i \cdot \boldsymbol{s}_j) \tag{3.10}$$

where

$$f(\boldsymbol{r}_i, \boldsymbol{r}_j) = - \sum_{k, k'} \frac{\langle \boldsymbol{k} | J(|\boldsymbol{r}_i - \boldsymbol{r}|) | \boldsymbol{k}' \rangle \langle \boldsymbol{k}' | J(|\boldsymbol{r}_j - \boldsymbol{r}|) | \boldsymbol{k} \rangle}{E_k - E_{k'}}. \tag{3.11}$$

According to (3.10), the indirect exchange leads to an interaction between the spins, which corresponds to the Heisenberg operator (3.7). The dependence of interaction on the distance, however, is quite different. Consider, for example, a strong localisation of the magnetic electrons, or nuclear spins with a Fermi interaction (2.75). The exchange integral can then be approximated by a function: $J(|\boldsymbol{r}_i - \boldsymbol{r}|) = C\delta(\boldsymbol{r}_i - \boldsymbol{r})$ and, if we also assume free electrons and change from a sum to an integral, we obtain from (3.11)

$$f(\boldsymbol{r}_i, \boldsymbol{r}_j) = \frac{2mC^2}{(2\pi)^6\hbar^2} \int d^3\boldsymbol{k}\, d^3\boldsymbol{k}'\, \frac{e^{i([\boldsymbol{k}'-\boldsymbol{k}]\cdot[\boldsymbol{r}_i-\boldsymbol{r}_j])}}{k^2-k'^2}.$$

Changing to polar coordinates, the integration over the angles can be carried out at once. Since the states inside the Fermi sphere are occupied, $k < k_0$ and $k' > k_0$ so that

$$f(\boldsymbol{r}_i, \boldsymbol{r}_j) = \frac{mC^2}{2\pi^4\hbar^2}\,\frac{k_0^2}{r_{ij}^2} \int_1^\infty u\,du \int_0^1 v\,dv\, \frac{\sin\alpha u\cdot\sin\alpha v}{u^2-v^2}$$

where $\alpha = k_0 r_{ij}$ and $r_{ij} = |\boldsymbol{r}_i - \boldsymbol{r}_j|$. The integral over u can be taken from $u = 0$ since the integral between 0 and 1 vanishes together with the integration over v, as the integrand is asymmetrical in u and v. The lower limit can then be extended to $-\infty$ and the integral can be calculated using the method of residues. At the points $u = \pm v$ we must take half the residue (principal value) and obtain

$$f(\boldsymbol{r}_i, \boldsymbol{r}_j) = \frac{k_0^2 C^2}{32\pi^4\zeta_0 r_{ij}^4}(\sin 2k_0 r_{ij} - 2k_0 r_{ij}\cos 2k_0 r_{ij}). \qquad (3.12)$$

Here ζ_0 is the Fermi energy at $T = 0$. Compared with the direct exchange interaction, the indirect exchange interaction decreases less rapidly with increasing distance of the atoms, so that we may expect the indirect exchange to play a decisive part in the case of metals such as the rare earths.

Although the Heisenberg operator is not well founded in the case of metals, we shall to begin with base our calculations upon it. The exchange integrals J_{ij} are considered to be parameters which must be determined experimentally.

Exchange coupling has a simple physical meaning; let us consider the Hamiltonian of two spins $\mathcal{H} = -2J(\boldsymbol{s}_1\cdot\boldsymbol{s}_2)$ and the corresponding time-dependent Schrödinger equation $\mathcal{H}\psi = i\hbar\dot{\psi}$. The general solution is a linear combination of the spin functions (3.5), i.e. we have

$$\psi(t) = c_0\chi_0 e^{-iE_0t/\hbar} + (c_1\chi_1 + c_2\chi_2 + c_3\chi_3)e^{-iEt/\hbar}.$$

E_0 is the energy of the singlet state, $E_0 = \frac{3}{2} J$, and E is the energy of the triplet state, $E = -\frac{1}{2} J$. The coefficients c_ν are determined from the initial conditions. At the time $t = 0$ the spin at point 1 is assumed to be parallel to the z-axis, and that at point 2 antiparallel to it, i.e. $\psi(0) = \alpha(1)\beta(2)$. Hence we obtain

$$c_0 = c_2 = 2^{-1/2}, \qquad c_1 = c_3 = 0,$$

so that

$$\psi(t) = [\alpha(1)\,\beta(2)\cos(Jt/\hbar) + i\beta(1)\,\alpha(2)\sin(Jt/\hbar)]\,\mathrm{e}^{-iJt/2\hbar}.$$

The expectation values of s_z at points 1 and 2 respectively, and at time t are thus given by

$$\overline{s_z(1)} = \tfrac{1}{2}\cos(2Jt/\hbar), \qquad \text{and} \qquad \overline{s_z(2)} = -\tfrac{1}{2}\cos(2Jt/\hbar),$$

respectively—that is, the spins are continuously exchanged between the two atoms. If therefore the magnetisation in a substance is varied locally, this perturbation does not remain localised but becomes spread over the entire substance by exchange coupling, in the same way as a local elastic deformation is expanded over a larger region by means of sound waves. In magnetic media the phenomenon corresponding to sound waves is called spin waves.

2. The Molecular-field Approximation

The Schrödinger equation with the Hamiltonian (3.7) cannot be solved exactly. The oldest and also simplest approximation is based on a linearization of the Hamiltonian, i.e. one of the angular momenta in (3.7) is replaced by its thermal mean.

For an isotropic ferromagnet in an external field H, which is parallel to the z-axis, we have

$$\mathscr{H} = g\mu_B H \sum_{i=1}^{N} s_{iz} - 2 \sum_{i>j} J_{ij}(\boldsymbol{s}_i \cdot \boldsymbol{s}_j).$$

If there were anisotropies, they could be taken into account in the spectroscopic splitting factor in the same way as in (2.81) and (2.82).

Because of the translational symmetry, the mean value of $\boldsymbol{s}_j$ will be independent of the lattice site and antiparallel to the external field. Therefore

$$\mathcal{H} = g\mu_B(H+H_W)\sum_{i=1}^{N} s_{iz} \tag{3.13}$$

and

$$g\mu_B H_W = -2\sum_{j=1}^{N} J_{ij}\bar{s}_{jz}. \tag{3.14}$$

In this approximation exchange coupling has the same effect as a magnetic field inside the medium. We call H_W the Weiss field or the molecular field. Because of the translational symmetry it is independent of the subscript i. This description of exchange interaction by a fictitious internal field was suggested by P. Weiss (1907), long before the era of quantum mechanics. This simple approximation permits a qualitatively correct description of the most important features of ferromagnetism, antiferromagnetism, and ferrimagnetism.

Since electron spin and magnetic moment are antiparallel to one another, the magnetisation is given by

$$M = -Ng\mu_B\bar{s}_{jz} = W^{-1}H_W. \tag{3.15}$$

W is called the Weiss factor.

It is an advantage of the linearisation that the initial problem with N spins can be reduced to a problem with one spin. We can therefore use the formulas derived in Section II.4 for the magnetisation of a paramagnetic ion. According to (3.13) and (2.51), we obtain for the magnetisation

$$M = Ng\mu_B sB_s(\beta g\mu_B s(H+H_W)). \tag{3.16}$$

where $\beta = 1/kT$, s is the atomic spin quantum number, and B_s is the Brillouin function defined in (2.48). For historical reasons we shall speak in the following only of the spin, since in the "classical" ferromagnetic metals of the iron group the orbital angular momentum is essentially quenched. This can be seen from both the susceptibility (Table 7) and from the g factors in paramagnetic resonance (Table 11). The oldest references to this fact stem from gyromagnetic experiments, the Einstein–de Haas effect or the Barnett effect.

It is well known that the Einstein–de Haas effect consists of the following: when the magnetisation of a sample is changed the angular momenta of the electrons in the sample must also be changed, which, because of the law of conservation of angular momentum, results in a rotation of the sample. The Barnett effect is the inverse phenomenon, i.e. magnetisation arises when the sample is rotated. The ratio between the change in magnetisation and the change in angular momentum yields the gyromagnetic factor. The g factor derived from it is not identical with the spectroscopic splitting factor. It will be denoted by g'.

The spectroscopic splitting factor g is obtained from the splitting of electron levels in a magnetic field. We consider the case of a strongly quenched orbital angular momentum, which is the only one we are here interested in. We also assume isotropic conditions. The ground state will then be split according to the spin quantum number M_s; disregarding fine structure, the splitting is given according to (2.81) and (2.82) by

$$E_{M_s} = g\mu_B H M_s,$$

where

$$g = (1/M_s) < M_s|2S_z+L_z|M_s > = 2(1-\zeta\Lambda),$$

and Λ is now isotropic. The expectation value must be formed with one of the states pertaining to the $(2S+1)$-fold degenerate ground state, which are slightly perturbed because of spin–orbit coupling.

The gyromagnetic factor g' of the magneto-mechanical experiments is given by the ratio between the magnetic moment produced by the field and the angular momentum caused by it (the similarly induced nuclear moment is so small that it can be neglected). Thus we have

$$g' = \frac{\langle M_s|2S_z+L_z|M_s\rangle}{\langle M_s|S_z+L_z|M_s\rangle} = \frac{2(1-\zeta\Lambda)}{1-2\zeta\Lambda}.$$

Hence we obtain the Kittel–Van Vleck relation

$$(g-1)(g'-1) = 1,$$

or, because of the quenching effect,

$$g-2 \cong 2-g'.$$

Resonance experiments carried out for ferromagnetic Fe, Co, and Ni yielded $g_{Fe} = 2{\cdot}10$, $g_{Co} = 2{\cdot}20$, and $g_{Ni} = 2{\cdot}21$; gyromagnetic experiments yielded

$$g'_{Fe} = 1{\cdot}92, \qquad g'_{Co} = 1{\cdot}85, \qquad g'_{Ni} = 1{\cdot}84$$

(Scott, 1962).

The definition (3.16) for the magnetisation has a non-trivial solution for $H = 0$, i.e. spontaneous magnetisation can arise. It vanishes at a certain temperature T_c, the Curie point. Near $T = T_c$ the Brillouin functions can be expanded

$$B_s(x) = \frac{s+1}{3s}x - \frac{(2s+1)^4-1}{(2s)^4}\frac{x^3}{45} + \dots \tag{3.17}$$

so that from (3.15) and (3.16) we obtain

$$M^2 = \frac{10Nk}{W}\frac{s(s+1)}{s^2+(s+1)^2}\frac{T^2}{T_c^2}(T_c - T). \tag{3.18}$$

Here

$$kT_c = N\frac{g^2\mu_B^2 s(s+1)W}{3}. \tag{3.19}$$

Equation (3.18) is a solution if $T < T_c$; above the Curie point the only solution that exists is $M = 0$.

Assuming the exchange interaction to decrease rapidly as the distance between the atoms increases, we can restrict ourselves to interactions between neighbouring atoms. Every atom has z nearest neighbours. Denoting by J the exchange integral between nearest neighbours, we have, according to (3.14), (3.15), and (3.19),

$$kT_c = \tfrac{2}{3}\, zJs(s+1). \tag{3.20}$$

Above the Curie point the magnetisation is proportional to the external field, $M = \chi H$. Together with (3.17) we obtain a Curie–Weiss law:

$$\chi = \frac{C}{T-T_c}; \qquad C = N\frac{g^2\mu_B^2 s(s+1)}{3k}. \tag{3.21}$$

We know from experience that, at temperatures considerably higher than the Curie temperature, χ^{-1} is a linear function of T. The so-called paramagnetic Curie temperature θ (obtained from the asymptotic behaviour of the reciprocal susceptibility) is always higher than the Curie temperature T_c. The difference may be as large as 50°C (cf. Table 12). The linear dependence of the reciprocal susceptibility on T and the coincidence of T_c and θ are typical for the molecular-field approximation.

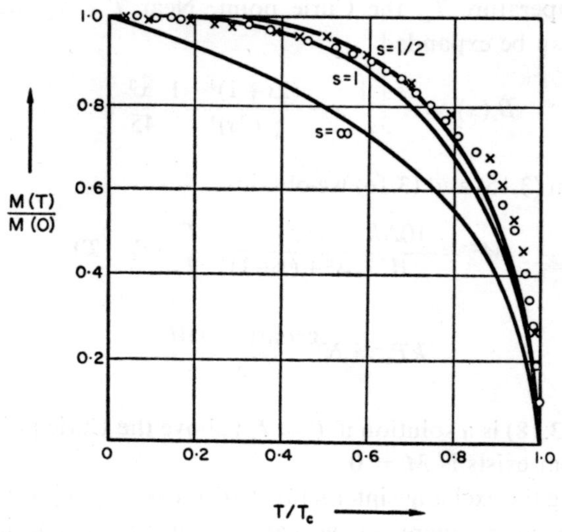

FIG. 18.

Relative magnetisation as a function of temperature. The solid lines represent the Brillouin functions for $s = \frac{1}{2}$, 1, and ∞. The experimental values given are those for nickel (○) and iron (×). (Landolt-Börnstein, 1962.)

At low temperatures the magnetisation tends exponentially to its peak value $Ng\mu_B s$, as the temperature is reduced (Fig. 18).

In the molecular-field approximation the mean energy of a ferromagnet is according to (3.13) and (3.15) by given

$$E = -MH - \tfrac{1}{2}WM^2.$$

The factor $\frac{1}{2}$ in the last term accounts for the fact that in M^2 the spin–spin interactions were counted twice. Near the Curie point the magnetisation, and thus also the energy, varies rapidly with temperature. This results in a rather large contribution to the specific heat. For $H = 0$ this magnetic contribution amounts to

$$C_m(T) = -\frac{1}{2}W\left(\frac{\partial M^2}{\partial T}\right)_{H=0}$$

Above the Curie point the spontaneous magnetisation vanishes so that $C_m = 0$ if $T > T_c$. A little below T_c we can apply (3.18) and obtain with (3.19)

$$C_m(T) = Nk\frac{5s(s+1)}{s^2+(s+1)^2}\frac{T^2}{T_c^2}\left(1+2\left(1-\frac{T_c}{T}\right)\right).$$

Below the Curie point the specific heat increases as the square of the temperature and drops at $T = T_c$ to zero, so that the curve shows a discontinuity

$$\Delta C_m = Nk\frac{5s(s+1)}{s^2+(s+1)^2}. \tag{3.22}$$

The discontinous behaviour of specific heat indicates that the transition from the ferromagnetic to the paramagnetic state is a second-order phase transition. Experimentally it can only be shown that above T_c the specific heat decreases more rapidly than it increases below T_c. Thus the discontinuity can be obtained only approximately from experiments. Table 12 shows that the agreement with eqn. (3.22) is poor.

TABLE 12

	T_c (°K)	Θ (°K)	p	p_{eff}	$\left(\frac{\Delta C}{Nk}\right)_{exp}$	$\left(\frac{\Delta C}{Nk}\right)_{MF}$
Fe	1043	1101	2·22	3·20	5	2
Co	1393	1411	1·71	3·15	2·2	2·2
Ni	631	649	0·605	1·61	1	1·5
Gd	289	302	7·1	7·97	3·5	3·4

In its first columns the table shows the ferromagnetic and paramagnetic Curie temperatures of Fe, Co, Ni and Gd. The next column contains the effective magnetic moments per atom in μ_B units as obtained from the saturation magnetisation at $T = 0$. Since all spins were assumed to be parallel, we have theoretically $p = g\cdot s$. The next column contains the effective moments determined from the susceptiblity above T_c,

$$3k(T-T_c)\chi = N\mu_B^2 p_{eff}^2.$$

The moments measured per atom proved to be neither integral nor half-integral multiples of $g\mu_B$ (Ni, Co). In the last two columns the experimental values of the change in C_m at T_c are compared with the values of eqn. (3.22). For Ni a value of $s = \frac{1}{2}$ was used.

The vanishing of the magnetic contribution to the energy (and all other thermodynamic quantities) above the Curie point is theoretically unsatisfactory, since for $T > T_c$ there will still exist a magnetic order of spins in smaller domains. This so-called short-range order is not encompassed by the molecular-field approximation since, as shown by (3.13), every spin will be oriented in the external and internal fields independently of its environment. The molecular-field theory describes only the long-range order in the system considered, that is, its magnetisation.

The molecular-field approximation can easily be applied to the case of antiferromagnetic coupling (Néel, 1932; Van Vleck, 1941). It is, however, necessary to know first the distribution of the magnetic moments in the crystal; a direct method yielding this distribution is connected with neutron experiments. Neutrons are scattered from the nuclear and atomic moments. A magnetically ordered structure manifests itself in the appearance of Bragg reflections. In ferromagnets the reflections will look similar to those obtained by means of X-rays, since the magnetic structure agrees with the lattice structure. In the case of antiferromagnets and ferrimagnets, the magnetic structure is of lower symmetry than the crystallographic structure, since the atomic moments at the lattice sites can be oriented in at least two different directions. This results in the appearance of additional reflections as compared with the X-ray patterns. The intensity of the reflections is a measure for the magnetisation, i.e. the magnetic moments of the atoms. Thus we obtain for MnO, for example, which has a rock-salt lattice, that similarly oriented magnetic moments are in planes whose normal coincides with the body diagonal. Adjacent planes have opposite spins. The simplest model for an antiferromagnet is obtained when we consider two identical mutually penetrating lattices whose atoms have opposite spins so that, without an external field, the total magnetisation vanishes. This is possible only if the exchange integrals for atoms in different sublattices are negative and exceed all other interactions in the sublattices.

When we first consider the case of sublattice magnetisations which are parallel or antiparallel to the external field, the Hamiltonian will in the molecular-field approximation be

$$\mathcal{H} = g\mu_B \sum_{i=1}^{N} (H+H_{Wi})s_{iz}. \tag{3.23}$$

The internal fields differ according to the sublattice;

$$g\mu_B H_{Wi} = -2 \sum_{j=1}^{N} J_{ij}\overline{s_{jz}}. \tag{3.24}$$

H_{Wi} is the internal field in the sublattice to which the spin with the subscript i pertains. If this sublattice contains N_i atoms, its magnetisation is given according to (2.51) by

$$M_i = N_i g\mu_B s B_s(\beta g\mu_B s(H+H_{Wi})). \tag{3.25}$$

In general we can restrict ourselves to the interaction of an atom with its nearest neighbours. The exchange interaction J with the z atoms belonging to the other sublattice is strongest. In addition to this, we have the interaction J' with the z' atoms of the same sublattice, which are neighbours of the atom considered. J' can be either positive or negative. According to (3.24), we shall then have for the two internal fields

$$\begin{aligned} H_{W1} &= -\frac{2}{g\mu_B}(zJ\overline{s_{2z}}+z'J'\overline{s_{1z}}) = W_{12}M_2+W_{11}M_1, \\ H_{W2} &= -\frac{2}{g\mu_B}(zJ\overline{s_{1z}}+z'J'\overline{s_{2z}}) = W_{12}M_1+W_{11}M_1. \end{aligned} \tag{3.26}$$

W_{11} and W_{12} are again Weiss factors. In the case of $H = 0$ the magnetisations of the two sublattices must be opposite and of equal magnitude: $M_1 = -M_2$, and $\overline{s_{1z}} = -\overline{s_{z2}}$. The definition equation of the spontaneous magnetisation of the sublattices, with $N_1 = N_2 = N/2$, can then be given with the help of (3.25) and (3.26):

$$M_1 = \tfrac{1}{2} N g\mu_B s B_s(\beta g\mu_B s(W_{11}-W_{12})M_1). \tag{3.27}$$

This relation has the same form as that for the spontaneous magnetisation of a ferromagnetic substance, (3.16). Thus the spontaneous magnetisation of the sublattices will vanish at the antiferromagnetic Curie point T_N, which is also called the Néel temperature. By analogy with (3.19) we have

$$kT_N = \tfrac{1}{6} N(g_B\mu)^2 s(s+1)(W_{11}-W_{12}) = -\tfrac{2}{3}\, s(s+1)\,(zJ-z'J'). \quad (3.28)$$

It is reasonable for this model to require that $J < 0$ and, according to (3.28), that $z\,|J| > z'\,|J'|$. This means that a possible ferromagnetic coupling between the atoms of a lattice must be weaker than the antiferromagnetic coupling between the sublattices. Above the Néel point M_1 and M_2 will be proportional to the external field. We can therefore expand the Brillouin function in (3.25) and, neglecting higher powers of H, we obtain

$$M_{1,\,2} = \tfrac{1}{6} N(g\mu_B)^2 s(s+1)\beta(H+H_{W1,\,2}).$$

Adding the magnetisations M_1 and M_2, we obtain for the total magnetisation

$$M = \frac{C}{T+\Theta} H = \chi H, \qquad C = N\frac{(g\mu_B)^2 s(s+1)}{3k} \quad (3.29)$$

with

$$k\Theta = -\tfrac{2}{3}\, s(s+1)(zJ+z'J'). \quad (3.30$$

According to (3.29), the susceptibility obeys a Curie–Weiss law; the exchange interaction between atoms of the same sublattice, however, results in a difference between the paramagnetic Néel point and T_N. In this model Θ, however, must be of the same order of magnitude as T_N (cf. Table 13). It is a characteristic property of antiferromagnets that, compared with (3.21), their Curie–Weiss law (3.29) has the opposite sign.

Below the Néel point the argument of the Brillouin function can no longer be considered small; we must then expand (2.25) with respect to the external field. It is easy to obtain

$$\chi_{||}(T) = N\frac{(g\mu_B)^2 s^2 B_s'(\alpha)}{k\left(T+\dfrac{3s}{s+1}\, B_s'(\alpha)\cdot\Theta\right)} \quad (3.31)$$

where $\alpha = \beta g\mu_B s(W_{11}-W_{12})M_{1,\,2}(H=0)$.

$B'(\alpha)$ is the derivative of the Brillouin function. Since B_s' is an even function of the argument, it is inessential which of the spontaneous magnetisations is used in the expression for α. The magnetisation $M_{1,2}(H=0)$ is obtained from (3.27). According to (3.31), the susceptibility drops exponentially to zero as the temperature rises. Equation (3.31) holds true only if the spins of the sublattices are parallel or antiparallel to the external field. (Above the Néel point this difference, of course, is of no interest, since then all spins can move freely.) If a field is applied at right angles to the preferred direction of the spontaneous magnetisations, then in a first approximation the sublattice magnetisations are rotated a little, so that their directions are no longer coincident with the preferred position; their absolute values, however, remain unchanged. The transverse susceptibility $\chi_\perp$ is therefore temperature-independent; however, as $\chi_\perp(T_N) = \chi(T_N)$, we have according to (3.29) for $T < T_N$

$$\chi_\perp = \frac{C}{T_N+\Theta} = N\frac{(g\mu_B)^2}{4z\,|J\,|^2}. \tag{3.32}$$

In the case of a powdered material the susceptibility will always be isotropic, so that

$$\chi_{\text{powder}}(T) = \tfrac{2}{3}\,\chi_\perp + \tfrac{1}{3}\chi_{||}(T);$$

in this case we have the remarkable relation

$$\chi_{\text{powder}}(0)/\chi_{\text{powder}}(T_N) = \tfrac{2}{3},$$

cf. Table 13. The susceptibility of an antiferromagnetic substance has according to our formulas a characteristic temperature dependence, as it has a maximum at the Néel point (cf. Fig. 19). At this point the specific heat displays the same anomaly as in the case of a ferromagnet.

The objections against the molecular-field approximations are the same as in the case of ferromagnetism. The model of two sublattices here considered again proves to be too simple. According to this model, the ratio Θ/T_N should not be much higher or much lower than 1. According to Table 13, however, this ratio may have high values. It is therefore necessary to consider models with several sublattices. Since

TABLE 13

	MnO	FeO	CoO	NiO	MnF_2	FeF_2	NiF_2
T_N (°K)	116	198	292	523	68	79	73
Θ (°K)	610	570	330	3000	113	117	115
Θ/T_N	5·3	2·9	1·1	5·7	1·7	1·5	1·6
$\chi_{powd}(0)/\chi_{powd}(T_N)$	0·69	0·78		0·67	0·72	0·72	

This table contains the Néel temperatures and the paramagnetic Néel temperatures Θ of several antiferromagnetic non-metals (Smart, 1963). The last column shows the ratio of the susceptibilities at 0°K and at T_N (Nagamiya *et al.*, 1955). According to the molecular-field approximation this ratio must be $\frac{2}{3}$.

$\Theta > T_N$, the second integral J' will also be negative, cf. (3.28) and (3.30); a more exact theoretical analysis of the experiments verifies this for most antiferromagnetic substances (Smart, 1963).

The anisotropy of antiferromagnetic substances generally has but a minor influence on the magnitude of the susceptibility. It is, however, responsible for the presence of a preferred direction in the crystal of the sublattice magnetisation so that, as in the case of ferromagnets, we

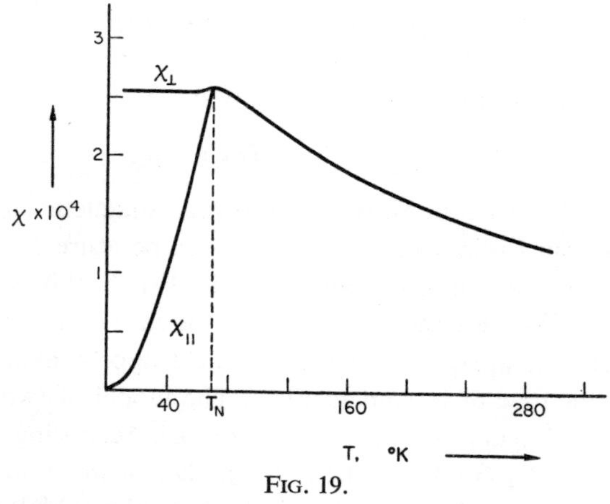

FIG. 19.

Susceptibility per gram of an antiferromagnet with one axis of symmetry, MnF_2, after Foner (1963).

can speak of directions of easy or difficult magnetisation. The anisotropy gives rise to a remarkable effect in antiferromagnetic substances. If a magnetic field is applied parallel to the preferred direction of magnetisation, which we assume to be stabilised by the anisotropy energy K with respect to the transverse direction, we obtain for the spin energy $-\frac{1}{2}\chi_{||}H^2 - K$. If the external field is perpendicular to the axis of magnetisation, the spin energy will be $-\frac{1}{2}\chi_{\perp}H^2$. Since below the Néel point we have $\chi_{\perp} > \chi_{||}$, it is for an antiferromagnetic substance above a certain field strength H_c more favourable to orient the magnetic moments transversely to the field if they had been parallel below H_c. The reversal of the spin (spin flop) obviously takes place when

$$\tfrac{1}{2}\chi_{\perp}H^2 = -\tfrac{1}{2}\chi_{||}H^2 - K,$$

i.e. at a field strength

$$H_c = \sqrt{\frac{2K}{\chi_{\perp} - \chi_{||}}}\,. \tag{3.33}$$

This spin flop is a measure for the anisotropy energy; it has been observed in MnF_2 and $CuCl_2 \cdot 2H_2O$, for instance.

Hitherto we always assumed the sublattice spins to be antiparallel, if no external field was applied. Actually this is a consequence of the translational invariance of the lattice. It can be shown that, if the position of a lattice site of a sublattice cannot be reached by a translation from another lattice site of the other sublattice (that is, we need translation plus, e.g. rotation), spin–orbit coupling gives rise to an additional term of the form $(\boldsymbol{d}\cdot[\boldsymbol{s}_1 \wedge \boldsymbol{s}_2])$ in the effective spin operator; here $\boldsymbol{s}_1$ and $\boldsymbol{s}_2$ denote the spins of different sublattices, and $\boldsymbol{d}$ is a constant vector (Moriya, 1963). The additional term is smallest if the spins are perpendicular to one another, which results in a slight mutual rotation of the directions of magnetisation of the sublattices. This has a similar effect as weak ferromagnetism; the effective moment per atom will then amount to about 1/100 of the atomic moments.

Moreover, there exist antiferromagnets which, in their spin structure, differ essentially from the simple model of two sublattices, such as the rare-earth metals above gadolinium. At low temperatures they are ferromagnetic and become antiferromagnetic at higher temperatures (Belov *et al.*, 1964). Among them dysprosium is the simplest example. It crystallises in a hexagonal lattice. Neutron scattering experiments show that the spins lie in planes which are perpendicular to the hexagonal axis, so that in the antiferromagnetic state the angle between the directions of the spins in two neighbouring layers is constant (according to the temperature it amounts to 26·5–43·2°). In this way the metal will become antiferromagnetic over a larger region and will possess a spiral spin structure; for details see the articles by Bertaut (1963) and Turov (1968).

The appearance of spin structures and weak ferromagnetism indicates that the Heisenberg operator (3.7) is too simple; we have to replace (3.7) by a more general bilinear expression:

$$\mathscr{H} = -\sum_{i \neq j} (\boldsymbol{s}_i \cdot \boldsymbol{J}_{ij} \cdot \boldsymbol{s}_j);$$

here the products are tensorial products. An inhomogeneous spin structure can also be due to oscillating long-range exchange interactions of the type of a Ruderman–Kittel interaction (Bertaut, 1963; Turov, 1968; Mattis, 1965).

When the two-sublattice model considered above is generalised by assuming the two lattices to consist of different atoms, or of identical atoms but different spins and g factors, so that the total magnetisation is non-zero, we are led to the simplest model of ferrimagnetism according to Néel (1948). Below a certain temperature, which is again called the Néel temperature, ferrimagnetic substances show spontaneous magnetisation. In this respect they are closer to the ferromagnets. Since they must have a lattice structure, which has at least two non-equivalent lattice sites for the paramagnetic ions displaying antiferromagnetic coupling, their lattice structure is rather complicated (cf., for example, Smit and Wijn, 1959). Important groups among the ferrimagnets are the ferrites with spinel structure of the chemical formula $M\mathrm{Fe_2O_4}$, where M is a bivalent metal ion, and also the garnet structure ferrites such as yttrium–iron garnet (YIG) with the chemical formula $\mathrm{Y_3Fe_5O_{12}}$, or the rare-earth–iron garnets (MIG), $M_3\mathrm{Fe_5O_3}$.

In the molecular-field approximation ferrimagnetism is treated in the same way as antiferromagnetism. The sublattice magnetisation is described by (3.25), where the spin quantum number and g factor can be different sublattices. Thus we have

$$M_i = N_i g_i \mu_B s_i B_{s_i}(\beta g_i \mu_B s_i (H + H_{W_i})),$$

and, according to (3.24) and (3.26), we have for the internal fields

$$H_{W_i} = \sum_j W_{ij} M_j.$$

The subscripts i and j number the sublattices. The Weiss factors W_{ij} are symmetrical, i.e. $W_{ij} = W_{ji}$ (*actio* = *reactio*). The calculations for two sublattices are not given, as they are rather obvious. The following expressions are obtained for the Néel temperature and the paramagnetic susceptibility

$$T_N = \frac{C_1 W_{11} + C_2 W_{22}}{2} \pm \frac{1}{2}\sqrt{(C_1 W_{11} - C_2 W_{22})^2 + 4 C_1 C_2 W_{12}^2},$$

$$\chi(T) = \frac{(C_1 + C_2)T + C_1 C_2 (2W_{12} - W_{11} - W_{22})}{(T - C_1 W_{11})(T - C_2 W_{22}) - C_1 C_2 W_{12}^2}.$$

Here

$$C_1 = N_1 \frac{g_1^2 \mu_B^2 s_1(s_1+1)}{3k}, \qquad C_2 = N_2 \frac{g_2^2 \mu_B^2 s_2(s_2+1)}{3k}.$$

Thus $\chi^{-1}(T)$ is a linear function of the temperature only at high temperatures, at $T = T_N$ it becomes equal to zero; Fig. 20 shows schematically the temperature dependence of the ferrimagnetic susceptibility. Because of the antiferromagnetic coupling of the two sublattices, W_{12} is negative and, as to its absolute value, it is the largest of the three Weiss factors. The Weiss factors for the coupling within the two sublattices are different. As a consequence of this, the magnetisation is generally not a monotonic function of the temperature since the sublattice magnetisations depend on the temperature in different ways. It may happen that the magnetisations of the sublattices compensate one another at a certain temperature, the compensation temperature. This is especially observed in ferrites and garnet-type structures (MIG).

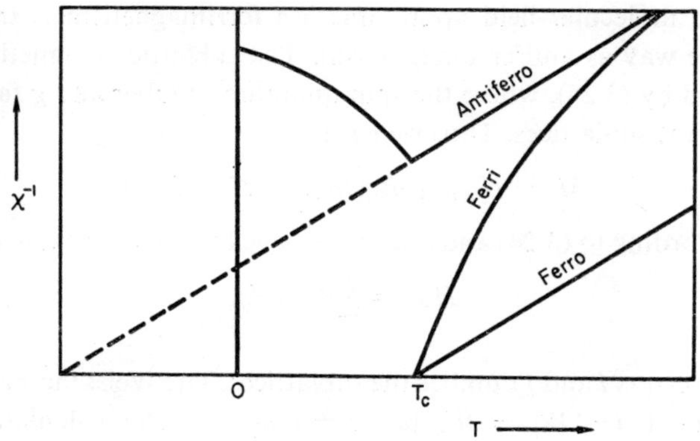

FIG. 20.

Schematic diagram of the temperature dependence of the reciprocal of the susceptibility for ferromagnetic, antiferromagnetic, and ferrimagnetic substances.

As in the case of antiferromagnetic substances, the spin structure of ferrites can be very complex.

3. Spin Waves

For a more detailed consideration of the Heisenberg operator (3.7) it is convenient to treat the problem differently in different temperature ranges, since the approximations necessary to solve the problem at low temperatures are different from those to be applied at high temperatures or near the Curie point.

The ground state of a ferromagnetic crystal is well known. When we ignore anisotropy effects, the magnetic moments at absolute zero are all parallel to the external field. The lowest states, lying first above the ground state, are called spin waves following Bloch (1930).

The Hamiltonian of a system of N spins in an external field H, which is parallel to the z-axis, is given according to (3.7) by

$$\mathcal{H} = g\mu_B H \sum_{i=1}^{N} s_{iz} - \sum_{i,j=1}^{N} J_{ij}(\boldsymbol{s}_i \cdot \boldsymbol{s}_j). \tag{3.34}$$

Using for brevity the symbol $s_i^{\pm} = s_{ix} \pm i s_{iy}$, the commutation relations of the spin components read

$$[s_i^+, s_j^-]_- = 2s_{iz}\delta_{i,j}, \qquad [s_{iz}, s_j^{\pm}]_- = \pm s_i^{\pm}\delta_{i,j}. \tag{3.35}$$

It is convenient to use Fourier-transformed quantities:

$$s_i^{\pm} = \sum_k s_k^{\pm} e^{\mp i(k\cdot r_i)}, \qquad s_{iz} = \sum_k s_k^z e^{-i(k\cdot r_i)},$$
$$J_{ij} \equiv J(|r_i - r_j|) = \frac{1}{N}\sum_k J_k\, e^{i(k\cdot r_i - r_j)}. \tag{3.36}$$

The vectors $\boldsymbol{k}$ are vectors in the reciprocal lattice and lie in the first Brillouin zone. There exist precisely N different $\boldsymbol{k}$ vectors. The transformations which are the inverse of (3.36) are:

$$s_k^{\pm} = \frac{1}{N}\sum_{i=1}^{N} s_i^{\pm}\, e^{\pm i(k\cdot r_i)},$$
$$s_k^z = \frac{1}{N}\sum_{i=1}^{N} s_{iz}\, e^{i(k\cdot r_i)}, \tag{3.37}$$
$$J_k = \sum_{i=1}^{N} J(r_i)\, e^{-i(k\cdot r_i)}.$$

Since the exchange integrals depend only on the distance and vanish if referred to one and the same atom, we have

$$J_k = J_{-k}, \qquad \sum_k J_k = 0. \tag{3.38}$$

The operators s_k^+ and s_k^- are Hermitean conjugate since s_i^+ and s_i^- are Hermitean conjugate. The commutation relations of the Fourier-transformed operators are according to (3.35) and (3.37):

$$[s_k^+, s_{k'}^-]_- = \frac{2}{N} s_{k-k'}^z, \qquad [s_k^z, s_{k'}^{\pm}]_- = \pm\frac{1}{N} s_{k'\pm k}^{\pm}. \tag{3.39}$$

The Hamiltonian (3.34) can now be written as follows:

$$\mathcal{H} = Ng\mu_B H s_0^z - N\sum_k J_k(s_k^z s_{-k}^z + s_k^+ s_k^-). \tag{3.40}$$

(It is worth mentioning that $\sum_{i=1}^{N} \exp i(\boldsymbol{k}\cdot\boldsymbol{r}_i) = N\delta_{k,0}$.)

In the ground state all moments are parallel to the field. When we denote the ground-state vector by $|0\rangle$, we can write

$$s_{iz}|0\rangle = -s|0\rangle \quad \text{and} \quad s_k^z|0\rangle = -s|0\rangle\,\delta_{k,0}. \tag{3.41}$$

In addition to this, because of (2.37), we have

$$s_i^-|0\rangle = 0 \quad \text{and} \quad s_k^-|0\rangle = 0. \tag{3.42}$$

The ground-state energy of (3.39) is therefore given by

$$E_0 = -Ng\mu_B sH - Ns^2 J_0. \tag{3.43}$$

It is easy to show that the states $s_k^+|0\rangle$ are eigenstates of the Hamiltonian. These states are called spin-wave states (more exactly: single-spin-wave states), as they correspond to a magnetisation wave in the spin system.

Using the commutation relations (3.39) we obtain for the commutator

$$[\mathscr{H}, s_k^+]_- = g\mu_B H s_k^+ + 2\sum_{k'}(J_{k'} - J_{k-k'})s_{k'}^+ s_{k-k'}^z. \tag{3.44}$$

Hence it follows immediately that the states $s_k^+|0\rangle$ are eigenstates:

$$\mathscr{H}s_k^+|0\rangle = s_k^+\mathscr{H}|0\rangle + [\mathscr{H}, s_k^+]_-|0\rangle = (E_0 + \varepsilon_k)s_k^+|0\rangle,$$

where

$$\varepsilon_k = g\mu_B H + 2s(J_0 - J_k). \tag{3.45}$$

The quantity ε_k is the excitation energy of the spin wave with the wave vector $\boldsymbol{k}$. The field-dependent term in ε_k indicates that the magnetic moment of the system, compared with the ground state, has been changed by $g\mu_B$ if it is in the spin-wave state (3.45) so that the angular momentum of the spin wave is equal to $\hbar$. Thus the spin waves are bosons, just like photons or phonons. The expectation value of the spin z component is the same for each lattice site in a spin-wave state and exceeds its value in the ground state by $\hbar/N$; this is easily verified. Thus a spin wave describes a collective excitation which involves all spins to the same degree. The operators s_k^+ and s_k^- act in ground state as field-theoretical creation and annihilation operators. It is thus reasonable to try to find such operators which have this property independently of the ground state, i.e. to change by means of a unitary transformation to operators a and a^+ which satisfy Bose commutation relations:

$$[a_i, a_j^+]_- = \delta_{ij}.$$

In this way we obtain the Holstein–Primakoff representation of the spin angular-momentum matrices

$$s_{iz} = -s + a_i^+ a_i,$$

$$s_i^+ = \sqrt{2s}\, a_i^+ \left(1 - \frac{a_i^+ a_i}{2s}\right)^{1/2},$$

$$s_i^- = \sqrt{2s} \left(1 - \frac{a_i^+ a_i}{2s}\right)^{1/2} a_i.$$

The meaning of $a_i^+ a_i$ is that it gives the number of spins reversed at point $\boldsymbol{r}_i$.

It is easy to show that the commutation relations (3.35) are satisfied if the a_i, a_i^+ satisfy Bose commutation relations. It is a disadvantage of this representation that the Bose operators act in an infinite-dimensional Hilbert space, while the spin operators act only in a $(2s+1)$-dimensional space, corresponding to the $2s+1$ different possibilities for the spin orientation. Owing to this fact, the roots in the above representation become imaginary if the eigenvalue of $a_i^+ a_i$ exceeds $2s$. It would therefore be necessary to provide the right-hand sides of the representation with projection operators which project into the "smaller" Hilbert space. This is, however, unnecessary at low temperatures, since the spins are only slightly excited and $a_i^+ a_i/2s$ can be considered a small quantity which can be used for series expansions (in the ground state $|0\rangle$ we have $a_i|0\rangle = 0$).

The zeroth order of this expansion corresponds precisely to the spin-wave approximation (see below) as can easily be verified. Moreover, the Holstein–Primakoff representation is suitable when the classical limiting case $s \to \infty$ is considered. For other details we refer to the book by Mattis (1965).

It is reasonable to describe states with several spin waves by a product of the form

$$\prod_k (s_k^+)^{n_k} |0\rangle. \tag{3.46}$$

Such a state contains n_k spin waves with wave vector k. It is easy to see that the state vectors formed in this way are not all linearly independent. The Hilbert space formed by the eigenvectors of the Hamiltonian (3.34) or (3.44) has the dimension $(2s+1)^N$, i.e. there exist $(2s+1)^N$ linearly independent state vectors since each spin has $2s+1$ eigenstates. The number of different $\boldsymbol{k}$ vectors is equal to the number N of spins; since the n_k can take arbitrary positive values, the state vectors formed according to (3.46) are not all linearly independent so that, for example they cannot be orthogonal with respect to one another. Moreover, these states are not eigenstates of (3.44), as one can easily verify by considering the example of two spin waves. This shows that the spin waves must interact with one another. In ferromagnetism the spin-wave approximation means that the states (3.46) are considered to be linearly independent, and, moreover, that they are considered eigenstates of the Hamiltonian. At low temperatures, when only a few spin waves are excited, this will be a good approximation.

In the spin-wave approximation, the mean number of spin-wave quanta (also called magnons) at a temperature T is given according to the Bose–Einstein statistics by

$$\bar{n}_k = \frac{1}{e^{\beta\varepsilon_k}-1}, \qquad \beta = \frac{1}{kT}. \tag{3.47}$$

Since a change of magnetisation by $g\mu_B$ is connected with each excited spin wave, the magnetisation of the system at a temperature T is given by

$$M(T) = M_0 - \sum_k g\mu_B \bar{n}_k \tag{3.48}$$

where $M_0 = Ng\mu_B s$. The mean energy is thus given by

$$E = E_0 + \sum_k \varepsilon_k \bar{n}_k. \tag{3.49}$$

In order to calculate the sums, we must make a definite choice for the exchange interaction. The simplest assumption is to assume only an interaction between nearest neighbours. Let z be the number of nearest neighbours and J the exchange integral between two neighbouring atoms; according to (3.37) we then have

$$J_0 = zJ, \qquad J_k = J\sum_{\delta} e^{-i(k\cdot\delta)}. \tag{3.50}$$

The sum is to be taken over the nearest neighbours of the atom considered; it depends on the lattice structure. When we restrict ourselves to cubic crystals, the position of the nearest neighbour of an atom placed at the origin of a simple cubic (s.c.) crystal is given by the following six vectors: $l(\pm 1, 0, 0)$, $l(0, \pm 1, 0)$, $l(0, 0, \pm 1)$. Here l is the length of an edge of the elementary cell. In the body-centred cubic (b.c.c.) crystal the positions of the eight nearest neighbours are given by the vectors $\frac{1}{2}l(\pm 1, \pm 1, \pm 1)$ and in the face-centred cubic (f.c.c.) crystal by the twelve vectors $\frac{1}{2}l(0, \pm 1, \pm 1)$, $\frac{1}{2}l(\pm 1, 0, \pm 1)$, $\frac{1}{2}l(\pm 1, \pm 1, 0)$. Thus we obtain

$$\begin{aligned} J_0 - J_k &= 2J(3 - \cos lk_x - \cos lk_y - \cos lk_z) \quad \text{s.c.,} \\ J_0 - J_k &= 8J\left(1 - \cos\frac{lk_x}{2}\cos\frac{lk_y}{2}\cos\frac{lk_z}{2}\right) \quad \text{b.c.c.,} \\ J_0 - J_k &= 4J\left(3 - \cos\frac{lk_x}{2}\cos\frac{lk_y}{2} - \cos\frac{lk_y}{2}\cos\frac{lk_z}{2}\right. \\ &\qquad \left. - \cos\frac{lk_x}{2}\cos\frac{lk_z}{2}\right) \quad \text{f.c.c.} \end{aligned} \tag{3.51}$$

The spin-wave energies are smallest for $k = 0$ and increase according to (3.51) almost quadratically with the wave vector. At low temperatures we can therefore restrict ourselves to long wavelengths and expand ε_k in terms of k; according to (3.51) we have

$$\varepsilon_k = g\mu_B H + 2sJl^2k^2 \tag{3.52}$$

for the three cubic lattices. Changing from a sum to an integral

$$\sum_k \to \frac{V}{(2\pi)^3}\int d^3\boldsymbol{k},$$

and using (3.52), we obtain for the magnetisation (3.48) at $H = 0$

$$M(T) = M_0 - g\mu_B \frac{V}{2\pi^2} \int_0^\infty \frac{k^2\, dk}{e^{2s\beta J l^2 k^2} - 1}. \tag{3.53}$$

Here we have already integrated over the angles. The upper limit of the integral over k can be extended to ∞, since at sufficiently low temperatures only the long wavelengths contribute considerably to the integral. We also see that in the one- or two-dimensional case the integral would be divergent, as we would then in the numerator of the integrand have a k^0 or k^1 while the denominator of the integrand is proportional to k^2 as $k \to 0$. This indicates that ferromagnetism can occur only in three-dimensions and not, for example, in layer lattices. This statement can also be given in a more exact form (Mermin and Wagner, 1966). The integral (3.53) can be evaluated:

$$\int_0^\infty \frac{k^2\, dk}{e^{\alpha k^2} - 1} = \sum_{n=1}^\infty \int_0^\infty k^2\, e^{-n\alpha k^2}\, dk = \frac{\sqrt{\pi}}{4} \sum_{n=1}^\infty \frac{1}{(n\alpha)^{3/2}} = \frac{\sqrt{\pi}}{4\alpha^{3/2}} \zeta\left(\frac{3}{2}\right).$$

Here ζ (3/2) is the Riemann zeta function of argument 3/2; it has the approximate value of 2·612. Thus we obtain for the magnetisation

$$M(T) = M_0\left(1 - \frac{1}{\nu \cdot s}\left(\frac{kT}{8\pi s J}\right)^{3/2} \zeta\left(\frac{3}{2}\right)\right). \tag{3.54}$$

The law was first derived by Bloch; ν is the number of atoms per elementary cell and is 1, 2, and 4 for the simple, body-centred, and face-centred cubic lattices, respectively. The quantities V, N, l, and ν are interrelated: $\nu \cdot V = N \cdot l^3$. Bloch's $T^{3/2}$-law is well satisfied particularly in metals with a high Curie point so that, compared with the molecular-field theory which predicts an exponential increase of magnetisation up to its maximum value, we can speak of satisfactory agreement. If we had taken into account higher powers of the wave vector in the spin-wave energy, we would have obtained correction terms to (3.54), whose temperature dependence goes as $T^{5/2}$, $T^{7/2}$, $T^{9/2}$, etc.

The energy of the spin wave is calculated in the same way. From (3.49) we obtain

$$E = E_0 + \frac{3N}{2v}\frac{(kT)^{5/2}}{(8\pi sJ)^{3/2}}\zeta\left(\frac{5}{2}\right).$$

From this we obtain the magnetic contribution to specific heat

$$C_{\mathrm{m}}(T) = \left(\frac{\partial E}{\partial T}\right)_{H=0} = Nk\frac{15}{4v}\left(\frac{kT}{8\pi sJ}\right)^{3/2}\zeta\left(\frac{5}{2}\right). \qquad (3.55)$$

where $\zeta(\frac{5}{2}) = 1{\cdot}341$. A direct proof for the spin-wave concept is again obtained from neutron experiments. The scattered neutrons contain an elastically scattered and an inelastically scattered component. The elastically scattered neutrons still possess their initial energies and have suffered Bragg reflections in the magnetic crystal. The inelastically scattered neutrons have changed their momenta and energies by exciting spin waves and sound waves. By means of the laws of energy and momentum conservation, the dispersion curve of the spin waves, i.e. the dependence of the spin-wave energy on the wave vector, can be checked experimentally by measuring these changes. According to Brockhouse and Sinclair (1960), measurements carried out with a Co–Fe alloy yielded a dispersion curve which is in good agreement with the quadratic function (3.52).

It was not until 1956 that the applicability of the spin-wave approximation at low temperatures was theoretically justified (Dyson, 1956). The most important result is the following: interaction between the spin waves gives rise to correction terms in the expression for the magnetisation; the largest of these terms is proportional to the fourth power of temperature so that it can be neglected against the main term ($\sim T^{3/2}$).

There exist several less correct but simpler ways to take the interaction into account. The simplest way is based upon a linearisation of the equations of motion of the spin operators or upon the double-time Green functions (Zubarev, 1960). The last approach we shall briefly discuss in a general form.

For two arbitrary operators A and B in the Heisenberg representation we define two different Green functions:

a retarded Green function

$$\langle\langle A(t);\ B\rangle\rangle_{\mathrm{r}} = -i\theta(t)\,\{\langle A(t)B\rangle - \eta\langle BA(t)\rangle\}, \tag{3.56}$$

and an advanced Green function

$$\langle\langle A(t); B\rangle\rangle_{\mathrm{a}} = i\theta(-t)\,\{\langle A(t)B\rangle - \eta\langle BA(t)\rangle\}.$$

As in (2.45a), $\theta(t)$ is the step function; the correlation functions $\langle A(t)B\rangle$ and $\langle BA(t)\rangle$ are the thermal averages of the two products evaluated using the density matrix; $B \equiv B(t = 0)$. The quantity η can take the values ± 1; the sign is chosen according to expediency. The equation of motion for the Green functions is obtained from that for the operator $A(t)$,

$$i\dot{A}(t) = [A, \mathscr{H}]_-$$

(here $\hbar = 1$). As $\dot{\theta}(t) = \delta(t)$, both Green functions satisfy the same equation of motion

$$i\langle\langle \dot{A}(t); B\rangle\rangle = \delta(t)\,\{\langle AB\rangle - \eta\langle BA\rangle\} + \langle\langle [A(t), \mathscr{H}]_-; B\rangle\rangle, \tag{3.57}$$

where the new Green function on the right-hand side is generally more complicated than the original one. This equation can be solved when the second Green function is simplified by means of physical arguments and reduced to well-known functions (linearisation), or when the equation of motion of this function is solved in an appropriate way. This method aims at calculating the correlation function considered by means of the Green function obtained in this way. In certain cases, however, the physical quantity sought is itself a Green function such as, for example, the general susceptibility (2.46a) introduced in Section II.4. We then change from the functions to their Fourier transforms:

$$\langle\langle A; B\rangle\rangle_{\mathrm{r,a}} = \int_{-\infty}^{+\infty} dt \langle\langle A(t); B\rangle\rangle_{\mathrm{r,a}}\, \mathrm{e}^{i\omega t}. \tag{3.58}$$

The Fourier transforms are originally functions of the real frequency ω. However, from the definition (3.56), i.e. from the fact that the retarded and advanced functions vanish for $t < 0$ and $t > 0$, respectively, it follows that the Fourier transforms of the two Green functions can be continued analytically to the upper and lower half-planes, respectively, of the complex frequency.

Similarly we change to the Fourier transforms of the correlation functions:

$$\begin{aligned} \langle BA(t)\rangle &= \frac{1}{2\pi}\int_{-\infty}^{+\infty} d\omega J(\omega)\, e^{-i\omega t}, \\ \langle A(t)B\rangle &= \frac{1}{2\pi}\int_{-\infty}^{+\infty} d\omega J'(\omega)\, e^{-i\omega t}. \end{aligned} \tag{3.59a}$$

The inverse transformations are:

$$\begin{aligned} J(\omega) &= \int_{-\infty}^{+\infty} dt \langle BA(t)\rangle\, e^{i\omega t}, \\ J'(\omega) &= \int_{-\infty}^{+\infty} dt \langle A(t)B\rangle\, e^{i\omega t}. \end{aligned} \tag{3.59b}$$

The functions $J(\omega)$ and $J'(\omega)$ are related. To show this, we have to extend the definition of an operator in the Heisenberg representation to complex times, as in Section II.4; for example,

$$A(t-i\beta) = e^{\beta\mathcal{H}} A(t)\, e^{-\beta\mathcal{H}}.$$

Because of the occurrence of the density matrix in the thermal average $\langle A(t)B\rangle$ it is obvious that this correlation function can be continued *analytically* into the lower plane of complex time as long as $\text{Im}\, t \geqslant -\beta$. Thus the average $\langle A(t-i\beta)B\rangle$ is the analytical continuation of $\langle A(t)B\rangle$ into the lower half-plane; since cyclic permutation is allowed under the trace in $\langle \ldots \rangle$, we obtain

$$\langle A(t-i\beta)B\rangle = \langle BA(t)\rangle,$$

and similarly (3.60)

$$\langle A(t)B\rangle = \langle BA(t+i\beta)\rangle.$$

Making use of this property we can write

$$J'(\omega) = \int_{-\infty}^{+\infty} dt \langle BA(t+i\beta)\rangle \, e^{i\omega t}$$
$$= \int_{C} dt \langle BA(t)\rangle \, e^{i\omega t} \, e^{\beta\omega}.$$

The path C is here parallel to the real axis at a distance $i\beta$. Since the continuation of the correlation functions is analytical, it is possible, according to Cauchy's integral theorem, also to integrate along the real axis—that is, we have

$$J'(\omega) = \int_{-\infty}^{+\infty} dt \langle BA(t)\rangle \, e^{i\omega t} \, e^{\beta\omega} = J(\omega) \, e^{\beta\omega}.$$

The spectral function $J(\omega)$ determines the two correlation functions. If, in addition to this, the two operators A and B are Hermitean conjugates with respect to one another ($A^+ = B$), the spectral density will be real:

$$J^*(\omega) = \int_{-\infty}^{+\infty} dt \langle B(t)A\rangle \, e^{-i\omega t} = \int_{-\infty}^{+\infty} dt \langle B(-t)A\rangle \, e^{i\omega t}$$
$$= \int_{-\infty}^{+\infty} dt \langle BA(t)\rangle \, e^{i\omega t} = J(\omega).$$

The relation linking the spectral density with the Green functions is obtained from the difference of the retarded and advanced functions at the point ω on the real axis. From (3.58) and (3.56) we obtain

$$\begin{aligned} &\langle\langle A; B\rangle\rangle_r - \langle\langle A; B\rangle\rangle_a \\ &\quad = -i \int_{-\infty}^{+\infty} dt \{\theta(t) + \theta(-t)\} \{\langle A(t)B\rangle - \eta\langle BA(t)\rangle\} \, e^{i\omega t} \qquad (3.61) \\ &\quad = -iJ(\omega) \, (e^{\beta\omega} - \eta), \end{aligned}$$

since $\theta(t) + \theta(-t) = 1$. Hence we can calculate at once the spectral density in the "Fermi case" ($\eta = +1$); in the "Bose case" ($\eta = -1$) this cannot be done in an unambiguous way as both $J(\omega)$ and $J(\omega) + c\delta(\omega)$ satisfy relation (3.61).

From (3.59a) we see that the constant factor of the δ function depends on the asymptotic behaviour at large times. If the correlation function vanishes at large times, the spectral density has no singularity at $\omega = 0$ (Lucas and Horwitz, 1969; Kwok and Schultz, 1969).

It is expedient to consider the two Green functions defined above as a *single* function in the complex frequency plane, a function which is discontinuous at the real axis where it has a branch point. Such an analytical function which is everywhere analytic except along the real axis, is determined unambigously by the discontinuity (jump) on the real axis if the Green function and the spectral density are assumed to tend to zero for large ω (Muschelischwili, 1965). For the Green functions used in physics this assumption is fulfilled (cf. Abrikosov *et al.*, 1965). It is easy to verify that the function

$$\langle\langle A; B\rangle\rangle = \frac{1}{2\pi}\int_{-\infty}^{+\infty} d\omega' \, \frac{J(\omega')\,(e^{\beta\omega'}-\eta)}{\omega-\omega'}$$

possesses the necessary analytical properties and discontinuity characteristics. When Im $\omega > 0$ it is therefore identical with the retarded Green function (and when Im $\omega < 0$ with the advanced Green function). As already mentioned, the spectral density is real in all physically important cases. We then obtain from the above formula a relation between the real and imaginary parts of a Green function on the real axis, known as the Kramers–Kronig relation for the susceptibility. For this purpose we use the well-known identity

$$\lim_{\varepsilon\to 0}\frac{1}{\omega-\omega'+i\varepsilon} = \frac{\mathrm{P}}{\omega-\omega'} - i\pi\,\delta(\omega-\omega') \qquad (\mathrm{P} = \text{principal value}),$$

and obtain for the retarded Green function on the real axis

$$\mathrm{Re}\,\langle\langle A; B\rangle\rangle_{\mathrm{r}} = \frac{1}{2\pi}\,\mathrm{P}\int_{-\infty}^{+\infty} \frac{J(\omega')\,(e^{\beta\omega'}-\eta)}{\omega-\omega'}\,d\omega',$$

$$\mathrm{Im}\,\langle\langle A; B\rangle\rangle_{\mathrm{r}} = -\tfrac{1}{2}J(\omega)\,(e^{\beta\omega}-\eta),$$

and therefore

$$\mathrm{Re}\,\langle\langle A; B\rangle\rangle_r = \frac{\mathrm{P}}{\pi}\int\limits_{-\infty}^{+\infty} \frac{\mathrm{Im}\,\langle\langle A; B\rangle\rangle_r}{\omega'-\omega}\,d\omega'.$$

When applied to the general dynamic susceptibility (2.45b), which is a retarded Green function, this Kramers–Kronig relation indicates that the real part is determined by the dissipative part of the susceptibility, i.e. the imaginary part.

As regards the application to ferromagnetism, the type of Green function to be used can be derived from the above considerations on spin waves. We may ask what happens with a spin wave between $t = 0$ (when it is produced in the state $|\varphi\rangle$) and the time t, i.e. we are interested in the projection of $s_k^+ |\varphi\rangle$ on $s_k^+(t)|\varphi\rangle$:

$$\langle s_k^+(t)\varphi\,|\,s_k^+\varphi\rangle = \langle\varphi\,|\,s_k^-(t)s_k^+\,|\varphi\rangle.$$

In this way we are led to consider the Green function $\langle\langle s_k^-(t);\, s_k^+\rangle\rangle$. Its equation of motion is obtained from (3.39) and (3.57) with $\eta = +1$:

$$i\langle\langle \dot{s}_k^-(t);\, s_k^+\rangle\rangle = -\frac{2}{N}\,\delta(t)\,\langle s_0^z\rangle + \langle\langle [s_k^-(t), \mathscr{H}]_-;\, s_k^+\rangle\rangle$$

($\hbar = 1$). Using the commutator analogous to (3.44) we have

$$i\langle\langle \dot{s}_k^-(t);\, s_k^+\rangle\rangle = -\frac{2}{N}\,\delta(t)\,\langle s_0^z\rangle + g\mu_B H\langle\langle s_k^-(t);\, s_k^+\rangle\rangle$$
$$+2\sum_{k'}(J_{k'} - J_{k-k'})\,\langle\langle s_{k-k'}^z(t)\, s_{k'}^-(t);\, s_k^+\rangle\rangle.$$

The simplest method to decouple the more complicated Green function on the right-hand side, i.e. to linearize the equation of motion for s_k^-, is based on the replacement of $s_{k-k'}^z(t)$ by its thermal average; this can be done since at low temperatures the spins are almost entirely parallel to one another. Since according to (3.37)

$$\langle s_{k-k'}^z\rangle = \frac{1}{N}\sum_{i=1}^{N}\langle s_{iz}\rangle\, e^{i([k-k']\cdot r_i)},$$

we have

$$\langle s^z_{k-k'}(t)\rangle = -s\sigma\,\delta_{k,k'} \quad \text{where} \quad \langle s_{iz}\rangle = -s\sigma,$$

because $\langle s_{iz}\rangle$ is independent both of position and of time. The quantity σ denotes the relative magnetisation.

In this way, after having changed to the Fourier transform, we can write the equation of motion in the so-called random-phase approximation (RPA):

$$(\omega-\tilde{\varepsilon}_k)\,\langle\langle s_k^-;\, s_k^+\rangle\rangle = \frac{2s}{N}\,\sigma, \tag{3.62}$$

where

$$\tilde{\varepsilon}_k = g\mu_B H + 2s\sigma(J_0 - J_k).$$

Comparison with (3.45) shows that the excitations of the spin system are obtained as the poles of the Green function and that here these poles lie on the real axis at $\omega = \tilde{\varepsilon}_k$. The undamped renormalised magnons obtained in this way are identical with the interaction-free magnons (3.45) as $T\to 0$ or $\sigma\to 1$. The spectral density is obtained from (3.61) and (3.62):

$$J(\omega) = 2\pi\,\frac{2s\sigma}{N}\,\bar{n}_k\,\delta(\omega-\tilde{\varepsilon}_k),$$

with

$$\bar{n}_k = \frac{1}{e^{\beta\tilde{\varepsilon}_k}-1}.$$

(The singular part of $J(\omega)$ must be vanishing since $\lim\,\langle s_k^-(t)s_k^+\rangle = 0$ if $k \neq 0$.) Thus we obtain for the simultaneous correlation function according to (3.59)

$$\langle s_k^+ s_k^-\rangle = \frac{2s\sigma}{N}\,\bar{n}_k\,.$$

Changing to coordinate space using (3.36) or (3.37) we find

$$\langle s_i^+ s_j^-\rangle = \frac{2s\sigma}{N}\sum_k \bar{n}_k\, e^{-i(k\cdot[r_i-r_j])}.$$

Because of the translational invariance of the spin system the expectation value with $i = j$ will be independent of position and we have

$$\langle s^+ s^- \rangle = \frac{2s\sigma}{N} \sum_k \bar{n}_k \,.$$

In order to obtain the magnetisation, we have to connect $\langle s^+ s^- \rangle$ with $\langle s_z \rangle$. For simplicity we shall only consider the $s = \frac{1}{2}$ case; we then obtain from $s^+s^- + s^-s^+ + 2s_z^2 = 2s(s+1)$ and $s_z^2 = \frac{1}{4}$ together with (3.35)

$$s^+ s^- = s_z + \tfrac{1}{2},$$

so that

$$\sigma = 1 - 2\langle s^+ s^- \rangle.$$

Substituting the above expression for $\langle s^+ s^- \rangle$ and using the relation $M(T) = M_0 \sigma$ we obtain for the magnetisation

$$M(T) = M_0 - 2M(T) \frac{1}{N} \sum_k \frac{1}{e^{\beta \bar{\varepsilon}_k} - 1} \,. \tag{3.63}$$

For the case of arbitrary spin quantum numbers we refer to the literature (Sauter, 1963; Callen, 1963; Tahir-Kheli and ter Haar, 1962; Tyablikov, 1969).

To calculate the magnetisation we proceed in the same way is in the case of free spin waves. Interactions give rise to a correction term $\sim T^3$ which does not appear in the exact calculation. Bogolyubov and Tyablikov (1959) were first to derive (3.63). The fact that the renormalised spin wave energies contain the temperature-dependent magnetisation represents the most essential difference between (3.63) and (3.48) (see, for example, Englert, 1960). Obviously one may expect this or a similar dependence owing to interactions since at higher temperatures the spin waves must be excited more easily so that it must be easier to change the signs of the spins.

Considering the rough linearisation, we can understand the discrepancy with Dyson's temperature dependence of magnetisation. In the low-temperature range Dyson's result is easy to obtain in another way by means of a simple approximation (Oguchi, 1960; Marshall and Murray, 1968). The linearisation, however, yields a magnetisation formula which can be applied in the whole temperature range, although we know that linearisation will be a poor approximation at high temperatures since the fluctuations of s_z are intensified as the temperature rises.

Near the Curie point σ is small and the power function in (3.63) can be expanded in a series. For $H = 0$ we obtain from it the Curie temperature

$$\beta_c = \frac{1}{kT_c} = \frac{2}{N} \sum_k \frac{1}{J_0 - J_k}. \tag{3.64}$$

In the case of arbitrary spin quantum numbers we have to put the factor $\frac{3}{4}\, s(s+1)$ in front of the sum. If we restrict ourselves to interactions between nearest neighbours, eqns. (3.51) will hold true for the three cubic lattice types; as above we change from sums to integrals. The integrals obtained can be reduced to elliptic functions (cf. Magnus and Oberhettinger, 1948) and we arrive at the result

$$\frac{2kT_c}{zJ} = \begin{cases} 0{\cdot}660 & \text{s.c.} \\ 0{\cdot}718 & \text{b.c.c.} \\ 0{\cdot}744 & \text{f.c.c.} \end{cases}$$

The corresponding numerical value would be equal to 1 if we had used the molecular-field approximation. Considering the spontaneous magnetisation we obtain from (3.63) the same behaviour as in the molecular-field approximation (3.18). There also exists an agreement between the asymptotic paramagnetic Curie point and the paramagnetic or ferromagnetic Curie point (3.20) of the molecular-field approximation, since the assumption of independent spin alignment in the magnetic field of the molecular-field approximation is correct as $T \to \infty$. Thus (3.19) or (3.20) will always represent the correct paramagnetic Curie temperature for a Heisenberg-type ferromagnet. Since there is a considerable difference between the paramagnetic and the ferromagnetic (3.64) Curie temperatures, the reciprocal susceptibility will not be a linear function of the temperature as in the molecular-field approximation. After some calculations we obtain from (3.63)

$$\chi^{-1} \sim (T-T_c)^2$$

for the susceptibility near the Curie point. As to further applications of the Green function technique we refer to the literature (Tyablikov, 1969).

For ferromagnets the spin-wave approximation is a very good approximation, while for antiferromagnetic and ferrimagnetic substances it is subject to restrictions. On the one hand, a transfer of the spin-wave concept to antiferromagnetism is not obvious since, in contrast to the case of ferromagnetism, the antiferromagnetic ground state is unknown. If we apply the sublattice model to antiferromagnetism (a model suggested by neutron scattering experiments), one considers the magnetisation, i.e. the total spin of the sublattices, as a constant of motion. It is, however, obvious that it is not the sublattice spin, but the total spin of all sublattices, which commutes with the Hamiltonian. The perfectly ordered ground state with antiparallel spins in the sublattices will therefore not be the correct ground state. However, it can be shown that the deviations of the correct ground state from the ordered state are small, so that the sublattice model is a good approximation for considering the ground state and the excited states (Nagamiya *et al.*, 1955; van Kranendonk and Van Vleck, 1958; Keffer, 1966).

The simplest model for an antiferromagnet is that of two sublattices with antiferromagnetic coupling between nearest neighbours. In order to stabilise the ground state, we take into account the anisotropy energy, which is formally described in the Hamiltonian by an effective internal field H_A. If the external field is parallel to the internal field, the Hamiltonian will be

$$\mathscr{H} = -J\sum_{i,j}^{N/2} (\boldsymbol{s}_i \cdot \boldsymbol{s}_j) + g\mu_B(H+H_A)\sum_{i}^{N/2} s_{iz} + g\mu_B(H-H_A)\sum_{j}^{N/2} s_{jz} \quad (3.65)$$

where $J < 0$. The subscript i labels the spins of the first sublattice, j those of the second sublattice. In the exchange term of (3.65) summation is taken only over the nearest-neighbour atoms. The sign of the effective anisotropy field is chosen so that the first sublattice has its moments parallel to the z-axis, and the second sublattice has them in an antiparallel orientation. At $T = 0$, the field H_A is connected with the anisotropy energy K introduced above by the simple relation $K = Ng\mu_B sH_A$. The field H_A can be of the order of 10^2 to 10^3 oe. Again we introduce Fourier-transformed operators

$$\begin{aligned} s_i^{\pm} &= \sum_{k'} s_{1k}^{\pm} \, e^{\mp i(k\cdot r_i)}, \qquad s_{iz} = \sum_k s_{1k}^z \, e^{-i(k\cdot r_i)} \\ s_j^{\pm} &= \sum_k s_{2k}^{\pm} \, e^{\mp i(k\cdot r_j)}, \qquad s_{jz} = \sum_k s_{2k}^z \, e^{-i(k\cdot r_j)}. \end{aligned} \tag{3.66}$$

The Hamiltonian (3.65) can now be written in the form

$$\begin{aligned} \mathcal{H} = {} & \frac{N}{2} g\mu_B(H+H_A)s_{10}^z + \frac{N}{2} g\mu_B(H-H_A)s_{20}^z \\ & -N\sum_k J_k\left(s_{1k}^z s_{2-k}^z + \frac{1}{2} s_{1k}^+ s_{2k}^- + \frac{1}{2} s_{1k}^- s_{2k}^+\right). \end{aligned} \tag{3.67}$$

The quantity J_k is again defined by (3.50). The commutation relations are governed by (3.39), if N is replaced by $N/2$ for each sublattice.

The simplest way to determine the spin-wave energies is again based on the equations of motion. In order to obtain just the spin-wave frequencies we need not introduce the Green functions; it is sufficient to consider the suitably linearised equations of motion for s_k^+ or s_k^- of the two sublattices and to determine their eigen solutions (i.e. the spin waves). Using

$$[\mathcal{H}, s_{1k}^+]_- = g\mu_B(H+H_A)s_{1k}^+ + 2\sum_{k'} J_{k'}(s_{1k-k'}^z s_{2k'}^+ - s_{2k'}^z s_{1k-k'}^+),$$

$$[\mathcal{H}, s_{2k}^+]_- = g\mu_B(H-H_A)s_{2k}^+ + 2\sum_{k'} J_{k'}(s_{2k-k'}^z s_{1k'}^+ - s_{1k'}^z s_{2k-k'}^+)$$

we obtain the following equations of motion:

$$\begin{aligned} i\hbar\dot{s}_{1k}^+ &= -g\mu_B(H+H_A)s_{2k}^+ + 2\sum_{k'} J_{k'}(s_{2k'}^z s_{1k-k'}^+ - s_{1k-k'}^z s_{2k'}^+), \\ i\hbar\dot{s}_{2k}^+ &= -g\mu_B(H-H_A)s_{2k}^+ + 2\sum_{k'} J_{k'}(s_{1k'}^z s_{2k-k'}^+ - s_{2k-k'}^z s_{1k'}^+). \end{aligned} \tag{3.68}$$

At low temperatures this coupled system can again be linearised. Since the true ground state agrees approximately with the ordered state, we can put to a rough approximation (neglecting interactions between the spin waves):

$$s_{1k}^z = -s\delta_{k,0}, \qquad s_{2k}^z = s\delta_{k,0}.$$

This simplifies the equations of motion, which now read

$$\begin{aligned} i\hbar\dot{s}_{1k}^+ &= -g\mu_B(H+H_A)s_{1k}^+ + 2s(J_0 s_{1k}^+ + J_k s_{2k}^+), \\ i\hbar\dot{s}_{2k}^+ &= -g\mu_B(H-H_A)s_{2k}^+ - 2s(J_0 s_{2k}^+ + J_k s_{1k}^+). \end{aligned} \tag{3.69}$$

The equations of motion for s_{1k}^- and s_{2k}^- are obtained by changing to the Hermitean conjugates so that they can be dropped. The equations (3.69) are solved by means of the substitution $s_k^+(t) = s_k^+(0) \exp(i\varepsilon_k t/\hbar)$. From the condition that the determinant of the set of equations must vanish, we obtain the two excitation energies for the free spin waves in the antiferromagnetic:

$$\varepsilon = \sqrt{g\mu_B H_A(g\mu_B H_A + 4s|J_0|) + 4s^2(J_0^2 - J_k^2)} \pm g\mu_B H. \tag{3.70}$$

Thus there exist two spin-wave branches which are degenerate at $H = 0$. The energy has its minimum at $k = 0$ (long wavelengths).† The energy spectrum (3.70) differs essentially from the ferromagnetic spectrum (3.45). On the one hand, the field-strength dependence of the second branch, ε_{k-}, makes it possible that the spin-wave energies vanish. With $k = 0$ and $H_E = 2s|J_0|/g\mu_B$ the critical field strength is given by

$$H_c = \sqrt{H_A(H_A + 2H_E)}.$$

The anisotropy field H_A is of the order of about 10^3 oe and the "Weiss" field H_E is of the order of 10^6 oe, so that H_A can be neglected compared to H_E. This critical field H_c is identical with the critical field strength for spin flop (3.33), if $\chi_{||}$ is neglected compared to $\chi_\perp$ and if we put $K = Ng\mu_B sH_A$; this can be done if the temperatures are low enough. The vanishing of the spin-wave energy means an in-

† In sufficiently weak external fields.

stability of the system which, microscopically, results in the possible appearance of excitations of arbitrarily high order and vanishing energy. Of course, the spin-wave approximation is then no longer reasonable. Comparing this case with that of ferromagnets, we see that there the external field prevents the appearance of such an instability at $k = 0$ (see eqn. (3.45)).

The energy gap between the ground state and the minimum of (3.70) at $H = 0$ and $k = 0$ is a characteristic feature of the spin-wave spectrum of an antiferromagnetic substance. The energy gap amounts to

$$\varepsilon_0 = g\mu_B \sqrt{H_A(H_A+2H_E)}\,;$$

unlike for ferromagnets, where it depends only on the anisotropy field H_A, it is here depending on the product of the molecular field and the anisotropy field and therefore it may reach considerable values. This means that at sufficiently low temperatures, i.e. when $\beta\varepsilon_0 \ll 1$, the susceptibility, the magnetisation of the sublattices, and the specific heat depend exponentially on the temperature ($\sim \exp(-\beta\varepsilon_0)$). Measurements of the magnetisation through the frequency shift in the case of ^{19}F nuclear resonance in antiferromagnetic MnF_2 seem to prove this dependence (Jaccarino and Walker, 1959; Walker, 1963; cf. Section III.6).

The spin waves are again considered independent of one another in the sense of the spin-wave approximation. The average number of magnons with wave vector $\boldsymbol{k}$ will then be for the two branches given by

$$\bar{n}_{k+} = \frac{1}{e^{\beta\varepsilon_{k+}}-1}, \qquad \bar{n}_{k-} = \frac{1}{e^{\beta\varepsilon_{k-}}-1}, \tag{3.71}$$

and the average energy is

$$E = E_0 + \sum_{k} \varepsilon_{k+}\bar{n}_{k+} + \sum_{k-} \varepsilon_{k-}\bar{n}_{k-}\,. \tag{3.72}$$

E_0 is the energy of the ground state. The total magnetisation can be calculated as in the ferromagnetic case. A calculation similar to that which yielded (3.63) leads us to

$$M = g\mu_B \sum_k (\bar{n}_{k+} - \bar{n}_{k-}). \tag{3.73}$$

This is plausible since the excitation of a magnon with the energy $\varepsilon_{k\pm}$ according to (3.70) entails a change in magnetisation by $\pm g\mu_B$. Expanding (3.73) up to the term linear in H, we obtain for the susceptibility $\chi_{||}$:

$$\chi_{||} = 2\beta g^2 \mu_B^2 \sum_k \frac{e^{\beta\varepsilon_k}}{(e^{\beta\varepsilon_k} - 1)^2}, \tag{3.74}$$

where ε_k is the spin-wave energy at $H = 0$ according to (3.70). If $\beta\varepsilon \gg 1$, summation in (3.72) and (3.74) can be carried out easily, when we restrict ourselves to long wavelengths and ε_k is expanded in terms of k. In this way we obtain the exponential temperature dependence of $\chi_{||}$ and C_m. A more exact treatment of the integrals is complicated by the presence of the root in (3.70) (Eisele and Keffer, 1954; Joenk, 1962). At moderately low temperatures, when $\beta\varepsilon_0 \ll 1$, the anisotropy can be neglected. If we also restrict ourselves to the case of long wave lengths and consider cubic lattices at $H = 0$, we have, according to (3.51) and (3.70),

$$\varepsilon_k = 2s\sqrt{J_0^2 - J_k^2} \cong 2s|J|\sqrt{2zl}k. \tag{3.75}$$

Thus ε_k is a linear function of the magnitude of the wave vector, in contrast to the case of ferromagnetic substances (3.52). The spin waves in an antiferromagnetic substance will therefore behave similar to sound waves in an ordinary solid. In particular, the linear dependence results in a T^3-law for the specific heat. Using (3.75) we obtain for the average energy

$$E = E_0 + \frac{V}{\pi^2} \frac{(kT)^4}{(2s|J|\sqrt{2zl})^3} \int_0^\infty \frac{x^2\,dx}{e^x - 1}$$

$$= E_0 + V \frac{6}{\pi^2} \frac{(kT)^4}{(2s|J|\sqrt{2zl})^3} \cdot \zeta(4).$$

$\zeta(4)$ is the Riemann zeta function of argument 4; $\zeta(4) = \pi^4/90$. From this we obtain for the specific heat

$$C_m(T) = \left(\frac{\partial E}{\partial T}\right)_{H=0} = \frac{V}{l^3} k \frac{4\pi^2}{15} \frac{(kT)^3}{(2s\,|J|\,\sqrt{2z})^3}. \qquad (3.76)$$

The corresponding calculation for the susceptibility yields

$$\chi_{||}(T) = \frac{V}{l^3} \frac{g^2 \mu_B^2 (kT)^2}{3(2s\,|J|\,\sqrt{2z})^3}. \qquad (3.77)$$

Here V is the sublattice volume. For two face-centred cubic lattices (NaCl structure) which penetrate one another we have $V/l^3 = N$ and $z = 6$, for two mutually penetrating simple cubic sublattices (CsCl structure) $V/l^3 = N/2$ and $z = 8$. In contrast to the molecular-field approximation (3.31), the spin-wave approximation yields the result that the specific heat and the susceptibility vary as powers of the temperature T. Equations (3.76) and (3.77) should apply to a temperature range of $\varepsilon_0 \ll kT \ll |J|$; experimentally eqn. (3.77) is partly satisfied (MnF_2). In the spin-wave approximation no important deviation of the transverse susceptibility $\chi_\perp$ from the temperature-independent value (3.32) is obtained (cf. Kubo, 1952).

No difficulties arise when the methods of the spin-wave approximation are applied to ferrimagnetism. When we again assume only an antiferromagnetic interaction between the nearest-neighbour atoms of two different sublattices, the linearised equations of motion (3.69) hold true, with the only exception that we have different internal fields H_{A1} and H_{A2}, different g factors, g_1, g_2, and different spin quantum numbers, s_1, s_2, for the two sublattices. For the spin-wave excitation energy we then obtain

$$\begin{aligned} \varepsilon_{k\pm} &= \tfrac{1}{2}\sqrt{\{(g_1-g_2)\mu_B H + (g_1 H_{A_1} + g_2 H_{A_2})\mu_B + 2\,|J_0|\,(s_1+s_2)\}^2 - 16 s_1 s_2 J_k^2} \\ &\quad \pm \tfrac{1}{2}\{(g_1+g_2)\mu_B H + (g_1 H_{A_1} - g_2 H_{A_2})\mu_B + 2\,|J_0|\,(s_2 - s_1)\}. \end{aligned} \qquad (3.78)$$

Thus we again have two spin wave branches which are non-degenerate for $H = 0$. The energies have their minima at $k = 0$. Compared with antiferromagnetism, the external field and the anisotropy fields play a relatively small role. Neglecting them, we obtain from (3.78)

$$\varepsilon_{k\pm} = \sqrt{J_0^2(s_1-s_2)^2+4s_1s_2(J_0^2-J_k^2)} \pm |J_0|(s_2-s_1). \qquad (3.79)$$

As $k \to 0$ the energy of one of the two branches tends to the value $2|J_0|(s_2-s_1)$, $(s_2 > s_1)$; this means that at low temperatures this branch is usually not excited. The second branch tends to zero proportional to the square of k (as $k \to 0$). According to (3.79) and (3.51) we have

$$\varepsilon_k = \frac{4s_1s_2}{s_2-s_1}|J|lk^2 \qquad (3.80)$$

in the case of cubic structures and long wave lengths. Thus the ferrimagnet will at low temperatures behave like a ferromagnet with an exchange interaction which is changed according to (3.80). For the magnetisation and the specific heat we have in particular a $T^{3/2}$-law.

As regards more detailed representations of the spin-wave approximation, we refer the reader to the literature (Nagamiya, Yosida, and Kubo, 1955; van Kranendonk and Van Vleck, 1958; Keffer, 1966; Walker, 1963).

The problem of interactions between the spin waves in antiferromagnetic and ferrimagnetic substances has been dealt with only sporadically (Lines, 1964; Brooks Harris, 1969).

4. High Temperatures

At high temperatures the interaction between the spin waves will play a decisive role. The ideas upon which the spin-wave theory is based, which assume undamped excited states which behave like quasi-particles of an infinitely long lifetime, are certainly wrong (Marshall and Murray, 1968; Brooks Harris, 1968). The treatment of spin-wave interactions which we discussed above is therefore reasonable only at low temperatures, and its extrapolation up to the Curie point and beyond is hardly justified. Nonetheless, the Curie points calculated in this approximation are in relatively good agreement with the values obtained in another way; see Table 14.

There does not exist a method of describing ferromagnetism near the Curie point which is as good as the spin-wave approximation. It is possible, however, to apply Landau's theory of second-order phase transitions (Landau and Lifshitz, 1965) to magnetic transformations. It is a phenomenological theory and in certain respects equivalent to the molecular-field approximation. On the other hand, with the exception of certain special cases, it is only applicable to the temperature range near the transition point and, owing to its phenomenological nature, it is independent of the model used to describe magnetism. The theory is based upon the following assumptions: in a magnetic transition a solid with a high symmetry (paramagnetic) changes to one with a lower symmetry (ferromagnetic); the state of lower symmetry can be described by one or several symmetry-violating order parameters (spontaneous magnetisation). According to Landau it is a decisive criterion for second-order phase transitions that the order parameter is a continuous function of temperature vanishing at the transition point T_c. In contrast to this, in a typical first-order phase transition (solid–gas) the distance between the molecules, i.e. the volume, varies discontinuously at T_c while the symmetry remains unchanged. Another example of a second-order phase transition is the transition from the normal to the superconducting state, where the pair potential plays the role of the order parameter. The thermodynamic potential, which is to describe the magnetic system, is the free enthalpy $G(p, T)$ as p and T are generally the quantities which are fixed. (If volume effects can be ignored, the free energy $F(V, T)$ can also be used.) Within the framework of Landau's theory the magnetisation is not considered to be a thermodynamic variable, but taken as a parameter occurring in the theory. For constant p and T, the quantity G will therefore be stationary, i.e. $\delta G = 0$ when one or several parameters are varied. In thermodynamic equilibrium G must therefore be an extremum, more exactly a minimum, with respect to the order parameters. This condition defines the parameters as functions of temperature, pressure, and field strength H. Since in second-order phase transitions the variables and the thermodynamic quantities of state are continuous, the potential can near the transition point be written as a power series in the small order parameters. In Landau's theory the potentials are thus assumed to be analytic functions of the parameters. For the sake of simplicity we shall consider an isotropic ferromagnet. The spontaneous magnetisation M will then be the order parameter, and we obtain for G

$$G(p, T; H = 0) = G_0(p, T) + a(p, T)M_s^2 + b(p, T)M_s^4 + \ldots,$$

or, if $H \neq 0$,

$$G(p, T; H) = G_0(p, T) + a(p, T)M^2 + b(p, T)M^4 - (\boldsymbol{M} \cdot \boldsymbol{H}) + \ldots .$$

Here $\boldsymbol{M}$ is the magnetisation, and G_0 is the free enthalpy of the paramagnetic system. As G is a scalar, the magnetisation cannot appear in odd powers except together with the field vector $\boldsymbol{H}$ in the form of a scalar product; the three components of $\boldsymbol{M}$ must appear in a symmetric form as we assumed the solid to be isotropic. The parameters $\boldsymbol{a}$ and b are phenomenological quantities and may depend on pressure and temperature. Near T_c the terms of higher order in M are small and can be neglected. In the case of weak ferromagnetism (Section III.5), i.e. $M \ll M_0$, the theory can also be extended beyond the high-temperature range since M remains small also at low temperatures.

For the ferromagnetic phase to be the stable phase when $T < T_c$ and the paramagnetic phase ($M_s = 0$) to be the stable state when $T > T_c$, the function $a(p, T)$ must possess the following properties:

$$a(p, T) = \begin{cases} > 0, & \text{when} \quad T > T_c; \\ = 0, & \text{when} \quad T = T_c; \\ < 0, & \text{when} \quad T < T_c. \end{cases}$$

The function $b(p, T)$, however, must be positive as otherwise, at higher values of M_s, the paramagnetic phase would become unstable when $T > T_c$. Since the quantity b is not a critical function of the temperature, we may consider b to be constant near T_c.

It is easy to verify that, with this choice of signs, G will have the necessary minimum properties:

$$\left(\frac{\partial G}{\partial M}\right)_{p,\,T} = 0 \quad \text{or} \quad a(p,T)M + 2bM^3 - H = 0,$$

and

$$\left(\frac{\partial^2 G}{\partial M^2}\right)_{p,\,T} > 0 \quad \text{or} \quad a(p,T) + 6bM^2 > 0.$$

Here we make use of the fact that in an isotropic system $\boldsymbol{M}$ is parallel to $\boldsymbol{H}$. When $H = 0$ we obtain from these relations the following two solutions for the spontaneous magnetisation:

$$M_s = 0, \qquad \text{when} \quad T > T_c;$$

$$M_s^2 = -\frac{a(p,T)}{2b} \qquad \text{when} \quad T < T_c.$$

In the case of $H \neq 0$ we put $M = M_s + \chi H$ and obtain from $\partial G/\partial M = 0$ (taking into account only the terms linear in H):

$$\chi = \frac{1}{2a(p,T)}, \qquad \text{when} \quad T > T_c;$$

$$\chi = -\frac{1}{4a(p,T)}, \qquad \text{when} \quad T < T_c.$$

Thus the two phenomenological functions a and b can be expressed in terms of measurable quantities; in the case of, say, $T < T_c$ the potential G will be given by

$$G(p,T;H) = G_0(p,T) - \frac{1}{4\chi}M^2 + \frac{1}{8\chi M_s^2}M^4 - (\boldsymbol{M}\cdot\boldsymbol{H}).$$

According to Landau, the function $a(p,T)$ near the Curie point T_c can be expanded in a power series of $T - T_c$; breaking off after the first term, we have

$$a(p,T) = \alpha(p)(T - T_c); \qquad \alpha(p) > 0.$$

Thus we have for M_s and χ the same temperature dependence as in the molecular-field approximation: (3.18) and (3.21), respectively.

As is well known, the specific heat is obtained from the entropy $S = -(\partial G/\partial T)_p$:

$$S(p, T; H) = S_0(p, T) - \alpha M^2, \quad \text{and}$$

$$C_m(T) = \left(\frac{\partial S}{\partial T}\right)_{p,\,H=0} = C_{m0} + \begin{cases} 0 & \text{when} \quad T > T_c; \\ \dfrac{\alpha^2}{2b} & \text{when} \quad T < T_c. \end{cases}$$

Here S_0 is the entropy and C_{m0} the specific heat of the paramagnetic phase. The discontinuity of the specific heat at T_c is analogous to (3.22). This kind of singularity is characteristic of Landau's theory, but not of second-order phase transitions, as we can see from the two-dimensional Ising model with its logarithmic divergence.

The fact that Landau's theory agrees with the molecular-field approximation at $T \cong T_c$ is not accidental: both approximations describe the ordered phase by a macroscopic quantity (the magnetisation), and neglect both fluctuations which are important especially at the critical point, as well as microscopic effects such as short-range order, etc. On the other hand, the fact that the theory has no microscopic features makes it easy to understand its applicability to both the Heisenberg type of ferromagnetism of non-conducting substances (Section III.2) and to the band ferromagnetism of metals (Section III.5).

Effects due to anisotropy and magnetostriction (Bjelov, 1961; Wohlfarth, 1969) are easy to account for in Landau's theory.

Let us consider the simplest anisotropic case of a ferromagnet with uniaxial anisotropy; we obtain at once the following expansion for the free enthalpy:

$$G(p, T; H) = G_0(p, T) + a_1 M_\perp^2 + a_2 M_z^2 + \\ + b_1 M_\perp^4 + b_2 M_\perp^2 M_z^2 + b_3 M_z^4 - (\boldsymbol{M} \cdot \boldsymbol{H}).$$

Here M_z and $M_\perp = \sqrt{Mx_x^2 + My_y^2}$ are the two order parameters. The form of G is obtained from the requirement for G to be invariant under all transformations of the symmetry group of the crystal formed by the magnetic moments (Landau and Lifshitz, 1965; Dzyaloshinskii, 1958). This is similar to the case of the symmetry properties of the crystal potential in Section II.5, with the only exception that here the symmetries refer to the magnetic moments (axial vectors) and there they concern the charges (scalars). The minimum property of G,

$$\frac{\partial G}{\partial M_\perp} = 0, \qquad \frac{\partial G}{\partial M_z} = 0,$$

and
$$\frac{\partial^2 G}{\partial M_\perp^2} > 0, \quad \frac{\partial^2 G}{\partial M_z^2} > 0, \quad \frac{\partial^2 G}{\partial M_\perp^2}\cdot\frac{\partial^2 G}{\partial M_z^2} - \left(\frac{\partial^2 G}{\partial M_1\,\partial M_z}\right)^2 > 0,$$

becomes for $H = 0$

$$(a_1 + b_2 M_z^2 + 2b_1 M_\perp^2)^2 M_\perp = 0,$$
$$(a_2 + b_2 M_\perp^2 + 2b_3 M_z^2) M_z = 0.$$

In addition to this we have the inequalities. As above we again assume the positive quantities b_ν to be constant. Apart from the paramagnetic solution we have another two, ferromagnetic, solutions (a_1, $a_2 < 0$)

$$M_\perp = 0, \quad M_z^2 = -\frac{a_2(p, T)}{2b_3}, \quad \text{when} \quad T < T_c,$$

and

$$M_z = 0, \quad M_\perp^2 = -\frac{a_1(p, T)}{2b_1}, \quad \text{when} \quad T < T_c.$$

Substituting this in G we see that the first solution is the more stable one, if $a_2^2 b_1 > a_1^2 b_3$; the z-axis will then be the axis of easy magnetisation. The procedure to be applied in the case of antiferromagnetism or ferrimagnetism is obvious. The results obtained are the same as in the molecular-field approximation (see, for example, Borovik–Romanov, 1968).

It is, of course, impossible in this way to obtain the spiral structure mentioned above, which can be observed in the case of the rare-earth elements (Belov *et al.*, 1964; Bertaut, 1963; Turov, 1968), since the anisotropic ferromagnet considered is still translationally invariant. In order to account also for this case, we have to consider the thermodynamic potential as a function of the spatially variable magnetisation, which is invariant with respect to the elements of the symmetry group of the magnetic crystal. The local magnetisation is then obtained from the requirement that G be stationary under local variations of the magnetisation, i.e. we derive it from the demand that the functional derivative must vanish. This step would lead us beyond the molecular-field approximation.

We return now to the Heisenberg model; one of the simplest approximations is based on an omission of exchange interaction between the x and y components of the spin in the Heisenberg operator. The model described in this approximation is called the Ising model (Ising, 1925). It is simpler to handle, since the operators in the Hamiltonian of the Ising model commute with one another, while no characteristic features such as the spontaneous magnetisation, phase transitions, etc., are lost. When we restrict ourselves to interaction between nearest-neighbour atoms, the Hamiltonian reads

$$\mathscr{H}_{\text{Ising}} = g\mu_{\text{B}}H\sum_{i} s_{iz} - J\sum_{i,j} s_{iz}s_{jz}\,. \tag{3.81}$$

The problem described by this Hamiltonian was first solved exactly by Onsager (see Domb, 1960; Newell and Montroll, 1953) for a two-dimensional square lattice with $H = 0$. The most interesting result is the logarithmic divergence of specific heat at the Curie point: $C_{\text{m}}(T) = $ $= \text{const} \ln |T-T_{\text{c}}|$. This result is also true for other two-dimensional lattices (see Domb, 1960). In addition to this, the spontaneous magnetisation at the Curie point was found to be proportional to $(T-T_{\text{c}})^{1/8}$, whereas the molecular-field approximation according to (3.18) yields a proportionality with $(T-T_{\text{c}})^{1/2}$. The two-dimensional model with a magnetic field and the three-dimensional Ising model have so far not been solved exactly. Series expansions of the partition function, however, seem to show that the reciprocal susceptibility is not a linear function of T but is in the two-dimensional case proportional to $(T-T_{\text{c}})^{7/4}$ and in three dimensions proportional to $(T-T_{\text{c}})^{5/4}$ (Domb, 1960; Domb and Sykes, 1962).

Since the Ising model is a simple model for a system with a phase transition, it is important and interesting not only for ferromagnetism and antiferromagnetism. It looks reasonable to transfer the results of the Ising model to the Heisenberg models; the analogy between the two models, however, is not very deep, particularly at low temperatures where the components of the spin at right angles to the external field are of special importance (Section III.3).

An improved molecular-field approximation has to take into account the short-range order of the spins, which is completely ignored in the original version of this approximation. It is obviously a first step in this direction to account exactly for the interaction of a spin with its nearest neighbourhood, and to describe the interaction with the remainder of the lattice by means of an internal or molecular field. This approximation is called the Bethe–Peierls–Weiss (BPW) approximation. It was first applied by Bethe (1935) to order–disorder phenomena in alloys, and later by Peierls (1936) and Weiss (1948) to the Ising model and the Heisenberg model, respectively. For simplicity we shall here discuss it in connection with the Ising model.

Consider a system of atoms with $s = \frac{1}{2}$ with only nearest neighbour interactions. An arbitrarily chosen atom (subscript 0) interacts with its z nearest neighbours. The effect of the other atoms is accounted for, according to the BPW approximation, by an internal field H_1 acting on the outer atoms of the cluster of $z+1$ spins. The Hamiltonian of the cluster will then be given by

$$\mathcal{H}_{\mathrm{Cl}} = -2J \sum_{i=1}^{z} s_{0z} s_{iz} + g\mu_{\mathrm{B}} H \sum_{i=1}^{z} s_{iz} + g\mu_{\mathrm{B}} H_1 \sum_{i=1}^{z} s_{iz}. \tag{3.82}$$

Here we also assumed that the "outer atoms" are not nearest neighbours of one another as is, for example, the case in face-centred cubic lattices.

The partition function $Z = \mathrm{Tr}\exp(-\beta\mathcal{H})$ is easily obtained in this case; since $\exp(2\alpha s_z) = \cosh\alpha + 2s_z \sinh\alpha$, we have

$$\begin{aligned} Z_{\mathrm{Cl}} = \mathrm{Tr}\, e^{-\beta\mathcal{H}_{\mathrm{Cl}}} = e^{-\beta g\mu_{\mathrm{B}} H/2} \cosh^z \frac{\beta}{2}\left(J - g\mu_{\mathrm{B}}(H+H_1)\right) \\ + e^{\beta g\mu_{\mathrm{B}} H/2} \cosh^z \frac{\beta}{2}\left(J + g\mu_{\mathrm{B}}(H+H_1)\right). \end{aligned} \tag{3.83}$$

Since the lattice is translationally invariant, the central atom is physically not preferred over the outer atoms. The thermal average of the magnetic moment of the central atom must then be equal to the mean value for an outer atom. Thus, according to (2.45) we must have

$$\frac{1}{z+1}\,\frac{\partial \ln Z_{\mathrm{Cl}}}{\partial H} = \frac{1}{z}\,\frac{\partial \ln Z_{\mathrm{Cl}}}{\partial H_1}. \tag{3.84}$$

This consistency condition determines the internal field H_1. Using (3.83), we obtain from (3.84) after a few transformations

$$\mathrm{e}^{\beta J} = \frac{\sinh \frac{\beta}{2} g\mu_{\mathrm{B}}\left(H+\frac{z}{z-1}H_1\right)}{\sinh \frac{\beta}{2} g\mu_{\mathrm{B}}\left(H+\frac{z-2}{z-1}H_1\right)}. \tag{3.85}$$

At low temperatures the internal field H_1 is, according to (3.85), proportional to the exchange integral $g\mu_{\mathrm{B}}H_1 = (z-1)J$, and therefore the magnetisation and the specific heat will have the same temperature dependence as in the molecular-field approximation. Near the Curie point, H_1 will be small and (3.85) can be expanded in terms of H_1 $(H = 0)$, breaking off after the second-order term:

$$\mathrm{e}^{\beta J} = \frac{z}{z-2}+\frac{z}{3}\left(\frac{\beta}{2}\, g\mu_{\mathrm{B}}H_1\right)^2.$$

Thus H_1 can be non-zero only if

$$\mathrm{e}^{\beta J} > z/(z-2).$$

The Curie point is therefore defined by

$$\frac{kT_{\mathrm{c}}}{J} = \ln^{-1}\frac{z}{z-2}. \tag{3.86}$$

The magnetisation of the entire system can be calculated from the magnetisation of the cluster considered. According to (2.45), the average magnetic moment of a cluster atom is given by

$$\bar{m} = \frac{kT}{z+1}\,\frac{\partial \ln Z_{\mathrm{Cl}}}{\partial H},$$

so that, using (3.84) and (2.83), the magnetisation of the whole system is obtained as follows:

$$M = N\frac{kT}{z+1}\cdot\frac{\partial \ln Z_{\text{Cl}}}{\partial H} = N\frac{kT}{z}\,\frac{\partial \ln Z_{\text{Cl}}}{\partial H_1}$$
$$= M_0 \tanh \frac{\beta}{2} g\mu_{\text{B}}\left(H + \frac{z}{z-1} H_1\right).$$

Here $M_0 = \frac{1}{2}Ng\mu_{\text{B}}$. From this we can easily calculate the susceptibility; with (3.85), neglecting terms of higher order of H, we obtain

$$\chi = \frac{C}{T}\,\frac{1}{ze^{\beta J}+2-z}; \qquad C = N\frac{g^2\mu_{\text{B}}^2}{2k}.$$

We see that the susceptibility deviates essentially from the Curie–Weiss law, and only at high temperatures, where $\beta J \ll 1$, χ^{-1} becomes a linear function of temperature. The short-range order also becomes apparent in the energy and the specific heat of the system. In the Ising model the energy of the spin system with $H = 0$ is, according to (3.81), given by

$$E = -\frac{Nz}{2}\,2J\overline{s_{0z}s_{iz}} = -NJ\sum_{i=1}^{z}\overline{s_{0z}s_{iz}}.$$

Here $\frac{1}{2}Nz$ is the number of different, adjacent spin-pairs possible in the case considered. Of course the mean value $\overline{s_{0z}s_{iz}}$ is the same for each pair. Using the definition of the partition function, we have in the BPW approximation (3.82)

$$\sum_{i=1}^{z}\overline{s_{0z}s_{iz}} = kT\frac{\partial}{\partial J}\ln Z_{\text{Cl}},$$

so that

$$E = -NkTJ\frac{\partial}{\partial J}\ln Z_{\text{Cl}}.$$

Substituting (2.83) we obtain

$$E = -\frac{NzJ}{4}\,\frac{\cosh \beta g\mu_{\text{B}}H_1 - e^{-\beta J}}{\cosh \beta g\mu_{\text{B}}H_1 + e^{-\beta J}}.$$

At the Curie point ($H_1 = 0$) the energy does not drop to zero as in the molecular-field approximation but approaches this value asymptotically, as a power of the temperature. Accordingly, the specific heat at the Curie point does not drop discontinuously to zero but decreases continuously after the discontinuity at T_c. This is due to the magnetic short-range order.

It is also easy to apply this method to antiferromagnetism (Firgau, 1940).

In the Heisenberg model the BPW approximation is connected with essential mathematical difficulties; it does not yield anything qualitatively new for the range about the Curie point (Weiss, 1948). For a consideration of antiferromagnetism and ferrimagnetism we refer the reader to papers by Li (1951) and Smart (1956).

The difficulties of this method are connected with an exact treatment of a cluster of $z+1$ spins, which requires cumbersome numerical calculations. A serious error of the BPW method is due to the fact that at low temperatures the ferromagnetic state becomes unstable, and at a certain temperature the spontaneous magnetisation vanishes (anti-Curie temperature; Anderson 1950). Although this does not influence the applicability of the method at high temperatures, it shows that the consistency condition (3.84) is insufficient to remove the preferred status of the central spin of the cluster compared with the spins of its boundary.

The constant-coupling approximation by Kasteleijn and van Kranendonk (1956) is a much simpler method to treat ferromagnetism and antiferromagnetism at high temperatures. In this approximation the problem of N spins is reduced to a problem with two spins which, of course, is easier to deal with. The results agree almost completely with those obtained in the BPW approximation. We again start from the Heisenberg operator with nearest-neighbour interactions, which can also be written as a sum over two-particle operators:

$$\mathcal{H} = -2J \sum_{i<j} (\boldsymbol{s}_i \cdot \boldsymbol{s}_j) + g\mu_B H \sum_i s_{iz} = \sum_{i<j} \mathcal{H}_{ij}$$

where (3.87)

$$\mathcal{H}_{ij} = -2J(\boldsymbol{s}_i \cdot \boldsymbol{s}_j) + g\frac{\mu_B}{z}(s_{iz}+s_{jz}).$$

In the following all atoms are assumed to possess a spin $s = \frac{1}{2}$; a transition to higher spin quantum numbers entails no difficulties (Wagner, 1963). The thermodynamic properties of the system are given by the partition function Z, which determines the free energy F:

$$-\beta F = \ln Z(\beta) = \ln \mathrm{Tr}\, e^{-\beta \mathcal{H}}. \tag{3.88}$$

The energy is given by

$$E = -\frac{\partial}{\partial \beta} \ln Z. \tag{3.89}$$

It is convenient to calculate Z, F, and E for a given magnetisation, that is, to consider the magnetisation M a free parameter which is chosen such that the free energy becomes minimum with respect to M. It is well known that, when an infinitely large system ($N = \infty$) is considered, this method is free from errors, since the state with the magnetisation determined in this way is by far preponderant in the partition function Z. The relations (3.88) and (3.89) can be immediately applied with the only exception that F, Z, and E depend not only on β but also on $\sigma = M/M_0$ where $M_0 = \frac{1}{2} N g \mu_B$:

$$E(\beta, \sigma) = -\frac{\partial \ln Z(\beta, \sigma)}{\partial \beta}, \tag{3.90}$$

or integrated

$$\int_0^\beta E(x, \sigma)\, dx = \ln Z(0, \sigma) - \ln Z(\beta, \sigma).$$

$Z(0, \sigma) = \mathrm{Tr}_\sigma 1$; now the trace extends over all states with constant magnetisation σ. Thus $Z(0, \sigma)$ is equal to the number of states with constant σ; if N_+ magnetic moments are parallel to the external field and N_- moments are antiparallel to it, this number will be given by

$$Z(0, \sigma) = \frac{N!}{N_+!\, N_-!},$$

where $N_+ + N_- = N$ and $N_+ - N = \sigma N$. Applying Stirling's formula to the factorial expressions, we obtain

$$\ln Z(0, \sigma) = N \ln 2 - \tfrac{1}{2} N\{(1+\sigma) \ln (1+\sigma) + (1-\sigma) \ln (1-\sigma)\}. \tag{3.91}$$

The spin system described by the Heisenberg operator (3.87) is isotropic; every pair of neighbouring spins will therefore make the same contribution to the energy. With z nearest neighbours there exist $\frac{1}{2}Nz$ different pairs so that the energy is given by

$$E(\beta, \sigma) = \frac{Nz}{2} \frac{\mathrm{Tr}_\sigma \mathscr{H}_{12} e^{-\beta \mathscr{H}}}{Z(\beta, \sigma)} = \frac{Nz}{2} \frac{Z(\beta)}{Z(\beta, \sigma)} \mathrm{Tr}_\sigma \mathscr{H}_{12} \varrho. \quad (3.92)$$

Here ϱ is the density matrix (2.42) and $\mathscr{H}_{12}$ is the Hamiltonian of an arbitrarily chosen pair (3.87). We now introduce a density matrix for the two spins 1 and 2 so that

$$E(\beta, \sigma) = \tfrac{1}{2} Nz \, \mathrm{Tr} \, \varrho_{12} \mathscr{H}_{12}.$$

According to (3.92)

$$\varrho_{12} = \frac{Z(\beta)}{Z(\beta, \sigma)} \mathrm{Tr}'_\sigma \, \varrho. \quad (3.93)$$

The trace must be formed only with eigenvectors of $s_{3z} \ldots s_{Nz}$, i.e. it is not taken over the selected pair of spins. To the density matrix σ_{12} we attribute an effective Hamiltonian $\mathscr{H}_{\mathrm{eff}}$:

$$\varrho_{12} = \frac{1}{Z_{\mathrm{eff}}} \exp(-\beta \mathscr{H}_{\mathrm{eff}}), \qquad Z_{\mathrm{eff}} = \mathrm{Tr} \exp(-\beta \mathscr{H}_{\mathrm{eff}}).$$

This effective Hamiltonian is defined by (3.93). Because of the interaction of this spin pair with the other spins it is not identical with H_{12}; its form is easily obtained from (3.93). From the isotropy of the whole problem it follows that $\mathscr{H}_{\mathrm{eff}}$ must be symmetrical with respect to the two spins. From (3.93) we also see that the z component of the system's total energy must be a constant of motion, and that $\mathscr{H}_{\mathrm{eff}}$ must commute with $s_{1z}+s_{2z}$. For the same reason the density matrix ϱ_{12} will no longer depend on the magnetic field according to (3.93), so that $\mathscr{H}_{\mathrm{eff}}$ will also be independent of H. Thus we obtain for the general form of an effective Hamiltonian

$$\mathscr{H}_{\mathrm{eff}} = -2A_1(\boldsymbol{s}_1 \cdot \boldsymbol{s}_2) - 2A_2 s_{1z} s_{2z} + g\mu_B(s_{1z}+s_{2z})A_3. \quad (3.94)$$

The coefficients A_ν are functions of β and σ; A_3 has the character of a molecular field.

The three functions are not independent of one another, since all spin pairs make the same contribution to the total magnetisation. Therefore

$$\sigma = \overline{s_{1z}+s_{2z}} = \text{Tr}\, \varrho_{12}(s_{1z}+s_{2z}). \tag{3.95}$$

This consistency condition corresponds to condition (3.84) in the BPW approximation with the exception that now all spins are treated in the same way.

So far we have introduced no approximations. The constant-coupling approximation is based on the assumption that the functions A_1 and A_2 are assumed to be constant, and that $A_1 = J$ and $A_2 = 0$. In this way we fix the properties of the system. The magnetisation σ is obtained from the requirement $\partial F(\beta, \sigma)/\partial\sigma = 0$; together with the equations (3.88)–(3.94) (cf. Kasteleijn and van Kranendonk, 1956) it can be written in the form

$$\beta g\mu_B(zA_3-H) = -\frac{z-1}{N}\,\frac{\partial \ln Z(0,\sigma)}{\partial\sigma} = (z-1)\ln\frac{1+\sigma}{1-\sigma}. \tag{3.96}$$

At the Curie point σ can be considered to be a small quantity. Therefore, according to (3.96), we have for $H = 0$ (neglecting higher powers of σ):

$$z\beta g\mu_B A_3 = 2(z-1)\sigma. \tag{3.97}$$

The internal field A_3 in the neighbourhood of T_c will also be small. We can therefore expand (3.95) in terms of A_3 and obtain to a first approximation

$$\sigma = \frac{\text{Tr}\,(s_{1z}+s_{2z})^2 \exp\big(2\beta(\boldsymbol{s}_1\cdot\boldsymbol{s}_2)J\big)}{\text{Tr}\exp\big(2\beta J(\boldsymbol{s}_1\cdot\boldsymbol{s}_2)\big)}\cdot\beta g\mu_B A_3.$$

As $2(\boldsymbol{s}_1\cdot\boldsymbol{s}_2) = \frac{1}{2}$ for the singlet state and $-\frac{3}{2}$ for the triplet state we obtain

$$\sigma = \beta g\mu_B A_3\cdot\frac{2e^{\beta J/2}}{3e^{\beta J/2}+e^{-3/2\beta J}}.$$

Together with (3.97) we obtain a homogeneous equation for σ or for A_3 which, generally, has only the solution $\sigma = 0$; only at the Curie point has it a non-trivial solution ($\sigma \neq 0$). The Curie temperature is therefore given by

$$kT_c = 2J \ln^{-1} \frac{z}{z-4}. \tag{3.98}$$

As $z \to \infty$ we obtain from this equation the Curie temperature in the molecular-field approximation:

$$kT_c = \frac{zJ}{2}.$$

Compared with the molecular-field approximation, the Curie temperature is lower in this approximation (cf. Table 14). Taking the interaction more accurately into account results in an earlier breakdown of the magnetic ordering, and the reciprocal susceptibility decreases less rapidly with increasing temperature than would correspond to the Curie–Weiss law. The thermal average of $(\boldsymbol{s}_1 \cdot \boldsymbol{s}_2)$ represents a direct measure of the short-range order. In the molecular-field approximation this average is the product of the averages of the individual spins and thus proportional to the square of the magnetisation, since the spins are independent of one another. At the Curie point and for $H = 0$, the mean $\overline{(\boldsymbol{s}_1 \cdot \boldsymbol{s}_2)}$ will therefore vanish. In the constant-coupling approximation, however, we obtain at the Curie point $(\boldsymbol{s}_1 \cdot \boldsymbol{s}_2) = \frac{3}{4}(z-1)^{-1}$. Since the mean value of $(\boldsymbol{s}_1 \cdot \boldsymbol{s}_2)$ is proportional to the energy, $E = -NzJ(\boldsymbol{s}_1 \cdot \boldsymbol{s}_2)$, this means that, owing to the short-range order above the Curie point, the magnetic energy only decreases above this point.

Compared with the molecular-field approximation, the constant-coupling approximation gives smaller changes in the spontaneous magnetisation than in the susceptibility or the specific heat.

In some way the constant-coupling theory is the simplest cluster theory, since the cluster considered comprises only two spins. Changing to three or more spins in the cluster, the difficulties grow rapidly (Strieb *et al.*, 1963; Ballensiefen and Wagner, 1965).

Although the cluster theories represent an improvement compared with the molecular-field approximation, it is not possible to estimate the degree of approximation and the accuracy. This can only be done in a theory in which the partition function is expanded systematically in terms of a parameter. For a theory applicable in the whole temperature range, z^{-1} would be a reasonable parameter (z being the number of nearest-neighbour atoms). When we restrict ourselves to interactions between the nearest neighbours, an expansion in terms of z^{-1} corresponds physically to an expansion in terms of the reciprocal of the range of the exchange interaction $J(|\boldsymbol{r}_i - \boldsymbol{r}_j|)$, since the range determines the number of spins with which a particular spin interacts. The advantage of such an expansion is that we know the zeroth approximation, since with an infinitely long range ($z = \infty$) the molecularfield approximation becomes exact. The difficulties, however, are considerable and most investigations are restricted to the Ising model (Brout, 1960; Horwitz and Callen, 1961; Stinchcombe *et al.*, 1963).

Above the Curie point the natural expansion parameter is the reciprocal temperature, more exactly, βJ. An expansion of the partition function leads us to a power series for the susceptibility:

$$\chi = N \cdot \tfrac{1}{3}\beta(g\mu_B)^2 s(s+1) \cdot \sum_{n=0}^{\infty} a_n(\beta J)^n. \tag{3.99}$$

Since at high temperatures the Curie–Weiss law is correct, the first two coefficients are known: $a_0 = 1$, $a_1 = \frac{2}{3} z(s+1)$. The higher coefficients were calculated first by Opechowski (1937, 1938), and later by Brown and Luttinger (1955), Domb and Sykes (1957, 1962) and Rushbrooke and Wood (1958). As these calculations are very cumbersome, only a few a_n have become known (up to $n = 6$). Using these a_n one can try to determine the Curie point;† on the one hand, this can be done by solving the equation $\chi^{-1} = 0$, where χ is given by the series (3.99), which has been broken off after a certain n; Brown and Luttinger (1955) continued up to $n = 4$. On the other hand, for most crystal structures the a_n are monotonic functions of n. It is therefore reasonable to extrapolate from the known a_n to the higher ones, and thus to determine the radius of convergence of the series (3.99) by means of the quotient or root criterion (Domb and Sykes, 1957; Rushbrooke

† Here, as in the Ising model, eqn. (3.99) yields information on the behaviour of susceptibility at the Curie point; we obtain $\chi^{-1} \sim (T - T_c)^{4/3}$ (Domb and Sykes, 1962).

and Wood, 1958). Both methods yield about the same values for the Curie point (see Table 14); the values obtained, with the help of this extrapolation method can therefore be considered as the almost true Curie points. The same table also contains the Curie points as obtained according to the linearised spin-wave approximation discussed earlier. They differ from those of molecular-field approximation only by a constant for a given lattice structure (z), so that we can say that the linearisation (3.38) neglects the correlations between the spins in a similar way as the molecular-field approximation. An improved

TABLE 14

s	Br.–Lutt.	Rush.–Wood	Linearised spin-wave	Tahir-Kheli	BPW	Const.-coupl.
			$z = 8$			
$\frac{1}{2}$	0·510	0·650	0·718	0·650	0·728	0·721
1	0·733	0·708	0·718	0·718	0·818	0·811
$\frac{3}{2}$	0·771	0·735	0·718	0·740	0·836	0·836
2	0·787	0·747	0·718	0·752	0·847	0·847
$\frac{5}{2}$	0·795	0·754	0·718	0·758	—	0·852
3	0·800	0·758	0·718	0·763	—	0·855
			$z = 12$			
$\frac{1}{2}$	0·707	0·692	0·744	0·680	—	0·822
1	0·796	0·750	0·744	0·744	—	0·878
$\frac{3}{2}$	0·822	0·767	0·744	0·765	—	0·894
2	0·833	0·776	0·744	0·776	—	0·900
$\frac{5}{2}$	0·839	0·783	0·744	0·782	—	0·904
3	0·843	0·785	0·744	0·786	—	0·906

The table contains the Curie points for the Heisenberg model with nearest-neighbour interaction for two lattice structures, $z = 8$ and $z = 12$. Numerical values are given for $3kT_c(2zJs(s+1))^{-1}$ as obtained by means of the various methods: Brown and Luttinger (1955), Rushbrooke and Wood (1958), linearised spin-wave (Bogolyubov and Tyablikov, 1959; Tahir-Kheli and ter Haar, 1962), Tahir-Kheli (1963), BPW (Brown and Luttinger, 1955), and constant coupling (Kasteleijn and van Kranendonk, 1956; Wagner, 1963) (see text).

linearised spin-wave approximation (Tahir-Kheli, 1963) yields almost the same Curie temperatures as obtained by Rushbrooke and Wood. The Curie temperatures calculated by means of the cluster methods, however, are a little higher and virtually equal to one another. It is true, however, that approximations with larger clusters yield essentially better values (Ballensiefen and Wagner, 1965).

The Curie temperatures calculated can also be used to determine the exchange integrals from the Curie temperatures measured. If the Heisenberg model with its exchange interaction between the nearest neighbours is correct, these values should agree with those derived from the magnetic energy $|E| = NzJs^2$, or from the temperature dependence of the magnetisation or the specific heat at low temperatures, i.e., from Bloch's $T^{3/2}$-law. Table 15 shows that such an agreement does exist as to the order of magnitude, for iron, nickel, and gadolinium. For iron and nickel the discrepancies are of no interest as we know that for metals with *d*-electrons the Heisenberg model is but a poor approximation. For gadolinium the theory should be correct with the indirect exchange interaction (3.10). It is therefore reasonable to attribute the discrepancy in the case of gadolinium to the fact that the exchange interaction with remote atoms was neglected.

TABLE 15

	s	z	T_c (°K)	|E| (cal/mole)	J (eV) (from T_c)	J (eV) (from E)	J (eV) (from (3.54))
Fe	1·1	8	1043	1930±120	0·012	0·010	0·020
Ni	0·3	12	631	420±50	0·013	0·010	0·018
Gd	3·55	12	289	806±40	0·00025	0·00024	0·00014

The table contains the values of the exchange integrals for iron, nickel, and gadolinium as obtained from the Curie temperatures from Rushbrooke and Wood, Table 14, from the magnetic energy measured ($E = -NzJs^2$), and from the temperature dependence of the magnetisation at low temperatures (Hofmann *et al.*, 1956, 1960). For $s = \frac{7}{2}$ the Curie point was obtained by extrapolation. In the case of nickel, 60% of the atoms were assumed to have $s = \frac{1}{2}$ and 40% of the atoms to have $s = 0$; the mean value for Ni was thus 0·3 (cf. Section III.5).

TABLE 16

$T > T_c$	M.F.	Ising ($s = \frac{1}{2}$)		Heisenberg ($d = 3$)	
		$d = 2$	$d = 3$	$s = \frac{1}{2}$	$s = \infty$
α	0	0	0·13	$< 0{\cdot}2$	$< 0{\cdot}2$
γ	1	$1\frac{3}{4}$	1·25	1·43	1·35
$T = T_c$					
δ	3	15	5	5	5
$T < T_c$					
α'	0	0	0·066	—	—
β	$\frac{1}{2}$	$\frac{1}{8}$	0·31	$0{\cdot}3 < \beta <$	0·4
γ'	1	$1\frac{3}{4}$	1·31	—	—

The table contains the critical exponents as obtained in the molecular-field approximation (M.F.) for the two- and three-dimensional Ising models ($d = 2$ or 3) and for the Heisenberg model (cf. Fisher, 1967a).

We can also try to determine the analytical form of the susceptibility as $T \to T_c$ by means of this series expansion. As in the case of the Ising model, we may expect also for the Heisenberg model that the susceptibility is not an analytical function of $T - T_c$; we can therefore put

$$\chi(T) \sim \begin{cases} (T-T_c)^{-\gamma}, & T > T_c; \\ (T_c-T)^{-\gamma'}, & T < T_c, \end{cases}$$

provided the susceptibility exists when $T < T_c$. This is not obvious since, for example, we see from (3.63) that the susceptibility calculated in the RPA diverges.

The critical exponent γ can be obtained immediately from the series (3.99), from the asymptotical behaviour of the a_n as $n \to \infty$; it can also be obtained indirectly by improving the poor convergence of the series near T_c, for example, by representing (3.99) as the ratio of two polynomials in βJ (Padé approximation). As a rule this method yields a good convergence and makes it possible to determine γ rather accurately (cf. Fisher, 1967a).

An analogous method is applied when considering the magnetic contribution to the specific heat C_m at $H = 0$ and the spontaneous magnetisation $M_s(T)$; we put

$$C_m(H=0,T) \sim \begin{cases} (T-T_c)^{-\alpha}, & T > T_c; \\ (T_c-T)^{-\alpha'}, & T < T_c; \end{cases}$$

$$M_s(T) \sim (T_c-T)^{\beta}, \qquad T < T_c.$$

The logarithmic divergence of the specific heat in the two-dimensional Ising model corresponds to a critical exponent $\alpha = \alpha' = 0$, as every power diverges more strongly than the logarithm. Therefore the difference between a logarithmic divergence and a discontinuity of the type of the molecular-field approximation cannot be determined from the critical exponents.

Another critical exponent is obtained on the critical isotherm ($T = T_c$) where

$$M = \text{const} \cdot H^{1/\delta}, \qquad (M, H > 0).$$

In the molecular-field approximation using eqns. (3.17) and (3.19), we obtain $\delta = 3$ from (3.16). In Table 16 we have compiled the exact and approximate numerical values of the critical exponents for both the Ising and the Heisenberg models.

The experimental values are known only partly (Fisher, 1967a; Kadanoff *et al.*, 1967). For γ the values obtained are about 1·3, those for δ are about 4; α and α' prove to be small ($\leqslant 0{\cdot}15$). The critical exponent of the spontaneous magnetisation, β, is generally a function of temperature. In the temperature range of $T_c - T \simeq 10^{-2}T_c$, β lies between 0·35 and 0·4, whereas near the Curie point, $T_c - T \simeq 10^{-4}T_c$, β values of about 0·5 are obtained, a value which would correspond to the molecular field approximation. This might be due to the dipole interaction between the spins (Kadanoff *et al.*, 1967).

The critical exponents are not independent of one another. General thermodynamic considerations lead us to two inequalities for the exponents. Both are based on the assumption that entropy increases with temperature, i.e. on the assumption of positive specific heats and a stable system, that is, positive susceptibilities. In the case of liquids the latter assumption corresponds to that of positive compressibilities. For the adiabatic susceptibility we obtain in particular

$$\chi_{\mathrm{ad}}(T) = \left(\frac{\partial M}{\partial H}\right)_S > 0,$$

where S is the entropy. In order to use this fact, we express the adiabatic susceptibility in terms of the usual, isothermal susceptibility. In the case of liquids this corresponds to the following relation between adiabatic and isothermal compressibilities (Landau and Lifshitz, 1965, § 16):

$$\left(\frac{\partial V}{\partial p}\right)_S = \left(\frac{\partial V}{\partial p}\right)_T + \frac{T}{C_p}\left(\frac{\partial V}{\partial T}\right)_p^2.$$

When we consider magnetic systems, we see from (2.67) that we must replace the pressure p (an intensive quantity) by the field, $p \leftrightarrow -H$, and the volume (an extensive quantity) by the magnetisation, $V \leftrightarrow M$. The identity given above will then read

$$\chi_{\mathrm{ad}}(T) = \left(\frac{\partial M}{\partial H}\right)_S = \left(\frac{\partial M}{\partial H}\right)_T - \frac{T}{C_H}\left(\frac{\partial M}{\partial T}\right)_H^2;$$

C_H is the specific heat at constant field as defined in (2.67). Thus we have the inequality ($\chi_{\mathrm{ad}} > 0$):

$$T\left(\frac{\partial M}{\partial T}\right)_H^2 < C_H\left(\frac{\partial M}{\partial H}\right)_T = C_H\chi(T).$$

In the case of $H = 0$ and $T \to T_{\mathrm{c}}$ we have $M = M_{\mathrm{s}} \sim (T_{\mathrm{c}} - T)^{\beta}$ and $C_{H=0} = C_{\mathrm{m}}(H = 0, T) \sim (T_{\mathrm{c}} - T)^{-\alpha'}$.

Thus the inequality yields

$$(T_{\mathrm{c}} - T)^{\alpha' + 2(\beta - 1) + \gamma'} < \text{const.}$$

or

$$\alpha' + 2(\beta - 1) + \gamma' \geq 0. \tag{3.99a}$$

The validity of this inequality was first proved by Rushbrooke (1963). If the dominant singularities in the two susceptibilities do not compensate one another exactly, the inequality degenerates into an equation. This is, however, probably not the case, since the difference between χ_{ad} and $\chi(\chi \geqslant \chi_{ad})$ should be largest at the Curie point, since the coupling of a thermodynamic system (adiabatic or isothermal) should exert the strongest effect at T_c, because of the strong, long-range fluctuations.

At $T = T_c$ a similar inequality can be derived for α, β, and δ, which was first given by Griffiths (1965). Consider M and T to be thermodynamic variables, so that the free energy $F(M, T)$ becomes a stationary quantity, if M and T are constant:

$$dF = -S\,dT + H\,dM.$$

Hence we obtain for the entropy S and the field H

$$S = -\left(\frac{\partial F}{\partial T}\right)_M, \qquad H = \left(\frac{\partial F}{\partial M}\right)_T.$$

At $T < T_c$ the quantities S, H, and F are defined only for $M \geqslant M_s(T)$, where $M_s(T)$ is the spontaneous magnetisation at T. In particular, $H = H(M_s(T), T) = 0$. It is possible, however, to continue the equation of state, $H = H(M, T)$, into the range of $M < M_s(T)$, when we put there $H \equiv 0$. This leads to:

$$\left(\frac{\partial F}{\partial M}\right)_T \equiv 0, \qquad \left(\frac{\partial S}{\partial M}\right)_T \equiv 0 \qquad \text{when} \qquad M \leqslant M_s(T).$$

Therefore $F(M, T) = F(T)$ and $S(M, T) = S(T)$, independent of the magnetisation, when $M \leqslant M_s(T)$. Let us now consider the free energy at constant M and at two temperatures T_1 and T_2 ($T_2 > T_1$). As $S(M, T)$ is a monotonically increasing function of T, the following inequalities are valid:

$$F(M, T_2) - F(M, T_1) \begin{cases} < -(T_2 - T_1)S(M, T_1), \\ > -(T_2 - T_1)S(M, T_2). \end{cases}$$

We use the latter inequality for $M = M_s(T_2) \equiv M_2$ and the first one for $M = M_s(T_1) \equiv M_1$. Then

$$F(M_2, T_2) > F(M_2, T_1) - (T_2 - T_1)S(M_2, T_2),$$
$$F(M_1, T_2) < F(M_1, T_1) - (T_2 - T_1)S(M_1, T_1).$$

Considering what has been said earlier, subtraction yields

$$F(M_1, T_2) - F(T_2) < (T_2 - T_1)^2 \frac{S(T_2) - S(T_1)}{T_2 - T_1}.$$

Hence when $T_2 < T_c$ we can again derive the inequality (3.99a) (Fisher, 1967a). When $T_2 = T_c$ we put $M_1 = M(T_c, H)$; since M_1 becomes small as $T_1 \to T_c$, the left-hand side of the inequality tends to

$$HM(T_c, H) = M^{\delta+1}(T_c, H) = M_s^{\delta+1}(T_1, H = 0) \sim (T_c - T_1)^{\beta(\delta+1)}$$

because of the definition of dF; the right-hand side tends to

$$\frac{(T_c - T_1)^2}{T_1} C_{H=0}(T_1) \sim (T_c - T_1)^{2-\alpha'},$$

so that we arrive at the inequality

$$\alpha' + \beta(\delta + 1) \geq 2. \tag{3.99b}$$

In order to obtain additional information we have either to investigate special models for the magnetically ordered system or to make a phenomenological Ansatz for the equation of state, $H = H(M, T)$ (Widom, 1965; Fisher, 1967b; Kouvel and Comly, 1968).

A certain exception is made by the scaling laws, whose derivation is mainly based upon the long-range correlation between the spins which must appear in every ferromagnetic system (cf. Kadanoff, 1966; Kadanoff *et al.*, 1967). For this purpose we consider an Ising model with $s = \frac{1}{2}$ whose Hamiltonian according to (3.81) is given in a somewhat modified form by

$$\mathcal{H} = h \sum_i \sigma_{iz} - J \sum_{i,j} \sigma_{iz}\sigma_{jz}. \tag{3.81a}$$

The double sum is to be extended over nearest neighbours; the σ_{iz} are the z components of the Pauli matrices with eigenvalues ± 1. The interacting spin system is characterised by an order or correlation between the spins which acts over a large distance ξ; ξ is the so-called correlation length. The neighbourhood of a reference spin will therefore move in a more or less rigid connection with it, if the linear dimensions of this neighbourhood do not exceed the correlation length. This leads us to the assumption that the spin system can be described just as well by means of a system of clusters whose linear dimension La is smaller than ξ; within these limits L is freely variable; a is the distance between the atoms. "Just as well" means here that all macroscopic quantities of the two systems must be equal.

The Hamiltonian describing the equivalent system is then given by

$$\tilde{\mathscr{H}} = \tilde{h} \sum_k \tilde{\sigma}_{kz} - \tilde{J} \sum_{k,l} \tilde{\sigma}_{kz} \tilde{\sigma}_{lz}; \tag{3.81b}$$

$\tilde{h}$ is the field acting upon the clusters, $\tilde{J}$ is the interaction between neighbouring clusters and $\tilde{\sigma}_{kz}$ describes the two possibilities of adjustment of the kth cluster and is again a Pauli matrix.

The parameters of the cluster system are connected with those of the initial system (3.81). It is thus obvious that the size of the spin system (a is the distance between atoms) must be connected with that of the equivalent system (distance of the clusters $= La$) through $L : \tilde{l} = Ll$. As regards the connection between $\tilde{h}$ and h we assume $\tilde{h} = L^x h$, x being an unknown number; this Ansatz is reasonable since $\tilde{h} \to 0$ as $h \to 0$ and $\tilde{h} = h$ when $L = 1$. The connection between $\tilde{J}$ and J, however, is not of direct interest. Since the Curie temperature T_c depends on J, it is plausible to describe the new system by a temperature $\tilde{T}$ which is a function of T. As we assumed the macroscopic properties of both systems to be the same, $\tilde{T}_c = T_c$ is a necessary condition and it is therefore convenient to use the temperatures $t = T - T_c$ and $\tilde{t} = \tilde{T} - T_c$ which are reckoned from T_c. Following Kadanoff we put $\tilde{t} = L^y t$, y being an unknown number. This transformation obviously yields the correct relation since, on the one hand, for $L = 1$ the relation $\tilde{T} = T$ must be valid and, on the other hand, the temperature $\tilde{t}$ will be a linear function of t, if t is sufficiently low. Hence we already see that in this discussion we have to restrict ourselves to the immediate neighbourhood of the Curie point, which also becomes obvious when we consider the fact that $\tilde{T} \neq 0$ when $T = 0$, although, correctly, $\tilde{T}$ should be equal to zero. The transformation between l, h, t on the one hand, and $\tilde{l}$, $\tilde{h}$, $\tilde{t}$ on the other is thus a pure scale transformation

$$\tilde{l} = Ll, \qquad \tilde{h} = L^x h, \qquad \tilde{t} = L^y t.$$

It is convenient to consider the macroscopic thermodynamic potentials as functions of h and t or $\tilde{h}$ and $\tilde{t}$ and thus to discuss the free enthalpy. Owing to the equivalence of the two systems we have $G = \tilde{G}$ or, in the differential form, $dG = d\tilde{G}$ where

$$dG = -S\,dT - \mu\,dh,$$
$$d\tilde{G} = -\tilde{S}\,d\tilde{T} - \tilde{\mu}\,d\tilde{h},$$

where

$$\mu = \sum_i \langle \sigma_{iz} \rangle \qquad \text{and} \qquad \tilde{\mu} = \sum_k \langle \tilde{\sigma}_{kz} \rangle.$$

If T is kept constant we have

$$\mu\,dh = \tilde{\mu}\,d\tilde{h} = \tilde{\mu} L^x\,dh,$$

or $\mu = L^x \tilde{\mu}$.

Solving this equation for the contributions from the individual clusters, we obtain for the kth cluster

$$\mu_k = \sum_i \langle \sigma_{iz} \rangle = L^x \tilde{\mu}_k,$$

where the sum is to be taken over the spins in the kth cluster. According to the assumptions discussed above, the correlations in a cluster are so strong that each spin gives the same contribution: $\langle \sigma_{iz} \rangle = \sigma$. Since $\sum_i = L^d$ (d is the dimension of the system) we have

$$\sigma = L^{x-d} \tilde{\mu}_k,$$

where μ_k is independent of k by virtue of the translational invariance. Both quantities, σ and μ_k, have the meaning of a relative magnetisation; owing to the similarity of the two Hamiltonians (3.81a) and (3.81b) they must therefore have the same functional dependence on

h and t and $\tilde{h}$ and $\tilde{t}$:

$$\sigma = f(h, t), \qquad \tilde{\mu}_k = f(\tilde{h}, \tilde{t}).$$

Thus we obtain for f (with arbitrary L) the following equation:

$$f(h, t) = L^{x-d} f(\tilde{h}, \tilde{t}) = L^{x-d} f(L^x h, L^y t).$$

Let us consider the somewhat more general case

$$f(x_1, x_2) = \lambda^{-a} f(\lambda^b x_1, \lambda^c x_2)$$
$$= \lambda^{-a} f((\lambda x_1^{1/b})^b, (\lambda x_2^{1/c})^c).$$

With

$$x_1' = x_1^{1/b},\ x_2' = x_2^{1/c} \quad \text{and} \quad f(x_1'^b, x_2'^c) = \varphi(x_1', x_2')$$

we hence obtain for arbitrary λ:

$$\varphi(\lambda x_1', \lambda x_2') = \lambda^a \varphi(x_1', x_2').$$

Thus φ is a homogeneous function of the two variables x_1' and x_2' of the degree a. It therefore satisfies Euler's relation

$$a\varphi(x_1', x_2') = x_1' \frac{\partial \varphi}{\partial x_1'} + x_2' \frac{\partial \varphi}{\partial x_2'}.$$

Returning to the function f, this relation can be written in the following form:

$$af(x_1, x_2) = bx_1 \frac{\partial f}{\partial x_1} + cx_1 \frac{\partial f}{\partial x_2}.$$

Here obviously we have assumed f to be differentiable.

Euler's relation yields for the magnetisation

$$(d-x)\sigma(h, t) = xh \frac{\partial \sigma}{\partial h} + yt \frac{\partial \sigma}{\partial t},$$

so that, since $\sigma(h = 0) \sim (T_c - T)^\beta$ and $\sigma(t = 0) \sim h^{1/\delta}$, we obtain

$$d-x = y\beta \qquad \text{and} \qquad d-x = x/\delta.$$

In this way we fix x and y. When we also assume higher derivatives to exist, we have

$$(d-2x) \frac{\partial \sigma}{\partial h} = xh \frac{\partial^2 \sigma}{\partial h^2} + yt \frac{\partial^2 \sigma}{\partial h\, \partial t},$$

and therefore, with $h = 0$,

$$(d-2x)\left(\frac{\partial \sigma}{\partial h}\right)_{h=0} = yt\left(\frac{\partial^2 \sigma}{\partial t\, \partial h}\right)_{h=0}.$$

Since

$$\left(\frac{\partial \sigma}{\partial h}\right)_{h=0} \sim \chi(T) \sim (T_c - T)^{-\gamma'}$$

we obtain

$$\gamma' = \beta(\delta - 1).$$

In the case of a constant field we proceed analogously. Considering again the contribution of the kth cluster to the entropy, which, owing to translational invariance, is again independent of k, we obtain from

$$S\, dT = \tilde{S}\, d\tilde{T} = L^y \tilde{S}\, dT$$

the relation

$$S_k = L^y \tilde{s}_k = L^y \tilde{s}.$$

$\tilde{S}$ is the entropy per cluster. As all spins in the cluster are equivalent, they make the same contributions to the entropy S_k, i.e. $S_k = L^d s$. Therefore

$$s = L^{y-d} \tilde{s}.$$

Because of the similarity of the two Hamiltonians, the two entropies per spin or per cluster must have the same functional dependence on h and t and on $\tilde{h}$ and $\tilde{t}$:

$$s = g(h, t), \qquad \tilde{s} = g(\tilde{h}, \tilde{t}).$$

As above, we can hence obtain Euler's relation

$$(d-y)s(h, t) = xh\frac{\partial s}{\partial h} + yt\frac{\partial s}{\partial t}.$$

Differentiation with respect to t yields for $h = 0$:†

$$(d-2y)\left(\frac{\partial s}{\partial t}\right)_{h=0} = yt\left(\frac{\partial^2 s}{\partial t^2}\right)_{h=0}.$$

Since

$$\left(\frac{\partial s}{\partial t}\right)_{h=0} \sim C_{H=0}(T) \sim (T_c - T)^{-\alpha'}, \qquad (T < T_c)$$

† It is interesting to note that this differential equation has no solution which leads to a logarithmic divergence of the specific heat.

we obtain for the critical exponent α'

$$(d-2y) = -y\alpha', \qquad \text{or} \qquad \alpha'+\beta(\delta+1) = 2.$$

Thus we obtain a stricter form of inequality (3.99b). When we eliminate $\beta\delta$ from it by means of $\gamma' = \beta(\delta-1)$, we have

$$\alpha'+2(\beta-1)+\gamma' = 0,$$

which represents a stricter form of inequality (3.99a). From the existence of the higher derivatives we can also draw the conclusion that the critical exponents are symmetrical, i.e. that $\alpha = \alpha'$ and $\gamma = \gamma'$.

In this way the assumptions about the validity of the scale transformations lead us in a relatively general way (in the Ising model) to a series of relations between the critical exponents, so that, finally, there are only two exponents which are independent of one another.

Finally we shall briefly discuss the static spin correlation function in the neighbourhood of the critical point. It is directly connected with the neutron scattering cross-section (Van Hove, 1954; de Gennes, 1963) as well as with the susceptibility (cf. (2.46a)), and can therefore be subject to a direct experimental investigation. The static correlation function is defined as

$$g(\boldsymbol{r}) = \langle(\boldsymbol{s}_i \cdot \boldsymbol{s}_j)\rangle.$$

Owing to the translational invariance, the expectation value of the spin product depends only on the distance $\boldsymbol{r}$ of the two spins, and the two spin operators contain the same time.

Even if there exists a weak field, it is, near the Curie point, a good approximation to assume the expectation value of the three components to be equal,

$$\langle s_{ix}\, s_{jx}\rangle = \langle s_{iy}\, s_{jy}\rangle = \langle s_{iz}\, s_{jz}\rangle.$$

We can then express g in terms of s_i^+ and s_j^-:

$$g(\boldsymbol{r}) = \tfrac{3}{2}\langle s_i^+\, s_j^-\rangle.$$

In the RPA ($s = \frac{1}{2}$) we obtain for it from Section III.3

$$g(\boldsymbol{r}) = \frac{3}{2}\,\sigma\,\frac{1}{N}\sum_{\boldsymbol{k}} \frac{e^{i(\boldsymbol{k}\cdot\boldsymbol{r})}}{e^{\beta\bar{\varepsilon}_k}-1}\,.$$

$\tilde{\varepsilon}_k$ is the spin wave energy (3.62) and σ is the relative magnetisation. For large distances between the two spins only small k will contribute to the sum so that ε_k can be expanded in a series of k. In agreement with (3.52) we obtain from (3.62)

$$\tilde{\varepsilon}_k = g\mu_B H + \sigma J l^2 k^2,$$

so that, after changing to an integral, we obtain the following relation for a simple cubic crystal

$$\lim_{r\to\infty} g(r) = \frac{3\sigma}{2\beta}\cdot\frac{V}{N}\,\frac{1}{(2\pi)^3}\int d^3k\,\frac{e^{i(k\cdot r)}}{g\mu_B H + \sigma J l^2 k^2}\,.$$

In the case of $T > T_c$ we have $Ng\mu_B\sigma = 2\chi H$ and therefore

$$\lim_{r\to\infty} g(r) = \frac{3}{8}\,\frac{kT}{J}\,\frac{e^{-r/\xi}}{r} \qquad \text{with} \qquad \xi^2 = \frac{2Jl^2}{N(g\mu_B)^2}\chi.$$

ξ is the correlation length discussed in connection with the scaling laws. In the RPA it is $\sim \chi^{1/2}$, so that it diverges at the Curie point as $(T-T_c)^{\gamma/2}$. A strong increase of elastic scattering corresponds to this divergence of the correlation length; we call this critical scattering in analogy with the opalescence of liquids.

The close correlation between ξ and χ is typical for the RPA. In a more general consideration we define new critical exponents for ξ:

$$\xi \sim \begin{cases} (T-T_c)^{-\nu}, & T > T_c, \\ (T_c-T)^{-\nu'}, & T < T_c, \end{cases}$$

and put for $T = T_c$

$$g(r) = \frac{1}{r^{d-2+\eta}}\,.$$

The three new exponents can be reduced to α and γ by means of the scaling laws. We obtain (Kadanoff *et al.*, 1967)

$$\nu = \nu' = (2-\alpha)/d, \qquad \text{and} \qquad (2-\alpha)(2-\eta) = d\gamma.$$

Experimental investigations yield ν between 0·6 and 0·71; η, however, is a small quantity.

As to a more detailed discussion of the critical exponents we refer to the review articles by Fisher (1967a) and Kadanoff *et al.* (1967).

5. The Band Model

Hitherto we have considered ferromagnetism, antiferromagnetism and ferrimagnetism only within the framework of the Heisenberg model. It has already been mentioned that this model is based on the assumption of localised electrons (Section III.1). In the case of ferromagnetic metals with a *d* shell this will certainly be subject to restrictions. It is therefore not unexpected that we are led to difficulties when we apply the simple Heisenberg model to metals. A typical difficulty is due to the fact that in the model of localised electrons the mean magnetic moment per atom should be an integral multiple of μ_B, since in transition metals the orbital angular momentum is to a large extent quenched. From Table 12 we see that for nickel this is not correct. According to a suggestion by Van Vleck (1953), the Heisenberg model can be modified by assuming the atoms to possess different electron configurations, i.e. to be in different states of ionisation. In this model the number of possible states of ionisation is small, since the Coulomb interaction between the electrons prevents a strong accumulation or depletion of the electrons on an atom (minimum polarity model). Applied to nickel, this model means that about 40% of the Ni atoms have the $3d^{10}4s$ configuration and thus are non-magnetic, while about 60% have the $3d^{9}4s$ configuration. This results in an average moment of 0·6 μ_B per atom. Since the atoms interact with one another, the configurations are continuously exchanged between the atoms, i.e. the *d* electrons become mobile. This must occur within a relatively short time ($<10^{-13}$ sec), since neutron scattering experiments do not indicate the presence of different magnetic moments. In this way we are led from a model with localised electrons to a band model for the *d*-electrons, whose motions are strongly correlated because of the requirement of minimum polarity. So far no quantitative justification has been found for this model.

When we drop the restriction of only a few ionised states and consider the *d* electrons to be mobile and independent of one another, we are led to the band model of ferromagnetism, which was first discussed by Slater (1936) and Stoner (1936). The fact that the average magnetic moments are not always integral multiples of μ_B is easily explained by the band structure, when for the transition metals several bands ($s+d$) are assumed to overlap one another. The electron wave function will then be a mixture of s and d functions, so that the electron behaves partly as a d-electron and partly as an s electron (rather more d than s). Of course there must be an integral number of electrons in the s-band and in the d-band. The d-band (more exactly, the band with over whelming d-type character) is characterised by a high density of states, as indicated by the strong Pauli paramagnetism of non-ferromagnetic transition metals and the electronic heat at low temperatures (Fig. 14).

Moreover, within the framework of the band model it is easy to explain the magnetic properties of many alloys. A well-known example is the nickel–copper alloy. When for nickel we assume the $3d$-electrons of both spin directions to populate two identical bands which, because of the exchange interaction, are energetically shifted with respect to one another in such a way that one band is filled, then, owing to the amount of magnetisation at $T = 0$, there must be 0·6 holes per atom in the other band. When diamagnetic copper is alloyed to the nickel, the average magnetic moment per atom shows a decrease which is proportional to the copper concentration. It is obvious that copper transfers one of its surplus $4s$-electrons (compared with nickel) to the $3d$-band, which in this way is filled gradually.

This model agrees with the fact that at 60% Cu the alloy is no longer ferromagnetic. Similar conditions are observed for other alloys: except for small deviations and certain other exceptions, the average magnetic moment depends only on the number of electrons in the $3d$- and $4s$-bands (for iron-group alloys) and is independent of the other alloy component (Fig. 21).

This and other effects support a band model for the magnetism of transition metals (cf. Friedel, 1958; Mott, 1964).

A difficulty connected with the band model is the explanation of the appearance and the cause of ferromagnetism in the transition metals. There exist two opposite points of view. According to Slater (1936) and other authors, it is chiefly the exchange interaction between electrons in the same atomic cell which produces a parallel orientation of the spins: one calls this an intra-atomic exchange interaction. This is the

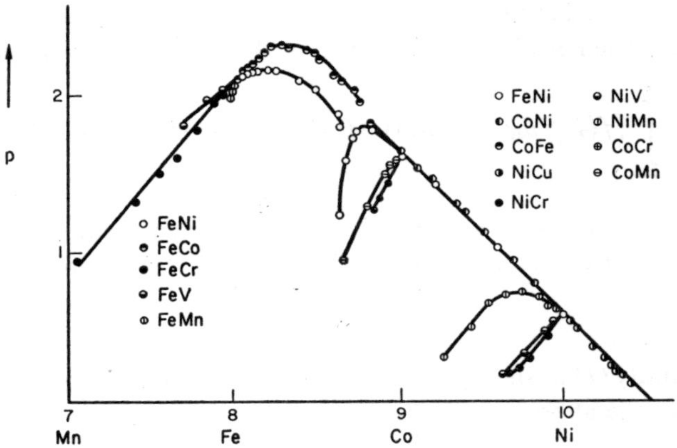

FIG. 21.

The mean magnetic moment p (in units of μ_B) per atom for binary alloys of the iron group as a function of the number of $3d+4s$ electrons (Slater–Pauling curve), after Crangle and Hallam (1963). The alloys on the right-hand side of the graph are face-centred cubic, those in the left-hand part are body-centred cubic.

same type of interaction which leads to Hund's rule and the permanent magnetic moments of the paramagnetic atoms (Section II.2). According to the other concept, the Coulomb interaction between electrons with antiparallel spins, i.e. the correlation energy, is the very cause of ferromagnetism. The problem has not been solved so far (cf. Mott, 1964).

The simplest band model for ferromagnetism is the collective electron model of Stoner (1938, 1939) and Wohlfarth (1949, 1951). In this model the electrons are treated as free electrons; the influence of the periodic lattice potential is described by an effective mass and the exchange interaction is described by a molecular field H_W which is parallel to the external field. The electrons are then governed by Fermi statistics. The magnetisation of the electrons is thus, according (2.2) and (2.3), given by

$$M = \mu_B \int_0^\infty \{f(E-\mu_B(H+H_W))-f(E+\mu_B(H+H_W))\} \frac{dZ}{dE} \, dE. \quad (3.100)$$

Here $f(E)$ is the Fermi distribution function and dZ/dE is the density of the states. The Fermi energy ζ is given by the total number of electrons:

$$N = \int_0^\infty \{f(E - \mu_B(H+H_W)) + f(E + \mu_B(H+H_W))\} \frac{dZ}{dE}\, dE. \quad (3.101)$$

The molecular field is proportional to the magnetisation. When we again introduce the relative magnetisation $\sigma = M/N\mu_B$, we put

$$\mu_B H_W = k\Theta\sigma. \quad (3.102)$$

The quantity Θ has the dimensions of a temperature. It is convenient to introduce the abbreviations

$$\begin{aligned} \alpha &= \beta\mu_B H_W, \quad & \xi &= \beta\zeta, \\ \alpha' &= \beta\mu_B H_W, \quad & x &= \beta E, \end{aligned} \qquad \xi_0 = \beta\zeta_0. \quad (3.103)$$

The parameter ζ_0 is given by (1.17). Writing

$$F_n(\eta) = \int_0^\infty \frac{x^n}{e^{x-\eta}+1}\, dx \quad (3.104)$$

and using the density of states (2.6) for the electrons, we can write the equations (3.100) and (3.101) in the form

$$\sigma = \frac{3}{4\xi_0^{3/2}} \{F_{1/2}(\xi+\alpha+\alpha') - F_{1/2}(\xi-\alpha-\alpha')\}, \quad (3.105)$$

$$1 = \frac{3}{4\xi_0^{3/2}} \{F_{1/2}(\xi+\alpha+\alpha') + F_{1/2}(\xi-\alpha-\alpha')\}. \quad (3.106)$$

We now have to show that these equations are suitable to describe ferromagnetism, i.e. to show that a spontaneous magnetisation results.

At low temperatures the arguments of the functions in (3.105) and (3.106) have high absolute values. First we assume the arguments to be positive, i.e. $\zeta > \mu_B H + k\Theta\sigma$. The function (3.104) can then be easily expanded (Wilson, 1958) and, neglecting higher powers of η^{-1}, we obtain

$$F_n(\eta) = \frac{1}{n+1}\eta^{n+1}\left(1+\frac{n(n+1)\pi^2}{6\eta^2}\right); \quad \eta \gg 1. \tag{3.107}$$

With this it becomes possible to eliminate ξ from eqns. (3.105) and (3.106). We obtain

$$\xi+\alpha+\alpha' = (1+\sigma)^{2/3}\xi_0 - \frac{\pi^2}{12}\cdot\frac{1}{\xi_0(1+\sigma)^{2/3}}, \tag{3.108}$$

$$\xi-\alpha-\alpha' = (1-\sigma)^{2/3}\xi_0 - \frac{\pi^2}{12}\cdot\frac{1}{\xi_0(1-\sigma)^{2/3}}. \tag{3.109}$$

The convergence of the last equation is connected with the condition $\sigma \ll 1$, which corresponds to the assumption of $\zeta > k\Theta+\mu_B H$. The parameter ξ is eliminated by subtraction and we obtain

$$2(\alpha+\alpha') = \xi_0\{(1+\sigma)^{2/3}-(1-\sigma)^{2/3}\} - \frac{\pi^2}{12\xi_0}\left\{\frac{1}{(1+\sigma)^{2/3}}-\frac{1}{(1-\sigma)^{2/3}}\right\}. \tag{3.110}$$

For $H = 0$ and $T = 0$ we obtain from this the equation from which the spontaneous magnetisation σ_0 can be determined:

$$\frac{2k\Theta}{\xi_0}\sigma_0 = (1+\sigma_0)^{2/3}-(1-\sigma_0)^{2/3}. \tag{3.111}$$

The equation has solutions with $\sigma_0 \neq 0$ for certain values of the parameter Θ. If Θ lies within the interval†

$$\frac{2}{3} \leqslant \frac{k\Theta}{\zeta_0} < 2^{-1/3}, \tag{3.112}$$

then, according to (3.111), the spontaneous magnetisation will lie in the interval $0 \leqslant \sigma_0 < 1$. Thus, a spontaneous magnetisation can arise in this model; however, we see that here the exchange interaction described by the molecular field is insufficient to parallelise all spins at

† We must expect a similar relation, yielding a certain proportionality between the exchange energy and the kinetic energy of the electrons, also from a more realistic model.

$T = 0$, if the parameter Θ lies in the interval (3.112). When $T \neq 0$, the magnetisation at low temperatures increases, according to (3.110), as the square of T. We obtain

$$\sigma = \sigma_0\left[1-\frac{3}{2}\left(\frac{3\pi kT}{\sigma_0\zeta_0}\right)^2\right].$$

In the case of strong exchange interactions, when the argument of $F_{1/2}(\xi-\alpha-\alpha')$ becomes negative ($\xi < k\Theta$), the function (3.104) is to a good approximation an exponential function:

$$F_n(\eta) = n!\,e^{\eta}; \qquad \eta < 0.$$

According to (3.105) and (3.104), the magnetisation will then depend exponentially on T (as in the molecular-field approximation), and we have $\sigma_0 = 1$. For the exchange parameter Θ we have

$$k\Theta \geqslant 2^{-1/3}\zeta_0.$$

The Curie temperature is determined from the vanishing of the spontaneous magnetisation. Expanding the functions in terms of α', we obtain from (3.104) and (3.105) an implicit equation to determine T_c

$$\frac{T_c}{\Theta} = \left(\frac{d\ln F_{1/2}(\xi)}{d\xi}\right)_{T=T_c} \quad \text{where} \quad F_{1/2}(\xi) = \frac{2}{3}\,\xi_0^{3/2}. \tag{3.113}$$

In order to obtain the susceptibility above the Curie point, we expand (3.105) and break off after the terms linear in H:

$$\chi^{-1} = \frac{1}{N\mu_B^2}\left\{\frac{F_{1/2}(\xi)}{F'_{1/2}(\xi)}kT - k\Theta\right\} = \chi_p^{-1}\frac{1-\dfrac{\Theta}{T}\dfrac{F'_{1/2}(\xi)}{F_{1/2}(\xi)}}{\xi_0^{-1/2}F'_{1/2}(\xi)} \tag{3.114}$$

where $F_{1/2}(\xi) = \frac{2}{3}\,\xi_0^{3/2}$ and $F'_{1/2}(\xi) = dF_{1/2}(\xi)/d\xi$; χ_p is the paramagnetic susceptibility of a free electron gas (2.1). In the classical limit when $\zeta_0 \ll kT$, the Fermi distribution changes to a Boltzmann distribution and therefore $F_{1/2}(\xi) \sim e^{\xi}$. We then obtain from (3.114) a Curie–Weiss law

$$\chi = \frac{N\mu_B^2}{k}\,\frac{1}{T-\Theta}.$$

In the case of paramagnetic metals, $k\Theta < \frac{2}{3}\zeta_0$, at medium and low temperatures $\zeta_0 \gg kT$ and thus $\xi \gg 1$. Using the expansion (3.107) we then obtain from (3.114)

$$\chi = \frac{\chi_{\mathrm{P}}}{1 - \dfrac{3}{2}\dfrac{k\Theta}{\zeta_0}}.$$

Taking exchange interactions into account through a molecular field corrects the susceptibility in the collective-electron model in the same way as the exchange energy (2.9) did in the free-electron model (cf. (2.10)).

The electron energy and specific heat are also easily calculated (Stoner, 1938, 1939). The magnetic contribution to the specific heat has a discontinuity at the Curie point and drops to zero, as is to be expected for a molecular-field approximation.

Antiferromagnetism can also be treated within the framework of the collective-electron model (Lidiard, 1954).

This model of band ferromagnetism is the simplest that is possible. Each electron aligns its spin in the effective magnetic field $H + H_{\mathrm{W}}$, independently of the other electrons. Thus the only possible excitations are due to spin flip processes; according to (3.102), the excitation energy at $H = 0$ is equal to $2\mu_{\mathrm{B}}H_{\mathrm{W}} = 2k\Theta\sigma$. This kind of excitations are called Stoner excitations. Because of them, the spontaneous magnetisation decreases as the square of the temperature. In this model, however, there are no collective excitations, i.e. spin waves. To demonstrate their presence in a relatively general form, it is expedient to use the Landau theory of Fermi fluids as developed in Section II.1. If there exist excitations of the type of spin waves, they must be solutions to the kinetic equation (2.8). With $\hbar = 1$ this equation reads

$$\begin{aligned}\frac{\partial \delta f}{\partial t} &+ \frac{1}{2}\left[\left(\frac{\partial \delta f}{\partial \boldsymbol{r}}\cdot\boldsymbol{v}\right) + \left(\boldsymbol{v}\cdot\frac{\partial \delta f}{\partial \boldsymbol{r}}\right)\right] \\ &- \frac{1}{2}\left[\left(\frac{\partial f_0}{\partial \boldsymbol{p}}\cdot\frac{\partial \delta\varepsilon}{\partial \boldsymbol{r}}\right) + \left(\frac{\partial \delta\varepsilon}{\partial \boldsymbol{r}}\cdot\frac{\partial f_0}{\partial \boldsymbol{p}}\right)\right] + i[\delta\varepsilon, f_0]_- + i[\varepsilon, \delta f]_- = 0,\end{aligned} \tag{2.8}$$

where

$$\delta\varepsilon(\boldsymbol{p}, \boldsymbol{s}, \boldsymbol{r}, t) = \mathrm{Tr}_{s'} \int \frac{d^3p'}{(2\pi)^3} \phi(\boldsymbol{p}, \boldsymbol{s}; \boldsymbol{p}', \boldsymbol{s}') \, \delta f(\boldsymbol{p}', \boldsymbol{s}', \boldsymbol{r}, t)$$

and

$$f_0 = (e^{\beta[\varepsilon(\boldsymbol{p}, \boldsymbol{s}) - \mu]} + 1)^{-1}.$$

In contrast to the paramagnetic case considered in Section II.1, the unperturbed quasi-particle energies $\varepsilon(\boldsymbol{p}, \boldsymbol{s})$ are here dependent on the orientation of the quasi-particle spin with respect to the magnetisation. We therefore put for $H = 0$

$$\varepsilon(\boldsymbol{p}, \boldsymbol{s}) = \varepsilon_0(\boldsymbol{p}) + \Delta(\boldsymbol{p})\,(\boldsymbol{\sigma}\cdot\boldsymbol{s}), \tag{3.115}$$

where $\boldsymbol{\sigma}$ is the unit vector in the direction of the magnetisation;

$$\boldsymbol{v} = \frac{\partial\varepsilon(\boldsymbol{p}, \boldsymbol{s})}{\partial\boldsymbol{p}}$$

is the velocity of a quasi-particle with momentum $\boldsymbol{p}$ and spin $\boldsymbol{s}$.

As in Section III.3, we have to find excitations which are periodic functions of space and time, i.e. we put

$$\delta f \sim \delta\varepsilon \sim e^{i(\boldsymbol{k}\cdot\boldsymbol{r}) - i\omega t}.$$

The kinetic equation for the Fourier components will then be

$$-\omega\,\delta f + \frac{1}{2}\left(\delta f(\boldsymbol{k}\cdot\boldsymbol{v}) + (\boldsymbol{k}\cdot\boldsymbol{v})\,\delta f\right)$$

$$-\frac{1}{2}\left(\left(\boldsymbol{k}\cdot\frac{\partial f_0}{\partial\boldsymbol{p}}\right)\delta\varepsilon + \delta\varepsilon\left(\boldsymbol{k}\cdot\frac{\partial f_0}{\partial\boldsymbol{p}}\right)\right) + [\delta\varepsilon, f_0]_- + [\varepsilon, \delta f]_- = 0.$$

Assuming the magnetisation to be parallel to the z-axis and choosing a representation for the spin matrices where s_z is diagonal, ε, $\boldsymbol{v}$, and f_0 will be diagonal matrices:

$$\varepsilon(\boldsymbol{p}, \boldsymbol{s}) = \varepsilon_0(\boldsymbol{p}) + \Delta(\boldsymbol{p}) s_z,$$

$$\boldsymbol{v}(\boldsymbol{p}, \boldsymbol{s}) = \boldsymbol{v}_0(\boldsymbol{p}) + \frac{\partial\Delta(\boldsymbol{p})}{\partial\boldsymbol{p}} s_z,$$

$$f_0(\boldsymbol{p}, \boldsymbol{s}) = f_1(\boldsymbol{p}) + f_2(\boldsymbol{p}) s_z,$$

where $f_1(\boldsymbol{p}) = \frac{1}{2}\mathrm{Tr}\, f_0(\boldsymbol{p}, \boldsymbol{s})$ and $2\mathrm{Tr}\, f_0(\boldsymbol{p}, \boldsymbol{s}) s_z = f_2(\boldsymbol{p})$, as $\mathrm{Tr}\, 1 = 2$ and $\mathrm{Tr}\, s_z^2 = \frac{1}{2}$; thus f_2 is a measure for the magnetisation of the system.

The quasi-particle spins will be predominantly opposite to the direction of the magnetisation; as in the Heisenberg model, the spin wave excitations are due to a slight deviation of the spin orientation from the preferred direction, while the particle density remains unchanged (Tr $\delta f = 0$), i.e. δf must have the form

$$\delta f(\boldsymbol{p}, \boldsymbol{s}) = \nu(\boldsymbol{p}) s^+; \qquad s^+ = s_x + i s_y.$$

Resolving the interaction Φ into φ_0 and φ_1 as in Section II.1, the relation assumed for δf yields the variation of energy

$$\delta\varepsilon(\boldsymbol{p}, \boldsymbol{s}) = \mathrm{Tr}_{s'} \int \frac{d^3p'}{(2\pi)^3} \big(\varphi_0(\boldsymbol{p}, \boldsymbol{p}') + \varphi_1(\boldsymbol{p}, \boldsymbol{p}')(\boldsymbol{s}\cdot\boldsymbol{s}')\big) \nu(\boldsymbol{p}') s'^+$$

$$= \frac{1}{2} \int \frac{d^3p'}{(2\pi)^3} \varphi_1(\boldsymbol{p}, \boldsymbol{p}')\, \nu(\boldsymbol{p}') s^+.$$

When forming the trace, the spin-independent component of the interaction drops out. Since s^+ commutes with the unit matrix and anticommutes with s_z, we obtain from the kinetic equation (3.115), with (3.35) and the Ansatz for δf, the following integral equation for $\nu(\boldsymbol{p})$:

$$\big(\omega - (\boldsymbol{k}\cdot\boldsymbol{v}_0(\boldsymbol{p})) - \Delta(\boldsymbol{p})\big)\, \nu(\boldsymbol{p}) + \frac{1}{2}\left(\boldsymbol{k}\cdot\frac{\partial f_1}{\partial \boldsymbol{p}}\right) \int \frac{d^3p'}{(2\pi)^3} \varphi_1(\boldsymbol{p}, \boldsymbol{p}')\, \nu(\boldsymbol{p}')$$
$$+ \frac{1}{2} f_2(\boldsymbol{p}) \int \frac{d^3p'}{(2\pi)^3} \varphi_1(\boldsymbol{p}, \boldsymbol{p}')\, \nu(\boldsymbol{p}') = 0. \qquad (3.116)$$

The eigenvalues of this equation, $\omega = \omega(k)$, must contain the spin wave frequencies we are looking for.

In the case of a zero external field one of the possible solutions to (3.116), must obviously always exist, owing to invariance: this is the solution corresponding to a rotation of the vector of magnetisation; this may be considered to be the limiting case of a spin wave with a very large wavelength ($k \to 0$) and zero excitation energy ($\omega = 0$).

The equation

$$\Delta(\boldsymbol{p})\, \nu(\boldsymbol{p}) = \frac{1}{2} f_2(\boldsymbol{p}) \int \frac{d^3p'}{(2\pi)^3} \varphi_1(\boldsymbol{p}, \boldsymbol{p}')\, \nu(\boldsymbol{p}')$$

will therefore possess a non-trivial solution if the magnetisation, i.e. f_2, is non-vanishing. Since $f_2(\boldsymbol{p})$ is proportional to the magnetisation of the quasi-particles with momentum $\boldsymbol{p}$,

$$\nu(\boldsymbol{p}) = \text{const}\, f_2(\boldsymbol{p})$$

must be this solution. Substituting it, we obtain

$$\Delta(\boldsymbol{p}) = \frac{1}{2}\int \frac{d^3\boldsymbol{p}'}{(2\pi)^3}\varphi_1(\boldsymbol{p}, \boldsymbol{p}')f_2(\boldsymbol{p}'). \tag{3.117}$$

This equation is thus some kind of consistency equation; it connects Δ with the exchange interaction φ_1 and—*cum grano salis*—with the magnetisation. It is decisive for the spin-wave dispersion, as it obviously implies that the spin-wave frequency $\omega(k)$ vanishes as $k \to 0$. This dependence of the spin-wave energy on the wave vector is therefore due to the invariance properties of the system. Such excitations or quasi-particles whose presence is connected with certain symmetries, more precisely, with symmetry-breaking parameters (spontaneous magnetisation), are also called Goldstone particles (cf., for example, Wagner, 1966).

Using the solution obtained for $k = 0$, eqn. (3.116) can be solved for $k \neq 0$ by means of an iteration method (Silin, 1958; Abrikosov and Dzyaloshinskii, 1959).

For the sake of simplicity we shall here consider only a case where the exchange interaction is independent of the momenta and $\varphi_1 = \text{const}$. According to eqn. (3.117) Δ will then be constant:

$$\Delta = \frac{\varphi_1}{2V}\frac{V}{(2\pi)^3}\int d^3\boldsymbol{p}\ \mathrm{Tr}_s\, f_0(\boldsymbol{p}, \boldsymbol{s})2s_z = -\frac{\varphi_1}{2V}(N_+ - N_-);$$

N_+ (N_-) is the number of quasi-particles whose moment is parallel (or antiparallel) to the direction of the magnetisation, and V is the volume (cf. Section II.1). It is obvious that $\varphi_1 < 0$. The eigenvalue equation is then obtained from (3.116) by dividing it by $(\omega - (\boldsymbol{k}\cdot\boldsymbol{v}_0) - \Delta)$ and integrating it over all momenta; in this way we obtain

$$1 = -\frac{\varphi_1}{2}\int \frac{d^3\boldsymbol{p}}{(2\pi)^3}\, \frac{f_2+\left(\boldsymbol{k}\cdot\frac{\partial f_1}{\partial \boldsymbol{p}}\right)}{\omega-(\boldsymbol{k}\cdot\boldsymbol{v}_0)-\Delta}\,. \tag{3.118}$$

This equation has solutions for all values of $\boldsymbol{k}$ and Δ. For small values of k it can be expanded as a power series in k and, using eqn. (3.117), we obtain as the lowest approximation

$$\omega = -\frac{\varphi_1}{6\Delta^2}k^2\int \frac{d^3\boldsymbol{p}}{(2\pi)^3}\, v_0^2(\boldsymbol{p})\left(f_2(\boldsymbol{p})-\Delta\frac{\partial f_1}{\partial \varepsilon_0}\right).$$

Thus, as in the case of the spin waves of a Heisenberg ferromagnet, for small propagation vectors the dispersion behaves as the square of k. It is obvious that in the expression in parentheses the second term is predominant, as otherwise the spin wave frequency would become negative; this would mean then that the assumed ferromagnetic ground state were not stable (cf. the analogous case for antiferromagnetism (3.70)).

From eqn. (3.116) we can also see that the Stoner excitations are contained as solution. We see that for $k = 0$, $\nu(\boldsymbol{p}) = a\delta(\boldsymbol{p}-\boldsymbol{p}_0)$ is a solution, since the last part in (3.116) can be made arbitrarily small after integrating over an arbitrarily small zone about p_0. The eigenvalue corresponding to this solution is clearly $\omega = \Delta(\boldsymbol{p}_0)$. If the exchange splitting Δ is independent of $\boldsymbol{p}$, Δ must correspond to the Stoner parameter, i.e. to $2k\Theta\sigma$ according to (3.102). When we assume both excitations to be independent of one another at low temperatures, the magnetisation will show both a $T^{3/2}$-dependence and a T^2-dependence.

Changing to a microscopic description of band ferromagnetism, we have to take into account a series of empirical facts. As a rule, metallic ferromagnets have two or several bands which are important for their physical properties, such as for example, the 3*d*- and the 4*s*-bands in the iron group. The second band, usually an *s* band, can be decisive for the presence of ferromagnetism or, more generally, for the presence of a magnetically ordered state (cf. the rare-earth metals, Section III.1).

Another fact is the high angular momentum of the electrons (d or f states) which are mainly responsible for the magnetism. The importance of this fact is not clear. In the following we shall disregard all effects of orbital angular momentum; all electrons are treated as s electrons.

In order to describe the electrons in a crystal we shall make use of the Bloch functions. They are solutions of the Schrödinger equation

$$\mathscr{H}_0\varphi_{kp} = \left(-\frac{\hbar^2}{2m}\nabla^2+V(\boldsymbol{r})\right)\varphi_{kp} = \varepsilon_{kp}\varphi_{kp}; \tag{3.119}$$

$V(\boldsymbol{r})$ is the periodic lattice potential, k stands for the wave vector $\boldsymbol{k}$ which lies in the first Brillouin zone, and the spin s, and p is the band index. When we take into account a constant external field H, we have, besides $\mathscr{H}_0$ and ε_{kp}, an extra term $g\mu_B(\boldsymbol{H}\cdot\boldsymbol{s})$. The Bloch functions are assumed to be orthonormalised:

$$\int \varphi^*_{kp}(\boldsymbol{r})\,\varphi_{k'p'}(\boldsymbol{r})\,d^3\boldsymbol{r} = \delta_{k,k'}\delta_{p,p'}.$$

Electrons of high mobility, which are not very well localised and thus possess a wide band, are described expediently by Bloch functions. For the d- and f-electrons it is often more convenient to work with Wannier functions $w_{ps}(\boldsymbol{r}-\boldsymbol{r}_i)$. They are linear combinations of Bloch functions:

$$w_{ps}(\boldsymbol{r}-\boldsymbol{r}_i) = \frac{1}{\sqrt{N}}\sum_k \varphi_{kp}(\boldsymbol{r})\,e^{-i(\boldsymbol{k}\cdot\boldsymbol{r}_i)},$$

and

$$\varphi_{kp}(\boldsymbol{r}) = \frac{1}{\sqrt{N}}\sum_i w_{ps}(\boldsymbol{r}-\boldsymbol{r}_i)\,e^{i(\boldsymbol{k}\cdot\boldsymbol{r}_i)}.$$

The $\boldsymbol{r}_i$ are the lattice sites of a Bravais lattice with N atoms. The Wannier functions $w_{ps}(\boldsymbol{r}-\boldsymbol{r}_i)$ have the property of being concentrated around the lattice sites $\boldsymbol{r}_i$, thus having a local character (Kittel, 1964); they are in fact the most strongly localised wave functions to be built up from the Bloch functions of a band. From the orthonormalisation of the Bloch functions it follows that the Wannier functions belonging to different lattice sites are orthogonal onto one another:

$$\int d^3r w_{ps}^*(\boldsymbol{r}-\boldsymbol{r}_i) w_{p's'}(\boldsymbol{r}-\boldsymbol{r}_j)$$
$$= \frac{1}{N} \sum_{k,k'} e^{i(k\cdot r_i)-i(k'\cdot r_j)} \int d^3r \varphi_{kp}^*(\boldsymbol{r}) \varphi_{k'p'}(\boldsymbol{r})$$
$$= \frac{1}{N} \sum_{k} e^{i(k\cdot [r_i-r_j])} \delta_{p,p'} \delta_{ss'} = \delta_{i,j} \delta_{p,p'} \delta_{s,s'}.$$

In order to take the Pauli principle for the electrons into account in a simple way, i.e. to avoid calculations involving the use of the cumbersome Slater determinants as in (3.9), it is expedient to formulate the problem using second quantisation methods (cf., for example, Abrikosov, Gorkov, Dzyaloshinskii, 1965; Nozières, 1964). We define mutually adjoint creation and annihilation operators for electrons at the point $\boldsymbol{r}$ with the spin s: $\psi_s^+(\boldsymbol{r})$ and $\psi_s(\boldsymbol{r})$. We can then write the Hamiltonian of the interacting electrons as follows:

$$\mathcal{H} = \sum_s \int d^3r \psi_s^+(\boldsymbol{r}) \mathcal{H}_0 \psi_s(\boldsymbol{r}) \tag{3.120}$$
$$+ \tfrac{1}{2} \sum_{s,s'} \int d^3r\, d^3r' \psi_s^+(\boldsymbol{r}) \psi_{s'}^+(\boldsymbol{r}') V(\boldsymbol{r}-\boldsymbol{r}') \psi_{s'}(\boldsymbol{r}') \psi_s(\boldsymbol{r}).$$

$\mathcal{H}_0$ is the single-particle operator (3.119) and $V(\boldsymbol{r}-\boldsymbol{r}')$ is the spin-independent Coulomb interaction. The field operators are governed by the anticommutation relations

$$[\psi_s(\boldsymbol{r}), \psi_{s'}^+(\boldsymbol{r}')]_+ = \delta_{s,s'} \delta(\boldsymbol{r}-\boldsymbol{r}'),$$
$$[\psi_s(\boldsymbol{r}), \psi_{s'}(\boldsymbol{r}')]_+ = [\psi_s^+(\boldsymbol{r}), \psi_{s'}^+(\boldsymbol{r}')]_+ = 0.$$

The operator $\psi_s^+(\boldsymbol{r})\psi_s(\boldsymbol{r})$ is a density operator for particles with spin s and

$$N_{\text{op}} = \sum_s \int d^3r \psi_s^+(\boldsymbol{r}) \psi_s(\boldsymbol{r}) \tag{3.121}$$

therefore corresponds to the operator of the total number of particles. Expanding the field operators in terms of the Bloch functions, which represent a complete system of functions, we obtain

$$\psi_s(\boldsymbol{r}) = \sum_{k,\,p} a_{kp}\varphi_{kp}(\boldsymbol{r}), \quad \psi_s^+(\boldsymbol{r}) = \sum_{k,\,p} a_{kp}^+\varphi_{kp}^*(\boldsymbol{r}),$$

$$a_{kp} = \int d^3\boldsymbol{r}\psi_s(\boldsymbol{r})\,\varphi_{kp}^*(\boldsymbol{r}), \quad a_{kp}^+ = \int d^3\boldsymbol{r}\psi_s^+(\boldsymbol{r})\,\varphi_{kp}(\boldsymbol{r}).$$

The a_{kp}^+ and a_{kp} are the creation and annihilation operators of electrons in the states determined by the quantum numbers k and p. Their commutation relations result immediately from those of the field operators and the orthonormalised nature of the Bloch functions:

$$[a_{kp}, a_{k'p'}^+]_+ = \delta_{k,\,k'}\delta_{p,\,p'}, \tag{3.122}$$
$$[a_{kp}, a_{k'p'}]_+ = [a_{kp}^+, a_{k'p'}^+]_+ = 0;$$

$n_{kp} = a_{kp}^+a_{kp}$ is the particle number operator for the state (k, p); $\sum_{k,p} n_{kp}$ is then again the operator (3.121) of the total number; this can be verified by substitution.

Using (3.119), the Hamiltonian (3.120) can be written in the Bloch representation

$$\mathscr{H} = \sum_{k,\,p} \varepsilon_{kp}a_{kp}^+a_{kp} + \tfrac{1}{2} \sum_{k_\nu,\,p_\nu,\,s,\,s'} V(1, 2; 3, 4)a_{k_1p_1}^+a_{k_2p_2}^+a_{k_3p_3}a_{k_4p_4}, \tag{3.123}$$

where $s_1 = s_4 = s$ and $s_2 = s_3 = s'$. The interaction matrix element V, when written out in full, is

$$V(1, 2; 3, 4) \tag{3.124}$$
$$= \int d^3\boldsymbol{r}\, d^3\boldsymbol{r}'\varphi_{k_1,\,s,\,p_1}^*(\boldsymbol{r})\,\varphi_{k_2,\,s',\,p_2}(\boldsymbol{r}')V(\boldsymbol{r}-\boldsymbol{r}')\,\varphi_{k_3,\,s',\,p_3}(\boldsymbol{r}')\,\varphi_{k_4,\,s,\,p_4}(\boldsymbol{r}).$$

Because of the translational symmetry the matrix element contains a Kronecker symbol $\delta_{k_1+k_2,\,k_3+k_4+g}$, where $\boldsymbol{g}$ is a reciprocal lattice vector. Since the momentum of the crystal electrons is $\hbar\boldsymbol{k}$ (mod $\hbar\boldsymbol{g}$), this means that the lattice momentum is conserved during interaction, i.e. the momentum of the particles annihilated must be equal to that of the particles created (Ziman, 1962).

In a similar way we can expand the field operators in terms of the Wannier functions

$$\psi_s(\boldsymbol{r}) = \sum_{i,p} a_{isp} w_{ps}(\boldsymbol{r}-\boldsymbol{r}_i), \qquad \psi_s^+(\boldsymbol{r}) = \sum_{i,p} a_{isp}^+ w_{ps}^*(\boldsymbol{r}-\boldsymbol{r}_i),$$

or

$$a_{isp} = \int d^3\boldsymbol{r}\psi_s(\boldsymbol{r})\, w_{ps}^*(\boldsymbol{r}-\boldsymbol{r}_i), \qquad a_{isp}^+ = \int d^3\boldsymbol{r}\psi_s^+(\boldsymbol{r})\, w_{ps}(\boldsymbol{r}-\boldsymbol{r}_i).$$

The creation operators a_{isp}^+ and the annihilation operators a_{isp} of particles at the point $\boldsymbol{r}_i$ obey the commutation relations

$$[a_{isp}, a_{js'p'}^+]_+ = \delta_{i,j}\delta_{s,s'}\delta_{p,p'}, \tag{3.125}$$
$$[a_{isp}, a_{js'p'}]_+ = [a_{isp}^+, a_{js'p'}^+]_+ = 0.$$

The relation between the a_{isp} and the a_{kp} is

$$a_{isp} = \frac{1}{\sqrt{N}}\sum_k a_{kp}\, e^{i(\boldsymbol{k}\cdot\boldsymbol{r}_i)}, \qquad a_{isp}^+ = \frac{1}{\sqrt{N}}\sum_k a_{kp}^+\, e^{-i(\boldsymbol{k}\cdot\boldsymbol{r}_i)}. \tag{3.126}$$

In the Wannier representation the Hamiltonian is

$$\mathcal{H} = \sum_{i,j,s,p} t_{ijsp} a_{isp}^+ a_{jsp} + \tfrac{1}{2} \sum_{i_\nu, p_\nu, s, s'} V(1, 2; 3, 4) a_{i_1sp_1}^+ a_{i_2s'p_2}^+ a_{i_3s'p_3} a_{i_4sp_4}. \tag{3.127}$$

Here we must take into account the fact that, according to (3.119), $\mathcal{H}_0$ is diagonal with respect to band index and spin so that

$$t_{ijps} = \int d^3 w_{ps}^*(\boldsymbol{r}-\boldsymbol{r}_i)\mathcal{H}_0 w_{ps}(\boldsymbol{r}-\boldsymbol{r}_j) = \frac{1}{N}\sum_k \varepsilon_{kp}\, e^{-i(\boldsymbol{k}\cdot[\boldsymbol{r}_i-\boldsymbol{r}_j])}. \tag{3.128}$$

The interaction matrix element in the Wannier representation is

$$V(1\ 2; 3, 4) \tag{3.129}$$
$$= \int d^3\boldsymbol{r}\, d^3\boldsymbol{r}' w_{p_1s}^*(\boldsymbol{r}-\boldsymbol{r}_{i_1})\, w_{p_2,s'}^*(\boldsymbol{r}'-\boldsymbol{r}_{i_2})\, V(\boldsymbol{r}-\boldsymbol{r}')\, w_{p_3s'}(\boldsymbol{r}'-\boldsymbol{r}_{i_3})\, w_{p_4s}(\boldsymbol{r}-\boldsymbol{r}_{i_4}).$$

If the lattice sites contained in it are far away from one another, the matrix element will be small because of the local character of the Wannier functions. Those matrix elements in which the Wannier functions refer to the same lattice site will therefore be largest. This part of the interaction is called the intra-atomic interaction. In the other part of the interaction at least two different lattice sites are contained, so that it is reasonable to call it the inter-atomic interaction. Compared with the first it is the smaller, as in final reckoning one is dealing with effective interactions and the inter-atomic interaction is certainly screened more strongly than intra-atomic interaction, and thus it is reduced. Among all possible inter-atomic interactions those between neighbouring lattice sites will be strongest and among them those where the integrand in (3.129) is positive, i.e. $i = i_1 = i_4$, $p_1 = p_4$ and $j = i_2 = i_3, p_2 = p_3$. This matrix element is given by

$$U_{ij} = \int d^3\boldsymbol{r}\, d^3\boldsymbol{r}'\, |w_{p_1s}(\boldsymbol{r}-\boldsymbol{r}_i)|^2\, |w_{p_2s'}(\boldsymbol{r}'-\boldsymbol{r}_j)|^2\, V(\boldsymbol{r}-\boldsymbol{r}').$$

It corresponds to the classical interaction of two charge densities which are concentrated around the lattice sites $\boldsymbol{r}_i$ and $\boldsymbol{r}_j$.

Of more interest for magnetism is the corresponding exchange term, i.e. the matrix element (3.129) with $i = i_1 = i_3$, $p_1 = p_3$ and $j = i_2 = i_4, p_2 = p_4$:

$$J_{ij}(p_1, p_2) = \int d^3\boldsymbol{r}\, d^3\boldsymbol{r}'\, w^*_{p_1}(\boldsymbol{r}-\boldsymbol{r}_i)\, w^*_{p_2}(\boldsymbol{r}'-\boldsymbol{r}_j)\, V(\boldsymbol{r}-\boldsymbol{r}')\, w_{p_1}(\boldsymbol{r}'-\boldsymbol{r}_i)\, w_{p_2}(\boldsymbol{r}-\boldsymbol{r}_j). \tag{3.130}$$

Its form is precisely that of the exchange integral (3.8) introduced above; it is independent of the spin quantum numbers s and s' since, because the spin-orbit coupling is neglected, w_{ps} is the product of an orbital function w_p and a spin function, α or β. The simplest way to show that (3.130) also leads us to an expression in (3.127) corresponding to the Heisenberg operator, when we introduce two-component spinor operators:

$$a_{ip} = \begin{pmatrix} a_{i\uparrow p} \\ a_{i\downarrow p} \end{pmatrix}, \quad a^+_{ip} = \begin{pmatrix} a^+_{i\uparrow p} \\ a^+_{i\downarrow p} \end{pmatrix}. \tag{3.131}$$

The arrows indicate the two possible directions of the angular momentum. The term in (3.127) corresponding to the exchange interaction can then be written as follows:

$$\frac{1}{2}\sum_{\substack{i,j,s,s'\\p_1,p_2}} J_{ij}(p_1,p_2)a^{+}_{isp_1}a_{js'p_2}a_{is'p_1}a_{jsp_2}$$

$$= -\frac{1}{4}\sum_{i,j,p_1,p_2} J_{ij}(p_1,p_2)(a^{+}_{ip_1}a_{ip_1})(a^{+}_{jp_2}a_{jp_2}) \tag{3.132}$$

$$-\sum_{i,j,p_1,p_2} J_{ij}(p_1,p_2)([a^{+}_{ip_1}s_i a_{ip_1}]\cdot[a^{+}_{jp_2}s_j a_{jp_2}]);$$

this is easily proved when the calculation is carried out. As usual, the s_i are the angular momentum matrices in terms of $\hbar$. The decomposition of (3.132), of course, means only that every interaction between electrons can be resolved into a spin-independent part (first term) and a spin-dependent part (last term); this was frequently used when considering Fermi liquids (Section II.1). The second term is thus the Hamiltonian (3.7) in second quantization.

Even if in the Hamiltonian (3.127) we took into account only the inter-atomic interaction (3.132), besides the stronger intra-atomic interaction, the problem to be solved would still be too complicated. Thus it seems reasonable (only at a first glance, however) to ignore all interatomic interactions (compare the detailed discussion by Herring, 1966).

When the intra-atomic interaction is further simplified by neglecting all band–band transitions, i.e. when we take into account only that part which refers to the band with the most strongly localised wave functions, the Hamiltonian (3.127) is resolved into the Hamiltonians of the various bands; with the exception of the narrowest band they all describe free electrons. When we drop the band index, we can write the Hamiltonian of the narrowest band as follows:

$$\mathcal{H} = \sum_{i,j} t_{ijs}a^{+}_{is}a_{js} + U\sum_{i} n_{i\uparrow}n_{i\downarrow}, \tag{3.133}$$

where

$$U = \int d^3r\, d^3r'\, |w(r)|^2 |w(r')|^2 V(r-r'),$$

and $n_{is} = a_{is}^+ a_{is}$. In the Bloch representation, however, we have

$$\mathcal{H} = \sum_{k,s} \varepsilon_{ks} a_{ks}^+ a_{ks} + \frac{U}{N} \sum_{k_\nu} a_{k_1\uparrow}^+ a_{k_4\uparrow} a_{k_2\downarrow}^+ a_{k_3\downarrow} \delta_{k_1+k_2,\, k_3+k_4+g}. \tag{3.133a}$$

The intra-atomic Coulomb interaction U results in a repulsion of two electrons with antiparallel spin at the same point; the electron system will therefore tend to align all electron spins in parallel positions, i.e. it will tend to become ferromagnetic in order to avoid such energetically unfavourable configurations.

If we accept that all assumptions made so far about the interaction are justified, the reduction of the problem to one with a single band was only possible because $\mathcal{H}_0$ was diagonal in the band index. This, however, would not be true if, for example, instead of another *d*-band besides the *s*-band we had only a localised *d*-state in the form of a paramagnetic impurity, such as is, for example, the case of a titanium atom in a copper lattice. Since the impurity problem is of both practical and theoretical importance, we shall briefly discuss it; for references we mention papers by Friedel (1958), Vonsovskij and Izjumov (1964), Mott (1964), and Marshall (1967). Repeating the steps of the calculation which led us from (3.120) to (3.133), we have first to expand the field operators in terms of the Bloch or Wannier functions of the *s*-band of the lattice and the *d*-function of the impurity:

$$\psi_s(\boldsymbol{r}) = \sum_i a_{is} w(\boldsymbol{r}-\boldsymbol{r}_i) + c_s w_{\boldsymbol{R}}(\boldsymbol{r}),$$

$$\psi_s^+(\boldsymbol{r}) = \sum_i a_{is}^+ w^*(\boldsymbol{r}-\boldsymbol{r}_i) + c_s^+ w_{\boldsymbol{R}}^*(\boldsymbol{r}).$$

Here $w_{\boldsymbol{R}}(\boldsymbol{r})$ is the atomic *d*-function of the impurity at the point $\boldsymbol{R}$. Apart from the matrix elements for the Wannier states of the *s*-band which we know already from (3.132) (without field)

$$t_{ij} = \int d^3r w^*(\boldsymbol{r}-\boldsymbol{r}_i)\mathcal{H}_0 w(\boldsymbol{r}-\boldsymbol{r}_j),$$

the matrix elements of $\mathcal{H}_0$ are the following:

$$E_d = \int d^3r w_{\boldsymbol{R}}^*(\boldsymbol{r})\mathcal{H}_0 w_{\boldsymbol{R}}(\boldsymbol{r}),$$

i.e. the expectation value of the d-state energy, and the mixing potential

$$V_{Ri} = V_{iR}^* = \int d^3r w_R^*(\mathbf{r}) \mathcal{H}_0 w(\mathbf{r}-\mathbf{r}_i).$$

It describes the mixing of s- and d-states or the transition of s-electrons to the d-state and vice versa.

When we take into account only the interaction for the d-state—the state which is most strongly localised—the Hamiltonian for this problem will be

$$\mathcal{H} = \sum_{i,j,s} t_{ij} a_{is}^+ a_{js} + \sum_{i,s} (V_{iR} a_{is}^+ c_s + V_{iR}^* c_s^+ a_{is}) + E_d \sum_s n_s + U n_\uparrow n_\downarrow,$$

where $n_s = c_s^+ c_s$.

When we use for the band electron the Bloch representation, we finally arrive at

$$\mathcal{H} = \sum_{k,s} \varepsilon_k a_{ks}^+ a_{ks} + \sum_{k,s} (V_k a_{ks}^+ c_s + V_k^* c_s^+ a_{ks}) + E_d \sum_s n_s + U n_\uparrow n_\downarrow; \qquad (3.134)$$

V_k is the Fourier transform of V_{iR}. This operator was first suggested by Anderson (1961) for studying the problem of the conditions under which an impurity described in this way has a localised magnetic moment, and when an alloy containing such impurities displays Langevin paramagnetism.

The intra-atomic energy U can be determined immediately when such impurities are investigated experimentally. In this way U values of about 5 eV are obtained for various atoms of the iron group contained in Cu (Klein and Heeger, 1966; Brewig *et al.*, 1969). This is about one-fourth of the Coulomb energies in free atoms, so that, even for the intra-atomic interaction, the screening effect of the band electrons is significant.

If E_d is large enough, i.e. if the band electrons can populate the d-level, and if U is large enough, so that it is energetically unfavourable to populate the d-level by two electrons, the impurity will contain only a single electron, thus behaving like an atom with the spin quantum number $s = \frac{1}{2}$. This was first shown explicitly by Schrieffer and Wolff (1966). In order to show this, we assume the mixing potential to be a small quantity, and transform the Hamiltonian (3.134) by means of a canonical transformation (i.e. we carry out a rotation in the Hilbert space of the *s* and *d* states), so that the transformed Hamiltonian $\tilde{\mathcal{H}}$ no longer contains the mixing potential in a linear form.

In the case of a transformation generated by $U = e^S$, the unitarity of U requires that $S^+ = -S$. The transformed Hamiltonian is then obtained from

$$\tilde{\mathcal{H}} = e^S H e^{-S} = \mathcal{H} + [S, \mathcal{H}]_- + \tfrac{1}{2}[S, [S, \mathcal{H}]_-]_- + \ldots,$$

when we expand it as a power series in S. We put $\mathcal{H} = \mathcal{H}_0 + \mathcal{H}_1$ so that $\mathcal{H}_1$ contains all terms with V_k. Under the assumption that S is a linear function of V_k (which can be checked) we have

$$\tilde{\mathcal{H}} = \mathcal{H}_0 + \mathcal{H}_1 + [S, \mathcal{H}_0]_- + [S, \mathcal{H}_1]_- + \tfrac{1}{2}[S, [S, \mathcal{H}_0]_-]_-,$$

where effects of third order in V_k are neglected.

The terms linear in V_k are eliminated from $\tilde{\mathcal{H}}$ if S satisfies the condition

$$\mathcal{H}_1 + [S, \mathcal{H}_0]_- = 0,$$

so that

$$\tilde{\mathcal{H}} = \mathcal{H}_0 + \tfrac{1}{2}[S, \mathcal{H}_1]_-.$$

Because of the constraint equation it is reasonable to write S in the following form:

$$S = \sum_{k,s} V_k \alpha_{ks} a_{ks}^+ c_s - \sum_{k,s} V_k^* \alpha_{ks}^+ c_s^+ a_{ks}$$

α_{ks} can also be an operator if it commutes only with $\mathcal{H}_0$. From the constraint equation for S it is easy to obtain

$$\alpha_{ks} = \frac{1}{\varepsilon_k - E_d - Un_{-s}} = \frac{n_{-s}}{\varepsilon_k - E_d - U} + \frac{1 - n_{-s}}{\varepsilon_k - E_d}$$

(it must be borne in mind that $n_s^2 = n_s$ and $n_s(1-n_s) = 0$). There are no problems as regards the calculation of the commutator in $\tilde{\mathcal{H}}$. Apart from terms of no interest we obtain for it

$$-\tfrac{1}{2} \sum_{k,k',s} J_{k,k'} \{\tfrac{1}{2}(a_{ks}^+ a_{k's} c_s^+ c_s - a_{ks}^+ a_{k's} c_{-s}^+ c_{-s}) + a_{ks}^+ a_{k'-s} c_{-s}^+ c_s\}.$$

Again changing to the spinor representation (3.131), we see that this corresponds to an exchange interaction between the band electrons and an $s = \frac{1}{2}$ impurity spin; we obtain

$$\tilde{\mathscr{H}} = \sum_{k, s} \varepsilon_k a_{ks}^+ a_{ks} - \tfrac{1}{2} \sum_{k, k'} J_{k, k'} (\{a_k^+ s_{k, k'} a_{k'}\} \cdot \{c^+ s c\}), \qquad (3.135)$$

with the exchange integral

$$J_{k, k'} = V_k V_{k'}^* \left(\frac{1}{\varepsilon_k - E_d - U} + \frac{1}{\varepsilon_{k'} - E_d - U} - \frac{1}{\varepsilon_k - E_d} - \frac{1}{\varepsilon_{k'} - E_d} \right).$$

When we assume, for example, V_k to depend only on the absolute magnitude of k, and the electron states near the Fermi energy ζ to be the only ones that are essential, we obtain for the exchange integral

$$J = 2 |V_{k_0}|^2 \frac{1}{(\zeta - E_d)(\zeta - E_d - U)}.$$

Since the band electrons are to populate the d-level, we must have $E_d < \zeta$. On the other hand, U should be so high that the second d-level with energy $E_d + U$ cannot be populated, i.e. $E_d + U > \zeta$. Hence we find $J < 0$. The band electrons are therefore antiferromagnetically coupled to the impurity spin.

The operator (3.135) is fundamental for the interaction between electrons and paramagnetic impurities. Using (3.135), Kondo (1964) succeeded in explaining the well-known resistivity minimum observed at low temperatures in alloys containing paramagnetic impurities.

Returning to the problem of band ferromagnetism as formulated in (3.137), we shall discuss within the framework of this model two questions, namely that of the conditions under which ferromagnetism arises and the problem of the spin waves. For this purpose we shall use the formalism of the double-time Green functions described in Section III.3. The problem first to be solved is to decide which Green function is the important one for the case considered. As we are ultimately interested in the number of electrons with spin s at a lattice site, $\langle n_s \rangle = \langle a_{is}^+ a_{is} \rangle$, or a state with a given momentum, $\langle n_{ks} \rangle = \langle a_{ks}^+ a_{ks} \rangle$, we are led to the single-particle Green function $\langle\langle a_{is}(t); a_{js}^+ \rangle\rangle$ or

$\langle\langle a_{ks}(t); a_{ks}^{+}\rangle\rangle$. They describe the time behaviour of a particle which enters the system at $t = 0$; see the analogous discussion in Section III.3 for spin waves. The poles of this Green function are therefore the single-particle excitation energies.

Unlike the operators dealt with in Section III.3, the creation and annihilation operators used in this case have no non-vanishing matrix elements in the Hilbert space of a system with fixed particle number N; this is an obvious consequence of the definition of these operators. It is thus necessary to change to the "larger" Hilbert space of systems with variable particle number but fixed chemical potential ζ or μ. This means at the same time that the statistical averages contained in the Green functions must be considered to be averages over a grand canonical ensemble and not over the canonical ensemble introduced in (2.42). It is a well-known fact that in the thermodynamic limit as $N \to \infty$ the results of a calculation are independent of whether this calculation was carried out within the framework of a canonical or a grand-canonical ensemble. When changing to the grand-canonical ensemble we have to replace everywhere the Hamiltonian $\mathcal{H}$ in Section III.3 by $\mathcal{H}-\mu N_{\text{op}}$, where N_{op} is the operator of the total particle number defined in (3.121)

$$N_{\text{op}} = \sum_{i,\,s} a_{is}^{+}a_{is} = \sum_{k,\,s} a_{ks}^{+}a_{ks}.$$

Therefore, instead of the Hamiltonian (3.137), we have to use in our calculations the Hamiltonian

$$\begin{aligned}\mathcal{H}-\mu N_{\text{op}} &= \sum_{k,\,s} (\varepsilon_{ks}-\mu)n_{ks} \qquad (3.136)\\ &\quad +\frac{U}{N}\sum_{k_\nu} a_{k_1\uparrow}^{+}a_{k_4\uparrow}a_{k_2\downarrow}^{+}a_{k_3\downarrow}\,\delta_{k_1+k_2,\,k_3+k_4+g}\\ &= \sum_{i,j,s} (t_{ijs}-\mu\,\delta_{ij})a_{is}^{+}a_{js}+U\sum_{i} n_{i\uparrow}n_{i\downarrow}.\end{aligned}$$

When we reckon, however, the single-particle energies ε_{ks} or t_{ii} from the chemical potential as zero (which will be done in the following), eqn. (3.136) is of the same form as (3.133) and all formulas for the Green functions in Section III.3 can be used literally in their original form.

According to (3.57), the equation of motion of the Green function $\langle\langle a_{is}(t);\, a_{js}^{+}\rangle\rangle$ with $\eta = -1$ and $\hbar = 1$ is

$$i\langle\langle \dot{a}_{is}(t);\, a_{js}^{+}\rangle\rangle = \delta_{ij} + \sum_{i_1} t_{ii_1 s}\langle\langle a_{i_1 s}(t);\, a_{js}^{+}\rangle\rangle + U\langle\langle a_{is}(t)\, n_{i-s}(t);\, a_{js}^{+}\rangle\rangle, \tag{3.137}$$

where we used (3.125) and (3.136).

The simplest approximation for the last Green function, which is a two-particle Green function, consists of a replacement of the operator n_{i-s} by its average $\langle n_{-s}\rangle$; and because of the translational symmetry of the system this operator is independent of the lattice site. This approximation is the Hartree–Fock approximation.

Changing to the Fourier transformation with respect to time, we obtain in this approximation from (3.137)

$$(\omega - U\langle n_{-s}\rangle)\,\langle\langle a_{is};\, a_{js}^{+}\rangle\rangle = \delta_{i,j} + \sum_{i_1} t_{ii_1 s}\langle\langle a_{i_1 s};\, a_{js}^{+}\rangle\rangle,$$

or, using (3.126) and (3.128)

$$(\omega - \varepsilon_{ks} - U\langle n_{-s}\rangle)\,\langle\langle a_{ks};\, a_{k's}^{+}\rangle\rangle = \delta_{k,k'}. \tag{3.138}$$

Compared with the unperturbed Bloch energies ε_{ks}, the single-particle energies in the Hartree–Fock approximation are thus only shifted by $U\langle n_{-s}\rangle$. The average number of electrons with momentum $\boldsymbol{k}$ and the spin s, $\langle n_{ks}\rangle$, is hence obtained from (3.59a) and (3.61). We obtain at once

$$\langle n_{ks}\rangle = f(\varepsilon_{ks} + U\langle n_{-s}\rangle),$$

where

$$f(\omega) = \frac{1}{e^{\beta\omega}+1}, \quad \beta = \frac{1}{kT},$$

or

$$\langle n_{ks}\rangle = f(\varepsilon_{ks} + \tfrac{1}{2}\, U\langle n_s + n_{-s}\rangle - \tfrac{1}{2}\, U\langle n_s - n_{-s}\rangle).$$

Hence we obtain for the magnetisation $M(g = 2)$

$$M = \mu_B \sum_{k} \langle n_{\downarrow k} - n_{k\uparrow}\rangle,$$

and for the total number of particles N_{el}

$$N_{el} = \sum_{k} \langle n_{k\downarrow} + n_{k\uparrow} \rangle.$$

When we change from the sums to integrals, these are the same expressions as in the Stoner model, i.e. the equations (3.100) and (3.101), when we use the following identities:

$$\mu = \zeta + \tfrac{1}{2}\, U\langle n_{\uparrow} + n_{\downarrow} \rangle = \zeta + \tfrac{1}{2}\, Uz,$$
$$\mu_B H_W = k\Theta\sigma = \tfrac{1}{2}\, U\langle n_{\downarrow} - n_{\uparrow} \rangle = \tfrac{1}{2}\, Uz\sigma.$$

Here $z = \langle n_{\downarrow} + n_{\uparrow} \rangle$ is the number of electrons per lattice site or per atom, and

$$\sigma = \frac{\langle n_{\downarrow} - n_{\uparrow} \rangle}{\langle n_{\downarrow} + n_{\uparrow} \rangle}$$

is the relative magnetisation. We then obtain for the Stoner parameter

$$k\Theta = \tfrac{1}{2}\, Uz.$$

The Stoner model is therefore identical with the Hartree–Fock approximation of the model described by (3.133). In particular, the same formula is obtained for the paramagnetic susceptibility

$$\chi = \frac{\chi_p}{1 - U\varrho(\zeta)}, \tag{3.139}$$

where χ_p is the susceptibility of the free electron gas defined by (2.4), and $\varrho(\zeta)$ is the density of states per atom, $Z'(\zeta)/N$ at the Fermi energy which, for free electrons, is given by (2.6). In the Hartree–Fock approximation the condition for ferromagnetism is therefore given by

$$U\varrho(\zeta) \geqslant 1. \tag{3.140}$$

This agrees formally with the condition resulting from the Landau theory, if $\bar{\varphi}_1$ in (2.10a) is chosen suitably. The similarity between Landau theory and the Hartree–Fock approximation observed is not accidental, since the Hamiltonian, i.e. the quasi-particle energy ε, which determines the motion of the quasi-particles, is a locally averaged quantity as in the Hartree–Fock approximation (cf. Section II.1). In the Landau theory, however, the quantities are renormalised quantities involving interaction effects, so that the similarity is rather a formal one and not due to the physical contents.

If we want to stay within the framework of the Hartree–Fock approximation it would not have been necessary to simplify the electron–electron interaction in (3.120) or (3.127) in such a way as was done here. For a detailed discussion and the pertinent literature, we refer the reader to a paper by Rajagopal *et al.* (1967).

From the Hartree–Fock criterion for ferromagnetism, (3.140), it might be deduced that a sufficiently strong intra-atomic interaction must always result in magnetic ordering. A consideration due to Kanamori (1963; see also Lang and Ehrenreich, 1968; Roth, 1969), shows that this is wrong; his consideration should apply to bands which contain but a few electrons. In this case, screening effects exerted by the other electrons on the interaction between two given electrons are less important than the continuous direct interaction or the continuous scattering of the two electrons by one another. Such a more accurate treatment of interaction effects will lead us to the necessity of replacing the bare interaction U in (3.140) by some U_{eff} corresponding to the T matrix for ordinary scattering processes (Messiah, 1960).

In order to calculate the effective interaction U_{eff}, we introduce an electron with momentum $\boldsymbol{k}$ and spin s in a band in which only one other electron with spin $-s$ is contained. We obtain the equation of motion of the Green function of the electron by subjecting (3.137) to a Fourier transformation, i.e. by a transition to Bloch functions

$$(\omega-\varepsilon_{ks})\langle\langle a_{ks};\, a_{ks}^{+}\rangle\rangle = 1+\frac{U}{N}\sum_{k_1,\,k_2}\langle\langle a_{k_1-s}^{+}a_{k_2-s}a_{k+k_1-k_2s};\, a_{ks}^{+}\rangle\rangle. \tag{3.141}$$

Here we assume all momentum vectors of the operators to lie in the first Brillouin zone; this can always be achieved by adding a suitable reciprocal lattice vector (see above). Since according to Section III.3 the

Green function $\langle\langle A(t), B\rangle\rangle$ is essentially a transition amplitude

$$\langle\psi\,|\,A(t)B\,|\,\psi\rangle = \langle A^{+}(t)\,\psi\,|\,B\psi\rangle,$$

where ψ describes the state of the system, eqn. (3.141) can also be interpreted in the following way: the variation with time of the transition amplitude of a particle with momentum $\boldsymbol{k}$ and spin s (corresponding to the single-particle Green function), $\langle a_{ks}^{+}(t)\,\psi\,|\,a_{ks}\psi\rangle$, is determined by the fact that the electron either returns to its initial state $(\boldsymbol{k}, s)$ like a free electron, or reaches the final state, $a_{k_3 s}^{+}a_{k_2 -s}^{+}a_{k_1 -s}\,|\,\psi\rangle$, $(\boldsymbol{k}_3 = \boldsymbol{k}+\boldsymbol{k}_1-\boldsymbol{k}_2)$ because of its interaction with other electrons. Because of the interaction the momentum of the electron was changed by $\boldsymbol{k}_1-\boldsymbol{k}_2$, and this change of momentum occurs via the annihilation of an electron (that is, the generation of a hole) in the state $(\boldsymbol{k}_1, -s)$, and the generation of an electron in the state $(\boldsymbol{k}_2, -s)$.

If we had considered U in (3.141) to be a small perturbation, we could have replaced the two-particle Green function in (3.141) by the corresponding free Green function:

$$\begin{aligned}\langle\langle a_{k_1 -s}^{+}a_{k_2 -s}a_{k+k_1-k_2 s};\, a_{ks}^{+}\rangle\rangle &= \delta_{k_1, k_2}\langle n_{k_1 -s}\rangle\langle\langle a_{ks};\, a_{ks}^{+}\rangle\rangle\\ &= \delta_{k_1, k_2}\langle n_{k_1 -s}\rangle\frac{1}{\omega-\varepsilon_{ks}},\end{aligned}$$

this can be easily verified by solving the equation of motion of this Green function for $U = 0$. In a first approximation in U we thus obtain from (3.141) for the single-particle Green function

$$(\omega-\varepsilon_{ks})\,\langle\langle a_{ks};\, a_{ks}^{+}\rangle\rangle = 1+\frac{1}{N}\sum_{k_1} U\langle n_{k_1 -s}\rangle\frac{1}{\omega-\varepsilon_{ks}}. \quad (3.142a)$$

When we want to consider the interaction in a more accurate way, we have to determine the two-particle Green function exactly; this can be done in the case of two electrons. It is easy to obtain the equation of motion of this Green function

$$\begin{aligned}(\omega+\varepsilon_{k_1 -s}-\varepsilon_{k_2 -s}-\varepsilon_{k+k_1-k_2 s})\,&\langle\langle a_{k_1 -s}^{+}a_{k_2 -s}a_{k+k_1-k_2 s};\, a_{ks}^{+}\rangle\rangle\\ &= \delta_{k_1, k_2}\langle n_{k_1 -s}\rangle+\frac{U}{N}\sum_{k_2'}\langle\langle a_{k_1 -s}^{+}a_{k_2' -s}a_{k+k_1-k_2' s};\, a_{ks}^{+}\rangle\rangle.\end{aligned} \quad (3.143)$$

Here we have made use of the fact that, because of the translational invariance of the system, or, what amounts to the same, because momentum is a constant of motion, we have

$$\langle a^{+}_{k_1-s} a_{k_2-s}\rangle = \delta_{k_1, k_2}\langle n_{k_1-s}\rangle;$$

higher Green functions of the type

$$\langle\langle a^{+}_{k_1} a^{+}_{k_2} a_{k_3} a_{k_4} a_{k_5}; a^{+}_{k}\rangle\rangle$$

have been omitted, since they obviously vanish in a band which, besides the electron with momentum $\boldsymbol{k}$, contains only one other electron. In this way we obtain from the above equation

$$\sum_{k_2} \langle\langle a^{+}_{k_1-s} a_{k_2-s} a_{k+k_1-k_2 s}; a^{+}_{ks}\rangle\rangle = \frac{\langle n_{k_1-s}\rangle}{1+UF(\boldsymbol{k}, \boldsymbol{k}_1)} \, \frac{1}{\omega-\varepsilon_{ks}},$$

where

$$F(\boldsymbol{k}, \boldsymbol{k}_1) = -\frac{1}{N}\sum_{k_2} \frac{1}{\omega+\varepsilon_{k_1-s}-\varepsilon_{k_2-s}-\varepsilon_{k+k_1-k_2 s}}.$$

Thus we obtain for the single-particle Green function exactly

$$(\omega-\varepsilon_{ks})\langle\langle a_{ks}; a^{+}_{ks}\rangle\rangle = 1+\frac{1}{N}\sum_{k_1} \frac{U\langle n_{k_1-s}\rangle}{1+UF(\boldsymbol{k}, \boldsymbol{k}_1)} \, \frac{1}{\omega-\varepsilon_{ks}}. \quad (3.142b)$$

In this Green function only frequencies near $\omega = \varepsilon_{ks}$ are important; it is thus possible to replace ω in $F(\boldsymbol{k}, \boldsymbol{k}_1)$ by $\omega = \varepsilon_{ks} \pm i\varepsilon$; the sign of the infinitesimal imaginary part $i\varepsilon$ depends on whether the Green function is an advanced or retarded function (cf. Section III.3). We then obtain for $F(\boldsymbol{k}, \boldsymbol{k}_1)$

$$F(\boldsymbol{k}, \boldsymbol{k}_1) = \frac{1}{N}\sum_{k_2, k_3} \frac{\delta_{k+k_1, k_2+k_3}}{\varepsilon_{k_3 s}+\varepsilon_{k_2-s}-\varepsilon_{k_1-s}-\varepsilon_{ks} \mp i\varepsilon}.$$

In this expression there is no dependence on the magnetic field so that the spin index can be omitted. Comparing (3.142a) with (3.142b), we see that the effective interaction U_{eff} must be assumed to be of the form

$$U_{\text{eff}}(\boldsymbol{k}, \boldsymbol{k}_1) = \frac{U}{1+UF(\boldsymbol{k}, \boldsymbol{k}_1)}.$$

When we apply this result to the many-electron problem, we have, of course, an approximation, since in (3.143) all higher correlations higher than two-particle correlations are neglected. Certainly this is only correct for small electron densities. Moreover, in this case, because of the other electrons in the Fermi sea, the whole phase space is no longer available for the electron scattering process. According to the discussion of eqn. (3.141), the final states ($\boldsymbol{k}_3$, s) and ($\boldsymbol{k}_2$, $-s$) must be unoccupied, so that the electrons can be scattered into them. In the case of $T = 0$ this means that in $F(\boldsymbol{k}, \boldsymbol{k}_1)$ the sum over $\boldsymbol{k}_2$ and $\boldsymbol{k}_3$ is to be extended only over $k_{2,3} > k_0$, that is, over the unoccupied states. In addition to this, according to (3.142b) $k_1 < k_0$ so that F is given by

$$F(\boldsymbol{k}, \boldsymbol{k}_1) = \frac{1}{N} \sum_{\substack{k_2, k_3 \\ k_{2,3} > k_0}} \frac{\delta_{k+k_1, k_2+k_3}}{\varepsilon_{k_3} + \varepsilon_{k_2} - \varepsilon_{k_1} - \varepsilon_k \mp i\varepsilon}.$$

When we consider an electron at the Fermi level ($k = k_0$), the real part of the denominator is positive, so that the imaginary part can be put equal to zero; moreover, $F > 0$. In a very rough approximation F may be considered a constant which, since the denominator varies between zero and $2W$ (W is the band width), must be of the order of $1/W$. Thus we obtain

$$U_{\text{eff}} = \frac{U}{1 + UF}, \tag{3.144}$$

with $F = \text{const} \cdot W^{-1}$; the constant is of the order of 1. It is an important fact that in the limit as $U \to \infty$ the effective interaction is no longer divergent.

Returning to the Hartree–Fock approximation of the many-electron problem (3.137) and (3.138) and replacing U approximately by U_{eff}, we obtain instead of the Stoner criterion for the ferromagnetism (3.140) the inequality

$$U_{\text{eff}}\varrho(\zeta) \geqslant 1, \tag{3.145}$$

which, in this case, is not necessarily satisfied for sufficiently high U values.

As already mentioned, this approximation should be applicable to weakly populated bands or almost filled bands, i.e. low concentration of holes as is, for example, the case in nickel.

In the case of a very narrow band where the electrons are strongly localised we must proceed in another way. In (3.133a) or (3.137) we may then consider the t_{ij} with $i \neq j$ to be small, as these t_{ij} describe the migration of an electron from the lattice site $\boldsymbol{r}_j$ to the site $\boldsymbol{r}_i$. It is therefore possible in this case to apply perturbation calculus in terms of the $t_{ij}(i \neq j)$ or, what is the same, we can approximate the Green function connected with these t_{ij} in the equation of motion of the two-particle Green function (3.137). This problem has been discussed in detail by Hubbard (1963, 1964a, b).

Finally we shall discuss the spin wave problem within the framework of this model. According to the general derivation in the Landau theory of Fermi liquids their existence is undubitable, if the ferromagnetic state is stable.

It is most convenient to proceed as in the Heisenberg model: we calculate the Green function suitably modified for the band model, and interpret its poles as collective excitations of the system which also comprise the spin waves. The Green function considered in Section III.3 in connection with the Heisenberg model has a simple interpretation. A comparison of the retarded function (3.62) with Kubo's formula for the dynamic susceptibility (2.45b) shows that it is, apart from a factor, identical with the transverse dynamic susceptibility

$$\chi_{-+}(\boldsymbol{k}, \omega) = -N^2 g^2 \mu_B^2 \langle\langle s_k^- ; s_k^+ \rangle\rangle, \tag{3.146}$$

since $\boldsymbol{m}_i = -g\mu_B s_i$ ($\hbar = 1$). The factor N^2 stems from the Fourier transformation (3.37). It is physically reasonable that the spin waves appear as poles of the transverse susceptibility, since the response of a ferromagnetic system to a circularly polarised transverse magnetic field which varies both in space and in time will be such that its magnetisation follows this field, i.e. it will produce magnetisation waves in the form of spin waves.

In a band the spin flop operators $s_i^{\pm}$ are obviously given by

$$s_i^+ = a_{i\uparrow}^+ a_{i\downarrow}, \quad s_i^- = a_{i\downarrow}^+ a_{i\uparrow}. \tag{3.147}$$

The Fourier-transformed operators defined according to (3.37),

$$s_k^{\pm} = \frac{1}{N} \sum_{i=1}^{N} s_i^{\pm} \, e^{\pm i(k \cdot r_i)},$$

can be expressed according to (3.126) in terms of the creation and annihilation operators of Bloch electrons:

$$s_k^{+} = \frac{1}{N} \sum_{k'} a^{+}_{k+k'\uparrow} a_{k'\downarrow}, \tag{3.148}$$

$$s_k^{-} = \frac{1}{N} \sum_{k'} a^{+}_{k'\downarrow} a_{k+k'\uparrow}.$$

Instead of the susceptibility (3.146) it is convenient first to discuss the Green function

$$\langle\langle a^{+}_{k'\downarrow}(t) a_{k+k'\uparrow}(t); s_k^{+}\rangle\rangle,$$

from which the susceptibility χ_{-+} is obtained by summation over $\boldsymbol{k}'$. After a Fourier transformation with respect to time, we obtain from the equation of motion of the operators with the Hamiltonian (3.133a):

$$(\omega + \varepsilon_{k'\downarrow} - \varepsilon_{k+k'\uparrow}) \langle\langle a^{+}_{k'\downarrow} a_{k+k'\uparrow}; s_k^{+}\rangle\rangle = \frac{1}{N} \langle n_{k'\downarrow} - n_{k+k'\uparrow}\rangle \tag{3.149}$$

$$- \frac{U}{N} \sum_{k'_v} \langle\langle a^{+}_{k'_1\downarrow} a^{+}_{k'_2\uparrow} a_{k'_3\uparrow} a_{k+k'\uparrow}; s_k^{+}\rangle\rangle \, \delta_{k'_1+k'_2,\, k'+k'_3}$$

$$+ \frac{U}{N} \sum_{k'_v} \langle\langle a^{+}_{k'\downarrow} a^{+}_{k'_1\downarrow} a_{k'_2\downarrow} a_{k'_3\uparrow}; s_k^{+}\rangle\rangle \, \delta_{k'_2+k'_3,\, k+k'+k'_1}.$$

As before, the sum over $\boldsymbol{k}'_v$ is extended over the first Brillouin zone, and the momentum conservation indicated by the Kronecker symbols is again taken as mod $\boldsymbol{g}$, i.e. if the sum of two vectors lies outside the first Brillouin zone, a suitable reciprocal-lattice vector must be added to this sum, so that the resulting vector comes to lie in the first zone. Moreover, the spin index is given explicitly for the Bloch energies in order to stress the interpretation, although no external field is present.

In the case of free electrons, i.e. $U = 0$, we obtain from this the transverse susceptibility of a non-interacting electron system:

$$\chi_{-+}^{(0)}(\boldsymbol{k}, \omega) = -g^2\mu_{\mathrm{B}}^2 \sum_{k'} \frac{f(\varepsilon_{k'\downarrow}) - f(\varepsilon_{k+k'\uparrow})}{\omega + \varepsilon_{k'\downarrow} - \varepsilon_{k+k'\uparrow}}, \tag{3.150}$$

where f is the Fermi distribution. In isotropic systems and vanishing external field we have, by virtue of the symmetry properties,

$$\chi_{-+}^{(0)} = \chi_{xx}^{(0)} + \chi_{yy}^{(0)} = 2\chi_{zz}^{(0)} = 2\chi^{(0)}, \tag{3.151}$$

where $\chi^{(0)}$ is the dynamic susceptibility of the system. It is easy to see that (3.150) tends to twice the Pauli susceptibility (2.4) (with $g = 2$) as $\omega \to 0$ and $k \to 0$:

$$\lim_{k \to 0} \lim_{\omega \to 0} \chi_{-+}^{(0)}(\boldsymbol{k}, \omega) = 2\chi_{\mathrm{P}}.$$

We therefore define

$$\chi_{-+}^{(0)}(\boldsymbol{k}, \omega) = 2\chi_{\mathrm{P}} \cdot u(\boldsymbol{k}, \omega), \tag{3.152}$$

with

$$u(\boldsymbol{k}, \omega) = -\frac{1}{N\varrho(\zeta)} \sum_{k'} \frac{f(\varepsilon_{k'\downarrow}) - f(\varepsilon_{k+k'\uparrow})}{\omega + \varepsilon_{k'\uparrow} - \varepsilon_{k+k'\uparrow}},$$

where $u(\boldsymbol{k}, \omega)$ is the so-called Lindhard function; $\varrho(\zeta)$ is again the density of states per atom at the Fermi energy.

The poles of the susceptibility (3.150) lie on the real axis. When we write (3.150) in the form

$$\chi_{-+}^{(0)}(\boldsymbol{k}, \omega) = -g^2\mu_{\mathrm{B}}^2 \sum_{k'} \frac{f(\varepsilon_{k'\downarrow})\left(1 - f(\varepsilon_{k+k'\uparrow})\right) - f(\varepsilon_{k+k'\uparrow})\left(1 - f(\varepsilon_{k'\downarrow})\right)}{\omega + \varepsilon_{k'\downarrow} - \varepsilon_{k+k'\uparrow}},$$

we see that the poles $\omega = \varepsilon_{k+k'\uparrow} - \varepsilon_{k'\downarrow}$ correspond to excitations in which an electron with spin down and momentum $\boldsymbol{k}'$ is annihilated, which is proportional to the population probability $f(\varepsilon_{k'\downarrow})$ in this state and an electron with spin up and momentum $\boldsymbol{k}+\boldsymbol{k}'$ is produced, which is proportional to the probability $1 - f(\varepsilon_{k+k'\uparrow})$ for this state to be unoccupied. In addition to this we also have the opposite process. Thus we have electron–hole pair excitations which are, so to speak, Stoner excitations with zero magnetisation.

The dissipative part of the dynamic susceptibility, which determines the absorption of a transverse field, is given by

$$\operatorname{Im} \chi_{-}^{(0)}(\boldsymbol{k}, \omega) = \pi g^2 \mu_B^2 \sum_{k'} \left(f(\varepsilon_{k'\downarrow}) - f(\varepsilon_{k+k'\uparrow})\right) \delta(\omega + \varepsilon_{k'\downarrow} - \varepsilon_{k+k'\uparrow}), \tag{3.153}$$

since $\chi_{-}^{(0)}$ is a retarded function. This shows that the absorption of a field, which varies in space and time takes place via electron–hole pair excitation; thus the imaginary part must have maxima at the frequencies which correspond to excitations of the system; this is a plausible conclusion. In the limiting case of long wave lengths ($\boldsymbol{k} \to 0$) we obtain from (3.153)

$$\lim_{k \to 0} \operatorname{Im} \chi_{-+}^{(0)}(\boldsymbol{k}, \omega) = -\pi g^2 \mu_B^2 \sum_{k'} \frac{\partial f}{\partial \varepsilon_{k'}} \left(\boldsymbol{k} \cdot \frac{\partial \varepsilon_{k'}}{\partial \boldsymbol{k}'}\right) \delta\left(\omega - \left(\boldsymbol{k} \cdot \frac{\partial \varepsilon_{k'}}{\partial \boldsymbol{k}'}\right)\right).$$

The integral is easily calculated for $T = 0$, when we change from a sum to an integral. Using the Fermi velocity $\boldsymbol{v}_0 = (\nabla_k \varepsilon_k)_{k=k_0}$ we obtain

$$\lim_{k \to 0} \operatorname{Im} \chi_{-+}^{(0)}(\boldsymbol{k}, \omega) = \begin{cases} \pi \dfrac{|\omega|}{k v_0} \chi_P & \text{when} \quad |\omega| < k v_0, \\ 0 & \text{when} \quad |\omega| > k v_0. \end{cases} \tag{3.154}$$

The vanishing of absorption in the latter case is based on the fact that in this range in an excitation the energy and momentum conservation laws cannot be satisfied at the same time.

According to (3.149), taking the electron–electron interaction into account yields more complicated Green functions. The simplest method to approximate them is a Hartree–Fock procedure which, just as the analogous approximation in the Heisenberg model, is again called Random Phase Approximation (RPA). In this approximation we put

$$\begin{aligned} \langle\langle a_{k_1'\downarrow}^{+} a_{k_2'\uparrow}^{+} a_{k_3'\uparrow} a_{k+k'\uparrow}; s_k^{+} \rangle\rangle &= \langle a_{k_2'\uparrow}^{+} a_{k_3'\uparrow} \rangle \langle\langle a_{k_1'\downarrow}^{+} a_{k+k'\uparrow}; s_k^{+} \rangle\rangle \\ &\quad - \langle a_{k_2'\uparrow}^{+} a_{k+k'\uparrow} \rangle \langle\langle a_{k_1'\downarrow}^{+} a_{k_3'\uparrow}; s_k^{+} \rangle\rangle = \langle n_{k_2'\uparrow} \rangle \langle\langle a_{k_1'\downarrow}^{+} a_{k+k'\uparrow}; s_k^{+} \rangle\rangle \, \delta_{k_2', k_3'} \\ &\quad - \langle n_{k+k'\uparrow} \rangle \langle\langle a_{k_1'\downarrow}^{+} a_{k_3'\uparrow}; s_k^{+} \rangle\rangle \, \delta_{k_2, k+k'}. \end{aligned}$$

The final equation is again obtained using the momentum conservation law. Just as in the Hartree–Fock method, in this approximation all possible combinations of operators are taken out of the Green function as thermal averages, taking into account the commutator properties of these operators. However, it cannot be said beforehand how good this method is.

When we treat the other Green function in (3.149) in the same way, we obtain from (3.149):

$$(\omega+\tilde{\varepsilon}_{k'\downarrow}-\tilde{\varepsilon}_{k+k'\uparrow})\langle\langle a^+_{k'\downarrow}a_{k+k'\uparrow};\,s^+_k\rangle\rangle = \frac{1}{N}\langle n_{k'\downarrow}-n_{k+k'\uparrow}\rangle$$

$$-\frac{U}{N}\langle n_{k'\downarrow}-n_{k+k'\uparrow}\rangle\sum_{k'_1}\langle\langle a^+_{k'_1\downarrow}a_{k+k'_1\uparrow};\,s^+_k\rangle\rangle.$$

The $\tilde{\varepsilon}_{ks}$ are the single-particle energies $\tilde{\varepsilon}_{ks} = \varepsilon_{ks}+U\langle n_{-s}\rangle$ obtained in the Hartree–Fock approximation (3.142), since

$$\langle n_s\rangle = \frac{1}{N}\sum_k\langle n_{k\,s}\rangle.$$

Using the relative magnetisation

$$\sigma = \frac{\langle n_\downarrow - n_\uparrow\rangle}{\langle n_\downarrow + n_\uparrow\rangle},$$

and (3.148) this equation can also be written in the form

$$(\omega+\varepsilon_{k'\downarrow}-\varepsilon_{k+k'\uparrow}-Uz\sigma)\langle\langle a^+_{k'\downarrow}a_{k+k'\uparrow};\,s^+_k\rangle\rangle$$

$$= \frac{1}{N}\langle n_{k'\downarrow}-n_{k+k'\uparrow}\rangle - U\langle n_{k'\downarrow}-n_{k+k'\uparrow}\rangle\langle\langle s^-_k;\,s^+_k\rangle\rangle.$$

Dividing by the first factor on the left-hand side and summing over $\boldsymbol{k}'$, we hence obtain $\langle\langle s^-_k;\,s^+_k\rangle\rangle$, and thus the transverse susceptibility in the RPA:

$$\chi^{\mathrm{RPA}}_{-+}(\boldsymbol{k},\omega) = -g^2\mu_{\mathrm{B}}^2\,\frac{\displaystyle\sum_{k'}\frac{\langle n_{k'\downarrow}\rangle-\langle n_{k+k'\uparrow}\rangle}{\omega+\varepsilon_{k'\downarrow}-\varepsilon_{k+k'\uparrow}-Uz\sigma}}{\displaystyle 1+\frac{U}{N}\sum_{k'}\frac{\langle n_{k'\downarrow}\rangle-\langle n_{k+k'\uparrow}\rangle}{\omega+\varepsilon_{k'\downarrow}-\varepsilon_{k+k'\uparrow}-Uz\sigma}}. \quad (3.155)$$

In the case of $U = 0$ and $\langle n_k\rangle = f(\varepsilon_k)$ this expression denotes the free-electron susceptibility (3.150). The poles of susceptibility are given by

$$1 = -\frac{U}{N}\sum_{k'}\frac{\langle n_{k'-k/2\downarrow}\rangle - \langle n_{k'+k/2\uparrow}\rangle}{\omega + \varepsilon_{k'-k/2\downarrow} - \varepsilon_{k'+k/2\uparrow} - Uz\sigma}. \tag{3.156}$$

This equation has two types of solution:

(a) *Stoner excitations*

In the limiting case as $N \to \infty$ these solutions lie at points where one of the denominators in the N summands of (3.156) vanishes, i.e. at

$$\omega = \varepsilon_{k'+k/2\uparrow} - \varepsilon_{k'-k/2\downarrow} + Uz\sigma.$$

Repeating the discussion made in connection with (3.152), we see that we are concerned with electron–hole pair excitations, in which one electron from the sub-band with spin $\downarrow$ is transferred to that with spin $\uparrow$ (thus leaving a hole in the first band). This process requires (besides the kinetic-energy difference) the exchange energy $Uz\sigma$, which is needed to turn the spin. Just as in the collective electron model or in the Hartree–Fock approximation the energy $Uz\sigma$ acts like an internal field.

(b) *Spin waves*

When we restrict ourselves to long waves, we can write eqn. (3.156) in the form

$$1 = -\frac{U}{N}\sum_{k'}\frac{\langle n_{k'\downarrow} - n_{k'\uparrow}\rangle \frac{1}{2}(\boldsymbol{k}\cdot\nabla_{k'})\langle n_{k'\downarrow} + n_{k'\uparrow}\rangle}{\omega - (\boldsymbol{k}\cdot\boldsymbol{v}(\boldsymbol{k}')) - Uz\sigma},$$

where $\boldsymbol{v}(\boldsymbol{k}') = \nabla_{k'}\varepsilon_{k'} = \partial\varepsilon_{k'}/\partial\boldsymbol{k}'$. This is formally the same to determine the spin-wave energies as occurs in the Landau theory of Fermi liquids if we put $\nabla = Uz\sigma$ and $\varphi_1 = 2UN/V$. Therefore, in the same way as there, we obtain for the spin wave frequency

$$\omega = Dk^2, \tag{3.157}$$

where

$$D = \frac{1}{6n\sigma} \int \frac{d^3k'}{(2\pi)^3} \left\{ \nabla^2_{k'} \varepsilon_{k'} \langle n_{k'\downarrow} + n_{k'\uparrow} \rangle - \frac{2}{Uz\sigma} (\nabla_{k'} \varepsilon_{k'})^2 \langle n_{k'\downarrow} - n_{k'\uparrow} \rangle \right\}.$$

Here we have replaced the sum by an integral and integrated once by parts; n is the electron density; D, as in the Heisenberg model (3.102), depends on the temperature. For low temperatures, however, this dependence is weak in the Heisenberg model, while it is relatively important in the band model (cf., for example, Mathon and Wohlfarth, 1968). Of course, $D > 0$. As already mentioned above, this is the condition for the stability of the ferromagnetic state. The model here considered need not necessarily satisfy this condition, i.e. it is not obvious whether ferromagnetism can exist within the framework of the Hamiltonian (3.133) (Nagaoka, 1966; Edwards, 1968).

According to (3.156) the spin waves are generally damped. When we split the right-hand side of (3.156) into a real and an imaginary part, the latter is given by

$$\pi i \frac{U}{N} \sum_{k'} \langle n_{k'-k/2\downarrow} - n_{k'+k/2\uparrow} \rangle \, \delta(\omega + \varepsilon_{k'-k/2\downarrow} - \varepsilon_{k'+k/2\uparrow} - Uz\sigma).$$

The formal similarity of this expression and the imaginary part of the free-electron transverse susceptibility (3.153) shows that the spin waves are damped by Stoner excitations. For small k and thus small $\omega = Dk^2$ the damping is zero, i.e. the spin waves are undamped in the RPA, if $k < k_{\text{max}}$. This reveals the fact that a spin wave with small momentum and low frequency cannot disintegrate into an electron–hole pair since the energy conservation law cannot be satisfied.

As in the case of the Hartree–Fock approximation, we may ask whether a more accurate treatment of the interaction U gives rise to essential corrections to (3.157), apart from the fact that U must be replaced by U_{eff}. However, this seems not to be the case (Edwards, 1967).

Within the framework of the RPA more general interactions can also be taken into account (Rajagopal *et al.*, 1967), such as either the interatomic interactions in a band (Englert and Antonoff, 1964), or interband interactions (Yamada and Shimizu, 1967). The qualitative result, i.e. the fact that the spin wave dispersion as $k \to 0$ behaves as the square of the wave vector, of course remains unchanged since, as shown in the discussion of the Landau theory, the spin waves exist by virtue of the symmetries of the ferromagnet and not by virtue of details of the interaction. For a detailed treatment of band ferromagnetism, we refer the reader to a paper by Herring (1966).

It is worth mentioning that even if the interaction is insufficient to give rise to a ferromagnetically ordered phase, the susceptibility (3.155) may show sharp peaks or, more accurately, the imaginary part has a more or less marked maximum at a certain frequency. In agreement with what has been said so far, this point must be considered to be a damped collective excitation of the system, i.e. a spin fluctuation wave below the ferromagnetic instability, $U\varrho \lessdot 1$, according to (3.139). These excitations are called paramagnons (Berk and Schrieffer, 1966; Doniach and Engelsberg, 1966; Schrieffer, 1968). In order to show this we put $\sigma = 0$ in (3.155) and replace the occupation numbers $\langle n_k \rangle$ by the Fermi functions. Since, because of $\sigma = 0$, the symmetry relations (3.151) hold true, we obtain in the RPA according to (3.155) and (3.152) the following expression for the dynamic susceptibility:

$$\chi^{\mathrm{RPA}}(\boldsymbol{k}, \omega) = \chi_{\mathrm{p}} \frac{u(\boldsymbol{k}, \omega)}{1 - U\varrho u(\boldsymbol{k}, \omega)}; \qquad (3.158)$$

χ_{p} is the Pauli susceptibility, ϱ is the density of states per atom at the Fermi energy, and $u(\boldsymbol{k}, \omega)$ is the Lindhard function introduced in (3.152). Since $\lim_{k \to 0} \lim_{\omega \to 0} u(k, \omega) = 1$, in this case the dynamic susceptibility in RPA becomes identical with that of the Hartree–Fock approximation (3.139). For small k and ω the Lindhard function can be easily expanded. We obtain

$$u(\boldsymbol{k}, \omega) = 1 + i\frac{\pi}{2}\frac{|\omega|}{kv_0} - \frac{k^2}{12k_0^2},$$

if $|\omega| < kv_0$. Otherwise the imaginary part vanishes according to

(3.154). For sufficiently small k the quadratic term in k can be neglected and we obtain for the imaginary part of the susceptibility from (3.158)

$$\operatorname{Im} \chi^{\mathrm{RPA}}(\boldsymbol{k}, \omega) = \chi_{\mathrm{P}} \frac{(2U\varrho - 1)\dfrac{\pi|\omega|}{2kv_0}}{(1-U\varrho)^2 + \left(U\varrho \dfrac{\pi\omega}{2kv_0}\right)^2}.$$

Near the ferromagnetic instability the maximum of the imaginary part is at

$$\omega = \frac{2v_0}{\pi U\varrho}(1-U\varrho)k \simeq \frac{2v_0}{\pi}(1-U\varrho)k.$$

The maximum of the imaginary part is the more marked the closer we are at the ferromagnetic instability, so that there we can speak of collective excitations, i.e. paramagnons. Their dispersion is the same as that of phonons. They can be measured immediately in inelastic scattering experiments with neutrons (just as when we prove the existence of spin waves); for this purpose we use substances which are almost ferromagnetic, such as, for example, palladium. They exert a more indirect influence on the effective masses of the electrons and thus on the specific heat (Berk and Schrieffer, 1966; Doniach and Engelsberg, 1966; Schrieffer, 1968).

6. Ferromagnetic Resonance

A high-frequency magnetic field perpendicular to a constant field can give rise to resonance absorption in ferromagnetic substances just as in the case of paramagnetic substances. The resonance frequency is given by the energy of the excited states, i.e. in the Heisenberg model by the energy of the spin-wave quanta (3.45). The Hamiltonian of the system (3.40), however, shows that an external, spatially uniform magnetic field is coupled only to the spin-wave with wave vector $k = 0$, i.e. with the magnetisation operator, so that only this wave can be directly excited. Thus, as in the paramagnetic case, the resonance frequency should be given by $\omega_0 = -\gamma H_0$, $\gamma = -g\mu_{\mathrm{B}}/\hbar$, H_0

being the constant external field. This relation, however, yields unreasonably high g factors. According to Kittel (1948); Van Vleck, (1950), this is due to the fact that the dipole interaction between the spins must be taken into account, which, in a ferromagnetic substance, causes the appearance of strong demagnetising fields. Because of the smallness of susceptibility they are insignificant for paramagnetism, when we disregard the very low temperatures ($< 1°K$) where χ becomes large as a consequence of the Curie law. When we restrict ourselves in the following to an ellipsoid whose axes are parallel to the coordinate axes, the demagnetising field is spatially homogeneous (cf., for example, Becker and Sauter, 1957). It is proportional to the magnetisation, so that in the ellipsoid the effective field is

$$H^i_{\text{eff}} = H_i - L_i M_i .$$

Here i denotes the components, H_i is the external field, and the L_i are the demagnetisation factors. For a sphere all L_i are equal and $L_i = 4\pi/3$; for a cylinder with its axis in the z-direction $L_z = 0$ and $L_x = L_y = 2\pi$.

For the magnetisation $\boldsymbol{M}$ the classically and quantum-mechanically correct equations of motion are the equations (2.73), when we disregard the relaxation times

$$\frac{d\boldsymbol{M}}{dt} = \gamma[\boldsymbol{M} \wedge \boldsymbol{H}_{\text{eff}}]. \tag{3.159}$$

In contrast to (2.73) this relation encompasses the demagnetising field which is contained in the effective field. In the case of a ferromagnetic sphere it is proportional to the magnetisation, so that it drops out of the equation of motion; the resonance frequency of the sphere is again given by $\omega_0 = -\gamma \boldsymbol{H}_0$.

In the other cases it is easy to solve the equation of motion, if the magnetisation M_z parallel to the constant field is assumed not to be influenced by the weak transverse field, and thus to be independent of time (no saturation effects). From (3.159) we then obtain the resonance frequency

$$\omega_0 = |\gamma| \sqrt{(H_0 + M_z(L_x - L_z))(H_0 + M_z(L_y - L_z))}. \tag{3.160}$$

Here we assume an external magnetic field of the form ($H_1 \sin \omega t$, $H_1 \cos \omega t$, H_0). Equation (3.160) is also the excitation frequency of the homogeneous spin wave. Of course, also the spin waves with $k \neq 0$ are modified by the dipole interaction. However, if the wave lengths are not too large, the exchange interaction is dominant according to (3.45) (Walker, 1963). The relaxation processes, and thus the line shape, can be described approximately by the Bloch equations (2.73). The most important relaxation process is based on the annihilation of the homogeneous spin wave ($k = 0$), and the simultaneous excitation of a spin wave with $k \neq 0$ by the inhomogeneities of the effective magnetic field (Haas and Callen, 1963).

In an antiferromagnetic substance the demagnetising field can again be neglected since the susceptibility is small. In the two-sublattices model the resonance frequencies are the two possible spin-wave frequencies with $k = 0$, i.e. according to (3.70),

$$\omega_0 = |\gamma| \left(\sqrt{H_A(H_A + 2H_E)} \pm H_0 \right). \tag{3.161}$$

This is true when the external field H_0 is parallel to the preferred direction of the two sublattices. When we express the two fields H_A and H_E in terms of the susceptibilities $\chi_\perp$ and $\chi_{||}$, the resonance condition (3.161) is experimentally verified. For $H = 0$ the two resonance frequencies (3.161) coincide, and ω_0 becomes a function of the sublattice magnetisation and thus of the temperature. In this way the antiferromagnetic resonance can be used to determine the temperature dependence of the anisotropy energy, if the magnetisation, which is proportional to H_E, is given (Fig. 22).

According to (3.78), ferrimagnetic substances have two resonance frequencies which are of completely different order of magnitude. The lower resonance frequency, which pertains to the acoustic branch, lies in the microwave band; the other resonance frequency pertaining to the optical branch lies in the infrared.

For a detailed discussion of antiferromagnetic and ferrimagnetic resonance we refer to an article by Foner (1963).

Nuclear resonance in ferromagnetic, antiferromagnetic, and ferrimagnetic substances differs considerably from that in diamagnetic materials. Thus, for example, in the case of ^{59}Co a strong frequency shift is observed which indicates a field strength at the nucleus of the order 10^5 Oe (Gossard and Portis, 1959, 1960). Such fields can be expected on the basis of the magnetisation, i.e. the electron polarisation in the unfilled shells. (Of course, these fields have nothing to do with

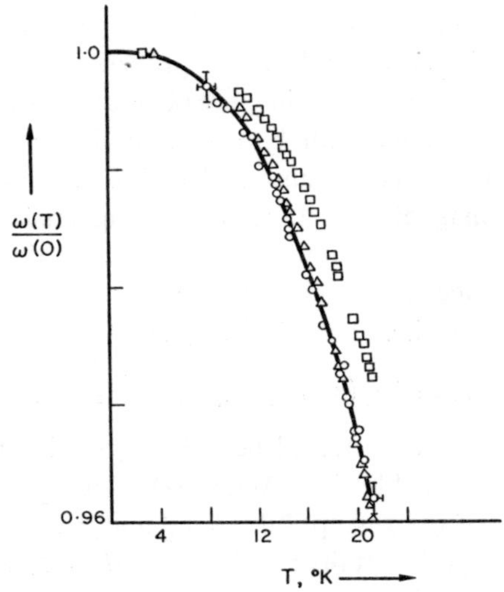

FIG. 22.

The graph shows the antiferromagnetic resonance frequency $\omega(T)/\omega(0)$ as a function of temperature for zero external field, after Johnson and Nethercot (1959), ○. The curve for the nuclear resonance frequency of ^{19}F in MnF_2, after Jaccarino and Walker (1959), □, yields the temperature dependence of the relative magnetisation. From the divergence of the curves there follows an antiferromagnetic resonance frequency which is proportional to $\sigma^{1\cdot45}$, △, i.e. according to (3.161), $H_A \sim \sigma^{1\cdot9}$ in agreement with theory (Oguchi, 1957).

the fictitious molecular field.) Fields of this order of magnitude are also obtained from the hyperfine structure of the paramagnetic resonance lines of ions which are brought into a magnetised medium (cf. Section II.6), the anisotropy of γ and β emission of the nuclei oriented in such fields, and the nuclear specific heats (Heer and Erikson, 1957).

The energy of a nuclear moment in a magnetic field $\boldsymbol{H}$ is given by $-g_{\text{nucl}}\mu_{\text{nucl}}(\boldsymbol{I}\cdot\boldsymbol{H})$; $\boldsymbol{I}$ is the nuclear angular momentum. The thermal average of the energy in a system of N nuclei is then equal to

$$E = -Ng_{\text{nucl}}\mu_{\text{nucl}} \frac{\text{Tr}\,(\boldsymbol{I}\cdot\boldsymbol{H})\, e^{\beta g_{\text{nucl}}\mu_{\text{nucl}}(\boldsymbol{I}\cdot\boldsymbol{H})}}{\text{Tr}\, e^{\beta g_{\text{nucl}}\mu_{\text{nucl}}(\boldsymbol{I}\cdot\boldsymbol{H})}} \cong -N\beta \frac{(g_{\text{nucl}}\mu_{\text{nucl}})^2}{2I+1}\,\text{Tr}\,(\boldsymbol{I}\cdot\boldsymbol{H})^2$$

$$= N\frac{(g_{\text{nucl}}\mu_{\text{nucl}})^2}{3kT}\, I(I+1)H^2,$$

provided $\beta\mu_{\text{nucl}} > 1$. Hence we obtain for the specific heat of nuclei in a magnetic field

$$C_{\text{nucl}} = \tfrac{1}{3} Nk(\beta g_{\text{nucl}}\mu_{\text{nucl}})^2\, I(I+1)H^2.$$

Because of its temperature dependence ($\sim T^{-2}$) this component exceeds at low temperatures ($< 1°$K) that of the electrons, of the lattice, or of the spin waves.

A final possibility for measuring nuclear fields is the measurement of the splitting of the recoil-free line in the Mössbauer effect, which arises as a consequence of the nuclear Zeeman effect. In this case the sign of the field with respect to magnetisation is determined at the same time (Hanna *et al.*, 1960); in most cases it is negative.

When we assume the greatest contribution to the magnetic fields at the nucleus to be made by the Fermi contact term of the hyperfine interaction, this field is given by

$$H_{\text{el}} = \frac{8\pi}{3}\mu_B(|\psi_+(0)|^2 - |\psi_-(0)|^2), \tag{3.162}$$

according to (2.75). Here $|\psi_\pm(0)|^2$ is the density of the s-electrons whose moments are parallel or antiparallel to the external field. The polarised, unfilled outer shells always yield in (3.162) a positive field at the nucleus. The main contribution to (3.162) must therefore be due to the inner s-shells which, obviously, are polarised by the exchange interaction with the outer electrons in such a way that the electrons of a shell which are parallel to the external field are more strongly drawn towards the periphery of the atom than the other electrons. The theoretical comprehension of these fields is very important as it permits conclusions as to the electron distribution with respect to their spins, and yields results about the electron–electron interaction. The calculation of the fields is complicated, since there are a series of effects which must be taken into account (Marshall, 1959; Watson and Freeman, 1961).

The field at a nucleus is proportional to the magnetisation and therefore temperature-dependent. When we assume the interaction

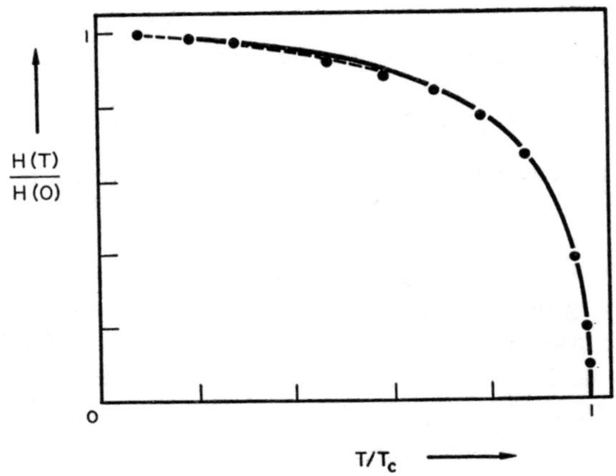

FIG. 23.

The graph shows the internal field in ^{57}Fe nuclei in ferromagnetic iron which is determined by the Mössbauer effect (Hanna, *et al.* 1962); $H(0) = 3{\cdot}38 \times 10^5$ Oe. For comparison we show the variation with temperature of the nuclear resonance frequency $\omega(T)/\omega(0)$ (dashed curve) and the relative magnetisation $\sigma(T)$ (solid line); the fact that the internal field vanishes at the Curie point does not mean that magnetic short-range order vanishes at $T > T_c$, but that short-range order exists only for a short time compared with the lifetime of the excited state of the ^{57}Fe nucleus (10^{-7} sec), or with the reciprocal Larmor frequency of the nucleus in the internal field.

between the electrons and between electrons and nucleus to be independent of T, the temperature dependence of the internal field should agree with that of magnetisation. This fact has been verified experimentally (Fig. 23).

References

ABRAGAM, A. (1962) *The Principles of Nuclear Magnetism* (Oxford University Press).

ABRAGAM, A. and PRYCE, M. H. L. (1951) *Proc. Roy. Soc.* A **205**, 135.

ABRIKOSOV, A. A. and DZYALOSHINSKII, I. E. (1959) *Sov. Phys. JETP* **8**, 535.

ABRIKOSOV, A. A., GORKOV, L. P. and DZYALOSKINSKII, I. E. (1965) *Methods of Quantum Field Theory in Statistical Physics* (Pergamon Press).

AKHIEZER, A. I. and BERESTEZKI, W. B. (1962) *Quantum-Electrodynamics* (J. Wiley, New York).

ANDERSON, P. W. (1950) *Phys. Rev.* **80**, 922.

ANDERSON, P. W. (1961) *Phys. Rev.* **124**, 41.

ANDERSON, P. W. (1963) *Solid State Phys.* **14**.

ANGUS, W. R. (1932) *Proc. Roy. Soc.* A **136**, 569.

AZBEL, M. I. and KANER, E. A. (1956) *Sov. Phys.—JETP* **3**, 772.

AZBEL, M. I. and KANER, E. A. (1957) *Sov. Phys.—JETP* **5**, 730.

BALLENSIEFEN, G. and WAGNER, D. (1965) *Phys. Status Solidi* **12**, 277.

BARDEEN, J. (1963) *Proc. of the 8th Intern. Conf. on Low Temperature Physics* (Butterworth, London).

BATES, L. F. (1961) *Modern Magnetism*, 4th ed. (Cambridge University Press).

BECKER, R. and DÖRING, W. (1939) *Ferromagnetismus* (Springer, Berlin).

BECKER, R. and SAUTER, F. (1957) *Theorie der Electrizität*, Vol. 1. 16th ed. (Teubner, Stuttgart).

BECKER, R. and SAUTER, F. (1959) *Theorie der Electrizität*, Vol. 2, 8th ed. (Teubner, Stuttgart).

BECQUEREL, J., DE HAAS, W. J., and VAN DEN HANDEL, J. (1938) *Physica*, **5**, 857.

BELOV, K. P., LEVITIN, R. Z. and NIKITIN, S. A. (1964) *Sov. Phys. Uspekhi* **7**, 179.

BERK, N. F. and SCHRIEFFER, J. R. (1966) *Phys. Rev. Lett.* **17**, 433.

BERTAUT, E. F. (1963) in *Magnetism*, Vol. 3 (Academic Press, New York).

BETHE, H. (1929) *Ann. Phys.* **3**, 133.

BETHE, H. (1935) *Proc. Roy. Soc.* A **150**, 552.

BETHE, H. A. and SALPETER, E. E. (1957) *Handbuch der Physik* XXV. (Springer).

BJELOV, K. P. (1961) *Magnetic Transitions* (Consultant Bureau, New York).

BLEANEY, B. and STEVENS, K. W. H. (1953) *Rep. Progr. Phys.* **26**, 108.

BLOCH, F. (1930) *Z. Phys.* **61**, 206.

BLOCH, F. (1946) *Phys. Rev.* **70**, 460.

BOGLE, G. S., COOKE, A. H. and WHITLEY, S. (1951) *Proc. Phys. Soc.* A **64**, 931.

BOGOLYUBOV, N. N. and TYABLIKOV, S. V. (1959) *Sov. Phys. Doklady* **4**. 607.

BOROVIK-ROMANOV, A. S. (1968) in *Elements of Theoretical Magnetism*, (Iliffe Books, London).

BOROVIK-ROMANOV, A. S. and KREINES, N. M. (1956) *Sov. Phys.—JETP* **2**, 657.

BOWERS, K. D. and OWEN, J. (1955) *Rep. Progr. Phys.* **28,** 304.
BOZORTH, R. M. (1951) *Ferromagnetism* (Van Norstrand, Princeton).
BREWIG, E., KIERSPE, W., SCHOTTE, U. and WAGNER, D. (1969) *J. Phys. Chem. Sol.* **30,** 483.
BROCKHOUSE, B. N. and SINCLAIR, R. N. (1960) *Phys. Rev.* **120,** 1638.
BROOKS HARRIS, A. (1968) *Phys. Rev.* **175,** 674.
BROOKS HARRIS, A. (1969) *Phys. Rev.* **183,** 486.
BROUT, R. (1960) *Phys. Rev.* **118,** 1009.
BROWN, H. A. and LUTTINGER, J. M. (1955) *Phys. Rev.* **100,** 685.
CALLEN, H. B. (1963) *Phys. Rev.* **130,** 890.
CASIMIR, H. G. B. (1961) *Magnetism and Very Low Temperatures,* (Dover Publ., New York).
CHIKAZUMI, S. (1964) *Physics of Magnetism* (New York).
CONDON, E. U. and SHORTLEY, G. H. (1959) *The Theory of Atomic Spectra,* (Cambridge Univ, Press).
CRANGLE, J. and HALLAM, G. C. (1963) *Proc. Roy. Soc.* A **272,** 119.
DENBIGH, J. S. and LOMER, W. M. (1963) *Proc. Phys. Soc.* **82,** 156.
DINGLE, R. B. (1952a) *Proc. Roy. Soc.* A **211,** 517.
DINGLE, R. B. (1952b) *Proc. Roy. Soc.* A**212,** 38.
DOMB, C. (1960) *Adv. Phys.* **9,** 149.
DOMB, C. and SYKES, M. F. (1957) *Proc. Roy. Soc.* A **240,** 214.
DOMB, C. and SYKES, M. F. (1962) *Phys. Rev.* **128,** 168.
DONIACH, S. and ENGELSBERG, S. (1966) *Phys. Rev. Lett.* **17,** 750.
DORFMAN, J. (1951) *Doklady Akad. SSSR* **81,** 765.
DRESSELHAUS, G., KIP, A. F. and KITTEL, C. (1955) *Phys. Rev.* **100,** 618.
DYSON, F. J. (1956) *Phys. Rev.* **102,** 1217; **102,** 1230.
DZYALOSHINSKII, I. E. (1958) *J. Phys. Chem. Sol.* **4,** 241.
EDMONDS, A. R. (1957. *Angular Momentum in Quantum Mechanics* (Princeton Univ. Press).
EDWARDS, D. M. (1967) *Proc. Roy. Soc. A* **300,** 373.
EDWARDS, D. M. (1968) *J. Appl. Phys.* **39,** 481.
EISELE, J. A. and KEFFER, F. (1954) *Phys. Rev.* **96,** 929.
ELLIOT, R. J. and STEVENS, K. W. H. (1952) *Proc. Roy. Soc.* A **215,** 437.
ELLIOT, R. J. and STEVENS, K. W. H. (1953) *Proc. Roy. Soc.* A **218,** 553; A **219,** 387.
ENGLERT, F. (1960) *Phys. Rev. Lett.* **5,** 102.
ENGLERT, F. and ANTONOFF, M. M. (1964) *Physica* **30,** 429.
FICK, E. and JOOS, G. (1957) *Handbuch der Physik* XVIII (Springer).
FIRGAU, U. (1940) *Ann. der Phys.* **40,** 295.
FISHER, M. E. (1967a) *Rep. Progr. Phys.* **30,** 615.
FISHER, M. E. (1967b) *J. Appl. Phys.* **38,** 981.
FONER, S. (1963) in *Magnetism,* Vol. 1 (Academic Press, New York).
FREEMAN, A. J., NESBET, R. K. and WATSON, R. E. (1962) *Phys. Rev.* **124,** 1978.
FREEMAN, A. J. and WATSON, R. E. (1961) *Phys. Rev.* **124,** 1439.
FRIEDEL, J. (1958) *Nuovo Cimento Suppl.* Vol. **7,** 287.
GENNES, P. G. DE (1963) in *Magnetism,* Vol. 3, ed by G. T. RADO and H. SUHL (Academic Press).

GOMBAS, P. (1956) *Handbuch der Physik* XXVI (Springer).
GORTER, C. J. (1947) *Paramagnetic Relaxation* (Elsevier Publ. Co., Amsterdam).
GORTER, C. J. (1957) *Progr. Low. Temp. Phys.* **2,** (North Holland).
GOSSARD, A. C. and PORTIS, A. M. (1959) *Phys. Rev. Lett.* **3,** 164.
GOSSARD, A. C. and PORTIS, A. M. (1960) *J. Appl. Phys.* **31,** 205.
GRIFFITHS, R. B. (1965) *Phys. Rev. Lett.* **14,** 623.
GUHA, B. C. (1951) *Proc. Roy. Soc.* A **206,** 353.
HAAR, D. TER (1961) *Rept. Progr. Phys.* **24,** 304.
HAAR, D. TER (1966) *Elements of Thermostatistics* (Holt, Rinehart and Winston).
HAAS, C. W. and CALLEN, H. B. (1963) in *Magnetism*, Vol. 1 (Academic Press, New York).
HAAS, W. J. DE and ALPHEN, P. M. VAN (1930) *Proc. Acad. Sci. Amsterdam* **33,** 1106.
HAM, F. S. (1962) *Phys. Rev.* **128,** 2524.
HANDEL, J. VAN DEN (1956) *Handbuch der Physik* XV (Springer).
HANNA, S. S., HEBERLE, J., LITTLEJOHN, C., PERLOW, G. J., PRESTON, R. S. and VINCENT, D. H. (1960) *Phys. Rev. Lett.* **4,** 177; 513.
HANNA, S. S., HEBERLE, J. and PRESTON, R. S. (1962) *Phys. Rev.* **128,** 2207.
HARTREE, D. R. (1928) *Proc. Cambr. Phil. Soc.* **24,** 89.
HARTREE, D. R. (1957) *The Calculation of Atomic Structures* (Chapman Hall, London).
HARTREE, D. R. and HARTREE, W. (1938) *Proc. Roy. Soc.* A **166,** 463.
HAVENS, G. G. (1933) *Phys. Rev.* **43,** 992.
HEER, C. V. and ERIKSON, R. A. (1957) *Phys. Rev.* **108,** 896.
HEINE, V. (1960) *Group Theory in Quantum Mechanics* (Pergamon Press).
HEISENBERG, W. (1928) *Z. Phys.* **49,** 619.
HENRY, W. E. (1952) *Phys. Rev.* **88,** 559.
HERRING, C. (1962) *Rev. Mod. Phys.* **34, 631**.
HERRING, C. (1966) in *Magnetism* 4, ed. by G. T. RADO and H. SUHL (Academic Press, New York).
HIRONE, T. (1935) *Scient. Rep. Tohoku Univ.* **24,** 264.
HOFMANN, J. A., PASKIN, A., TAUER, K. J. and WEISS, R. J. (1956) *J. Phys. Chem. Solids*, **1,** 45.
HOFMANN, J. A., PASKIN, A., TAUER, K. J. and WEISS, R. J. (1960) *J. Phys. Chem. Solids*, **15,** 187.
HORWITZ, G. and CALLEN, H. B. (1961) *Phys. Rev.* **124,** 1757.
HOVE, L. VAN (1954) *Phys. Rev.* **95,** 1374.
HUBBARD, J. (1963) *Proc. Roy. Soc.* A **276,** 238.
HUBBARD, J. (1964a) *Proc. Roy. Soc.* A **277,** 237.
HUBBARD, J. (1964b) *Proc. Roy. Soc.* A **281,** 401.
HYLLERAAS, E. A. (1929) *Z. Phys.* **54,** 347.
ISING, E. (1925) *Z. Phys.* **31,** 253.
IWANENKO, D. and SOKOLOV, A. (1953) *Klassische Feldtheorie* (Akademie Verlag, Berlin).
JACCARINO, V. and WALKER, L. R. (1959) *J. de Phys.* **20,** 341.
JAHN, H. A. and TELLER, E. (1937) *Proc. Roy. Soc.* A **161,** 220.

JENSEN, H. (1936) *Z. Phys.* **101**,, 141.
JOENK, R. J. (1962) *Phys. Rev.* **128**, 1634.
JOHNSON, F. M. and NETHERCOT, A. H. (1959) *Phys. Rev.* **114**, 705.
KADANOFF, L. P. (1966) *Proc.* 1966 *Midwest Conf.*, ed. by D. B. LICHTENBERG.
KADANOFF, L. P., GÖTZE, W., HAMBLEN, D., HECHT, R., LEWIS, E. A. S., PALCIAUSKAS, V. V., RAYL, M. and SWIFT, J. (1967) *Rev. Mod Phys.* **39**, 395.
KAHN, A. H. and FREDERIKSE, H. P. R. (1959) *Solid State Phys.* **9**, 257.
KANAMORI, J. (1963) *Progr. Theor. Phys.* **30**, 275; *Magnetism*, Vol. 1 (Academic Press, New York).
KASTELEIJN, P. W. and KRANENDONK, J. VAN (1956) *Physica* **22**, 317.
KASUYA, T. (1956) *Progr. Theor. Phys.* **16**, 45.
KEFFER, R. (1966) Spin Waves, *Handbuch der Physik* XVIII/2 (Springer).
KINOSHITA, T. (1957) *Phys. Rev.* **105**, 1490.
KIP, A. F., LANGENBERG, D. N. and MOORE, T. W. (1961) *Phys. Rev.* **124**, 359.
KITTEL, C. (1948) *Phys Rev.* **73**, 155.
KITTEL, C. and GALT, J. K. (1957) *Solid State Physics*, **3**, 439.
KITTEL, C. (1962) *Introduction to Solid State Physics*, 2nd ed. (J. WILEY, New York).
KITTEL, C. (1964) *Quantum Theory of Solids* (J. WILEY, New York).
KJELDAAS, T. and KOHN, W. (1956) *Phys. Rev.* **101, 66**.
KJELDAAS, T. and KOHN, W. (1957) *Phys. Rev.* **105**, 806.
KLEIN, A. P. and HEEGER, A. J. (1966) *Phys. Rev.* **144**, 458.
KLERK, D. DE (1956) *Handbuch der Physik* XV (Springer).
KNELLER, E. (1962) *Ferromagnetismus* (Springer, Berlin).
KNIGHT, W. D. (1956) *Solid State Physics* **2**, 93.
KOHN, W. (1954) *Phys. Rev.* **96**, 590.
KOHN, W. (1961) *Phys. Rev.* **123**, 1242.
KONDO, J. (1964) *Progr. Theor. Phys.* **32**, 37.
KOPFERMANN, H. (1956) *Kernmomente* 2. Aufl. (Akad. Verlagsges., Frankfurt am Main).
KOUVEL, J. S. and COMLY, J. B. (1968) *Phys. Rev. Lett.* **20**, 1237.
KRAMERS, H. A. (1930) *Proc. Kon. Ned. Weten.* **33**, 959.
KRANENDONK J. VAN and VAN VLECK, J. H. (1958) *Rev. Mod. Phys.* **30**, 1.
KUBO, R. (1952) *Phys. Rev.* **87**, 568.
KUBO, R. (1957) *J. Phys. Soc. (Japan)* **12**, 570.
KUBO, R. (1958) Lectures at the Summer Institute of Theoretical Physics, Vol. 1 Univ. of Colorado.
KWOK, P. C. and SCHULTZ, T. D. (1969) *J. Phys.* **C2**, 1196.
LANDAU, L. D. (1930) *Z. Phys.* **64**, 629.
LANDAU, L. D. (1956) *Sov. Phys.—JETP* **3**, 920.
LANDAU, L. D. (1957) *Sov. Phys.—JETP* **5**, 101.
LANDAU, L. D. and LIFSHITZ, E. M. (1958) *Quantum Mechanics* (Pergamon Press).
LANDAU, L. D. and LIFSHITZ, E. M. (1965) *Statistical Physics* (Pergamon Press).
LANDOLT–BÖRNSTEIN (1950) *Tabellen*, Bd. 1, Teil 1, 6. Aufl. (Springer).
LANDOLT–BÖRNSTEIN (1962) *Tabellen*, Bd. 2, Teil 9, 6. Aufl. (Springer).
LANG, N. D. and EHRENREICH H. (1968) *Phys. Rev.* **168**, 605.
LANGEVIN, P. (1905) *J. de Phys.* **4**, 678.

LAX, B. and MAVROIDES, J. G. (1960) *Solid State Phys.* **11**.
LAX, B. and WRIGHT G. B. (1960) *Phys. Rev. Lett.* **4,** 16.
LI, Y. Y. (1951) *Phys. Rev.* **84,** 721.
LIDIARD, A. B. (1954) *Proc. Roy. Soc.* A **224,** 161.
LIEB, E. and MATTIS, D. (1962) *Phys. Rev.* **125,** 164.
LIFSHITZ, E. M. and KOSEVITCH A. M. (1956) *Sov. Phys.—JETP* **2,** 636.
LINES, M. E. (1964) *Phys. Rev.* **135,** 1336.
LOW, W. (1960) *Solid State Physics* Suppl. Vol. 2 (Academic Press New York).
LUCAS, G. L. and HORWITZ, G. (1969) *J. Phys.* A **2,** 503.
LUDWIG, G. W. and WOODBURY, H. H. (1963) *Solid State Phys.* **13,** 223.
MAGNUS, W. and OBERHETTINGER, F. (1948) *Formeln und Sätze* 2nd ed. (Springer).
MARSHALL, W. (1959) *Phys. Rev.* **110,** 1280.
MARSHALL, W. (editor) (1967) *Proc. Intern. School Phys. "Enrico Fermi"* (Academic Press).
MARSHALL, W. and MURRAY, G. (1968) *J. Appl. Phys.* **39,** 380; *A. E. R. E. Report* 5879 (Harwell).
MATHON, J. and WOHLFARTH, E. P. (1968) *Proc. Roy. Soc.* A **302,** 409.
MATTIS, D. C. (1965) *The Theory of Magnetism* (Harper & Row, New York).
MERMIN, N. D. and WAGNER H. (1966) *Phys. Rev. Lett.* **17,** 1133.
MESSIAH, A. (1960) *Quantum Mechanics* (North Holland).
MEYER, H. and SMITH, P. L. (1959) *J. Phys. Chem. Solids* **9,** 285.
MORIYA, T. (1963) in *Magnetism*, Vol. 1 (Academic Press, New York).
MOTT, N. F. (1964) *Adv. Phys.* **13,** 325.
MOTT, N. F. and JONES, H. (1936) *The Theory of the Properties of Metals and Alloys.* (Oxford University Press)
MUSCHELISCHWILI, N. I. (1965) *Singuläre Integralgleichungen*, 31 (Akademie Verlag, Berlin).
MYERS, W. R. (1952) *Rev. Mod. Phys.* **24,** 15.
NAGAMIYA, T., YOSIDA, K. and KUBO, R. (1955) *Adv. Phys.* **4,** 1.
NAGAOKA, Y. (1966) *Phys. Rev.* **147,** 392.
NÉEL, L. (1932) *Ann. de Phys.* **17,** 64.
NÉEL, L. (1948) *Ann. de Phys.* **3,** 137.
NEWELL, G. F. and MONTROLL, E. W. (1953) *Rev. Mod. Phys.* **25,** 353.
NOZIÈRES, PH. (1964) *Theory of Interacting Fermi Systems* (Benjamin, New York).
OGUCHI, T. (1957) *Phys. Rev.* **111,** 1063.
OGUCHI, T. (1960) *Phys. Rev.* **117,** 117.
ONSAGER, L. (1952) *Phil. Mag.* **43,** 1006.
OPECHOWSKI, W. (1937) *Physica*, **4,** 181.
OPECHOWSKI, W. (1938) *Physica* **6,** 1112.
PAKE, G. E. (1962) *Paramagnetic Resonance* (Benjamin, New York).
PASCAL, A. (1920) *Jahrbuch der Radioakt und Elekt.* **17,** 184.
PAULI, W. (1920) *Z. Phys.* **2,** 201.
PAULING, L. (1927) *Proc. Roy. Soc.* A **114,** 181.
PEIERLS, R. (1933) *Z. Phys.* **81,** 186.
PEIERLS, R. (1936) *Proc. Roy. Soc.* A **154,** 207.
PEKERIS, C. L. (1958) *Phys. Rev.* **112,** 1649.

PEKERIS, C. L. (1959) *Phys. Rev.* **115,** 1216.
PINES, D. (1961) *The Many Body Problem* (Benjamin, New York).
PINES, D. and NOZIÈRES, PH. (1966) *The Theory of Quantum Liquids* (Benjamin, New York).
PIPPARD, A. B. (1960) *Rep. Progr. Phys.* **20,** 176.
PRYCE, M. H. L. (1950) *Proc. Phys. Soc.* A **63,** 25.
RAJAGOPAL, A. K., BROOKS, H. and RANAGANATHAN, N. P. (1967) *Nuovo Cimento, Suppl.* Vol. **5,** 807.
RICE, T. M. (1965) *Ann. Phys.* **31,** 100.
ROTH, L. M. (1969) *Phys. Rev.* **186,** 428.
RUDERMAN, M. A. and KITTEL, C. (1954) *Phys. Rev.* **96,** 99.
RUSHBROOKE, G. S. (1963) *J. Chem. Phys.* **39,** 842.
RUSHBROOKE, G. S. and WOOD, P. J. (1958) *Mol. Phys.* **1,** 257.
SACHS, M. (1963) *Solid State Theory* (McGraw Hill, New York).
SAMPSON, J. B. and SEITZ, F. (1940) *Phys. Rev.* **59,** 633.
SAUTER, F. (1963) *Ann. der Phys.* **11,** 190.
SCHLAPP, R. and PENNEY, W. G. (1932) *Phys. Rev.* **42,** 666.
SCHRIEFFER, J. R. (1968) *J. Appl. Phys.* **39,** 642.
SCHRIEFFER, J. R. and WOLFF, P. A. (1966) *Phys. Rev.* **149,** 491.
SCHUMACHER, R. T. and VEHSE, W. E. (1960) *Bull. Amer. Phys. Soc.* **4,** 296.
SCHUMACHER, R. T. and VEHSE, W. E. (1963) *J. Phys. Chem. Solids* **24,** 297.
SCOTT, G. G. (1962) *Rev. Mod. Phys.* **34,** 102.
SEITZ, F. (1940) *The Modern Theory of Solids* (McGraw Hill, New York).
SHIMIZU, M., TAKAHASHI, T. and KATSUKI, A. (1962) *J. Phys. Soc. Japan* **17,** 1740.
SHOENBERG, D. (1957) *Progr. Low Temp. Phys.* **2** (North Holland).
SHOENBERG, D. (1960) *The Fermi Surface* (J. Wiley, New York).
SILIN, V. P. (1958) *Sov. Phys.—JETP.* **6,** 945.
SILVERSTEIN, S. D. (1963) *Phys. Rev.* **130,** 1703.
SLATER, J. C. (1930) *Phys. Rev.* **36,** 57.
SLATER, J. C. (1936) *Phys. Rev.* **49,** 537.
SLATER, J. C. (1960) *Quantum Theory of Atomic Structure* (McGraw Hill, New York).
SLICHTER, CH. P. (1963) *Principles of Magnetic Resonance* (Harper & Row, New York).
SMART, J. S. (1956) *Phys. Rev.* **101,** 585.
SMART, J. S. (1963) in *Magnetism,* Vol. 3 (Academic Press, New York).
SMIT, J. and WIJN, H. P. J. (1959) *Ferrites* (Philips Library, Eindhoven).
SOBELMAN, I., I. (1972) *Introduction to the Theory of Atomic Spectra* (Pergamon, Oxford).
SPEDDING, F. H., LEGVOLD, S., DANE, A. H. and JENNINGS, L. B. (1957) *Progr. Low Temp. Phys.* **2.**
STEPHEN, M. J. (1962) *Proc. Roy. Soc.* A **265,** 215.
STEWART, A. L. (1963) *Adv. Phys.* **12,** 299.
STINCHCOMBE, R. B., HORWITZ, G., ENGLERT, F. and BROUT, R. (1963) *Phys. Rev.* **130,** 155.

STONER, E. C. (1929) *Proc. Leeds Phil. Soc.* **1,** 484.
STONER, E. C. (1936) *Proc. Roy. Soc.* A **154,** 656.
STONER, E. C. (1938) *Proc. Roy. Soc.* A **165,** 372.
STONER, E. C. (1939) *Proc Roy. Soc.* A **169,** 339.
STRIEB, B., CALLEN, H. B. and HORWITZ, G. (1963) *Phys. Rev.* **130,** 1798.
SYDORIAK, S. G. and ROBINSON, J. E. (1949) *Phys. Rev.* **75,** 118.
TAHIR-KHELI, R. A. (1963) *Phys. Rev.* **132,** 689.
TAHIR-KHELI, R. A. and HAAR, D. TER (1962) *Phys. Rev.* **127,** 88; 95.
TINKHAM, M., WEINSTEIN, R. and KIP, A. F. (1951) *Phys. Rev.* **84,** 848.
TUROV, E. A. (1968) *Elements of Theoretical Magnetism* (Iliffe Books, London).
TYABLIKOV, S. A. (1969) *Quantum Mechanical Methods in Magnetism* (Consultant Bureau, New York).
VAN VLECK, J. H. (1932a) *The Theory of Electric and Magnetic Susceptibilities* (Oxford University Press).
VAN VLECK, J. H. (1932b) *Phys. Rev.* **41,** 208.
VAN VLECK, J. H. (1941) *J. Chem. Phys.* **9,** 85.
VAN VLECK, J. H. (1945) *Rev. Mod. Phys.* **17,** 27.
VAN VLECK, J. H. (1950) *Phys. Rev.* **78,** 266.
VAN VLECK, J. H. (1953) *Rev. Mod. Phys.* **25,** 220.
VAN VLECK, J. H. (1962) *Rev. Mod. Phys.* **34,** 681.
VAN VLECK, J. H. and HEBB, M. H. (1934) *Phys. Rev.* **46,** 17.
VONSOVSKIJ, S. V. and IZJUMOV, J. A. (1964) *Fortschr. der Physik* **12,** 1.
WAGNER, D. (1963) *Physica* **29,** 803.
WAGNER, H. (1966) *Z. Phys.* **195,** 273.
WALKER, L. R. (1963) in *Magnetism*, Vol. 1 (Academic Press, New York).
WATSON, R. E. and FREEMAN, A. J. (1961) *Phys. Rev.* **123,** 2027.
WEISS, P. (1907) *J. de Phys.* **6,** 661.
WEISS, P. R. (1948) *Phys. Rev.* **74,** 1493.
WIDOM, B. (1965) *J. Chem. Phys.* **43,** 3898.
WILSON, A. H. (1958) *The Theory of Metals*, 2nd ed. (Cambridge University Press).
WOBSCHALL, D., GRAHAM, J. R. and MALONE, D. P. (1963) *Phys. Rev.* **131,** 1565.
WOHLFARTH, E. P. (1949) *Proc. Roy. Soc.* A **195,** 434.
WOHLFARTH, E. P. (1951) *Phil. Mag.* **42,** 374.
WOHLFARTH, E. P. (1969) *J. Phys.* C. **2,** 68.
YAFET, Y. Y. (1963) *Solid State Phys.* **14,** 2.
YAMADA, H. and SHIMIZU, M. (1967) *J. Phys. Soc. (Japan)* **22,** 1404.
YOST, W. J. (1940) *Phys. Rev.* **58,** 557.
ZIMAN, J. M. (1962) *Electrons and Phonons* (Oxford University Press).
ZUBAREV, D. N. (1960) *Sov. Phys.—Uspekhi* **3,** 320.

Index

OTHER TITLES IN THE SERIES IN NATURAL PHILOSOPHY

Vol. 38. Todorov—Analytical Properties of Feynman Diagrams in Quantum Field Theory
Vol. 39. Sitenko—Lectures in Scattering Theory
Vol. 40. Sobel'man—Introduction to the Theory of Atomic Spectra
Vol. 41. Armstrong & Nicholls—Emission, Absorption and Transfer of Radiation in Heated Atmospheres
Vol. 42. Brush—Kinetic Theory, Volume 3
Vol. 43. Bogolyubov—A Method for Studying Model Hamiltonians
Vol. 44. Tsytovich—An Introduction to the Theory of Plasma Turbulence
Vol. 45. Pathria—Statistical Mechanics
Vol. 46. Haug—Theoretical Solid State Physics, Volume 2
Vol. 47. Nieto—The Titius—Bode Law of Planetary Distances: Its History and Theory